Discrete Mathematics
A Concept-based Approach

Basavaraj S Anami
Principal
KLE Institute of Technology
Hubballi

Venkanna S Madalli
Professor
Department of MCA
KLE Institute of Technology
Hubballi

Universities Press

DISCRETE MATHEMATICS: A Concept-based Approach

UNIVERSITIES PRESS (INDIA) PRIVATE LIMITED

Registered office
3-6-747/1/A & 3-6-754/1, Himayatnagar, Hyderabad 500 029, Telangana, India
info@universitiespress.com; www.universitiespress.com

Distributed by
Orient Blackswan Private Limited

Registered office
3-6-752 Himayatnagar, Hyderabad 500 029, Telangana, India

Other offices
Bengaluru, Bhopal, Chennai, Guwahati, Hyderabad, Jaipur, Kolkata, Lucknow, Mumbai, New Delhi, Noida, Patna, Vijayawada

First published 2016

ISBN: 978-81-7371-999-8

Typeset in Albertina MT Std 10/12 by
Cameo Corporate Services, Tamil Nadu, India

Printed at SS Colour Impression Pvt. Ltd., Chennai 600 106

Published by
Universities Press (India) Private Limited
3-6-747/1/A & 3-6-754/1, Himayatnagar, Hyderabad 500 029, Telangana, India
e-mail: info@universitiespress.com

Dedicated to the
SAPTARSHIS of KLE Society, Belgaum, Karnataka,
on the eve of the Society's
Centenary Celebrations (1916–2016)

Contents

Foreword ix
Preface x
Acknowledgements xi
About the Authors xii

Chapter 1: Overview of Disciplines in Mathematics 1

1.1 Analytical Methods 1
1.2 Transforms 2
1.3 Numerical Methods 3
1.4 Discrete Mathematics 3
Summary 4

Chapter 2: Fundamentals of Mathematical Logic 5

2.1 Introduction 5
2.2 Symbolic Logic 5
2.3 Laws of Logic 11
2.4 Principle of Duality 12
2.5 Rule of Inference 15
2.6 Method of Resolution 21
2.7 Predicate Logic 23
Additional Solved Examples 27
Summary 29
Exercises 30

Chapter 3: Mathematical Induction 32

3.1 Introduction 32
3.2 Jokes and Paradoxes 45
Additional Solved Examples 46
Summary 50
Exercises 50

Chapter 4: Introduction to Set Theory 52

4.1 Sets and Set Definitions 52
4.2 Russell's Paradox 55
4.3 Subset 55
4.4 Power Set 56
4.5 Set Operations 57
4.6 Laws of Set Theory 58
4.7 Principle of Duality 59
4.8 Computer Representation of Sets 61
4.9 Venn Diagrams 62
4.10 Addition Principle 64

4.11 Sequences 67
4.12 Strings 68
4.13 Regular Expressions 69
Additional Solved Examples 71
Summary 73
Exercises 73

Chapter 5: Relations and Operations 76

5.1 Concept of Relations 76
5.2 Relations as Subsets of Cartesian Products 78
5.3 Properties of Binary Relations 79
5.4 Operations on Relations 85
5.5 Matrix Representation of Relations 88
5.6 Graphical Representation of Relations 92
5.7 Partially Ordered Set (POSET) 95
5.8 Lattices 100
5.9 Warshall's Algorithm to find Transitive Closure 104
Additional Solved Examples 107
Summary 108
Exercises 108

Chapter 6: General Functions and Growth Functions 111

6.1 Overview of Functions 111
6.2 Domain and Range of a Function 112
6.3 Different Type of Functions 115
6.4 Alternative Types of Functions 117
6.5 Inverse of an Element 119
6.6 Composite Function 121
6.7 Identity Function 126
6.8 Invertible Function 126
6.9 Growth Functions 128
6.10 Applications 133
6.11 Sorting 133
Additional Solved Examples 137
Summary 141
Exercises 141

Chapter 7: Algebraic Structures 143

7.1 History of Group Theory 143
7.2 Binary Operations and General Properties 145
7.3 Groups 148
7.4 Square Roots, Cube Roots and Fourth Roots of Unity 154
7.5 Properties of Groups 157
7.6 Cancellation Laws 159
7.7 Semi-groups and Monoids 160
7.8 Permutations 161
7.9 Symmetric Group 162

7.10 Geometrical Meanings of Six Permutations in S = {1, 2, 3} 166
7.11 Cyclic Groups 167
7.12 Congruence Modulo *n* 173
7.13 Cosets 174
7.14 Homomorphisms 176
7.15 Applications of Group Theory 184
Additional Solved Examples *185*
Summary *189*
Exercises *190*

Chapter 8: Coding Theory 191

8.1 Introduction 191
8.2 Communication Process 192
8.3 Encoders and Decoders 193
8.4 Decoding and Error Correction 195
8.5 Parity and Generator Matrices 196
8.6 Group Codes 203
8.7 Applications of Coding Theory 206
Additional Solved Examples *208*
Summary *214*
Exercises *214*

Chapter 9: Fundamentals of Counting Principles 217

9.1 Fundamentals of Counting 217
9.2 Permutations 220
9.3 Combinations 225
9.4 Pascal's Triangle 231
9.5 Combinations with Repetitions 232
Additional Solved Examples *233*
Summary *236*
Exercises *236*

Chapter 10: Introduction to Probability 238

10.1 Probability Models 238
10.2 Meaning of Probability 239
10.3 Sample Spaces and Events 240
10.4 Conditional Probability 243
10.5 Baye's Theorem 245
10.6 Random Variables 248
10.7 Bernoulli's Trials 248
10.8 Areas of Application 251
Additional Solved Examples *262*
Summary *266*
Exercises *266*

Chapter 11: Finite State Machine 269

11.1 Deterministic Finite Automata (DFA) 269

11.2 Non-Deterministic Finite Automata (NDFA) 274
11.3 NFA to DFA Conversion 281
11.4 NFA with ∈–transitions (∈–NFA) 285
11.5 Conversion from ∈–NFA to NFA 286
11.6 Conversion from ∈–NFA to DFA 287
11.7 Output Associated Finite Automata 289
11.8 Minimisation of Automata 294
11.9 Applications of Finite Automata 300
11.10 Solved Problems 301
Additional Solved Examples 305
Summary 309
Exercises 309

Chapter 12: Introduction to Recurrence Relations 311

12.1 Concept of Recurrence Relation 311
12.2 General Solution 313
12.3 More Applications of Recurrence 315
12.4 Non-homogeneous Recurrence Relations 316
Additional Solved Examples 320
Summary 326
Exercises 326

Chapter 13: Introduction to Graph Theory 327

13.1 History of Graph Theory 327
13.2 Definitions and Basic Properties 328
13.3 Graph Isomorphism 333
13.4 Walk, Paths, Cycles and Circuits 333
13.5 Hamiltonian Cycles 335
13.6 Planar Graphs 335
13.7 Detection of Planarity 336
13.8 Euler's Formula (Euler's Polyhedral Formula) 337
13.9 Graph Colouring 339
13.10 Basics of Trees 341
13.11 Directed, Weighted and Directed Acyclic Graphs 343
13.12 Certain Applications of Graph Theory 351
Additional Solved Examples 355
Summary 356
Exercises 356

Bibliography 359
Index 361

Foreword

I am happy to write the foreword for the book *Discrete Mathematics – A Concept-based Approach* authored by Dr Basavaraj S Anami and Dr Venkanna S Madalli. As seasoned academicians who have taught this subject for many years, the two authors have transposed their experience in teaching to a good reading material for undergraduate programmes in computer science and postgraduate programme in MCA.

The book runs into more than 350 pages and consists of 13 well-written chapters. The topics of each chapter are treated in a concise manner with clear definitions of terms and illustrative examples that make the concepts explicit and connected to the real world. End-of-chapter summary and exercises are provided to enable easy recapitulation of the topics discussed. The subject has been treated in a lucid language that would be definitely appealing to the students. The theorems are kept to the minimum with an aim to make students comfortable with abstract mathematical concepts. The flow from one chapter to another is logical. Students' interests and their study points are also considered.

The topics on mathematical logic, mathematical induction, set theory, relations, functions, algebraic systems, coding theory and finite state machines are dealt with in detail, as are the theories on probability, recurrence and counting, which are presented with an introduction to the origin of these concepts. The book draws diverse examples from the field of computer science and also from other related areas to drive home the themes and bring out the abstract concepts in a manner that is appealing to the students.

I am sure the book will serve the needs of students who take up the subject of discrete mathematics at the undergraduate and postgraduate levels. I appreciate the effort taken by the authors to write this student friendly book.

Dr K Chidananda Gowda
Former Vice Chancellor
Kuvempu University,
Shimoga, Karnataka, India.

Preface

Discrete mathematics is taught to students of mathematics and computer science at the freshman level. This book is an outcome of the authors' experience in teaching discrete mathematics for the last 15 years at the undergraduate and postgraduate levels in distinguished engineering colleges. The content and sequencing of the topics in this book reflect the order and manner in which the authors approached the subject in their classes. That the topics have been put together by two authors, each specialised in a distinct field – while one is an accomplished researcher in mathematics, the other has done intense work in computer science – makes the book unique in that it explores the mathematical foundations of the subject while also extrapolating its applicability in computer science. Its contents have been decided based on student feedback. It focuses on the applications of discrete mathematical concepts to real-life scenarios and makes the subject appealing to the student. The flow of the topics is gradual and designed to lead the students step by step from the first principles to the advanced topics. The topics of the book are well-supported by appropriate examples from computer science to showcase the application of discrete mathematics in the field of computers. The topics are so sequenced that they can be studied independently. The book can be used as a foundation course for studying advanced mathematical concepts. However, the prerequisite assumed of a reader for doing so is no more than conversance with high-school mathematics.

The book is a self-contained and comprehensive study material with adaptable chapters that the reader can choose to read according to his or her convenience. The writing style is direct and illustrative, with abstractions kept to the minimum throughout the book. A brief background of the advent of most of the topics is succinctly described in the book. At the end of every chapter is a brief summary and a fair number of exercises. The book elaborates on the elements of discrete mathematics in 13 chapters. Chapter 1 gives an introduction to discrete mathematics. Mathematical logic is dealt with in Chapter 2. The concept of mathematical induction is introduced in Chapter 3. Set theory is covered in Chapter 4 while Chapter 5 focuses on relations and operations on relations. Functions are discussed in Chapter 6. Chapter 7 gives an account of algebraic systems. Chapter 8 delves into the essentials of coding theory while Chapter 9 explores the principle of counting. Chapter 10 introduces the concept of probability. Chapter 11 elucidates finite state machines while Chapter 12 is a brief discussion on recurrence relations. Chapter 13, the last chapter of the book, is an introduction to graph theory – a sub-discipline of mathematics that has several applications in computer science, sociology and other physical sciences. An extensive bibliography is given at the end for the benefit of students who may be interested to pursue any given topic further. PowerPoint presentations on the pivotal points of each chapter, solutions to the chapter-end exercises, and solved university questions can be accessed online at
www.universitiespress.com/basavarajsanami/discretemathematics

This book, with its contributive study-aids, will be a useful tool for studying discrete mathematics.

Basavaraj S Anami
Venkanna S Madalli

Acknowledgements

The blessings of His Holiness Shri Sidharood Swamiji, Hubballi, and Shri Bilur Gurubasava Swamiji, Bilur, have enabled us to write this book, *Discrete Mathematics – A Concept-based Approach*, for the freshmen of computer science programme and get it published by Universities Press Pvt. Ltd, Hyderabad. It is indeed a coincidence that the KLE Society is celebrating its centenary year in 2015–2016. We dedicate this book to the Saptarshis, who founded the KLE Society in 1916 and also express our deep sense of gratitude to the people who, directly or indirectly, helped us in bringing out this book.

We thank the honourable chairman and architect of the KLE Society, Dr Prabhakar Kore, for his constant support and encouragement. We are obliged to Shri S C Metgud and Shri Shankaranna Munavalli, director of the KLE Society and members of the governing body of KLE Institute of Technology. We thank Dr Ashok Shettar, Principal, BVB College of Engineering and Technology, Hubballi, for his timely help. We are indebted to Shri Arun Khannur for his friendly words of advice at the time of writing this book. We also thank all our friends for their best wishes and support to make this endeavour a success.

We are indebted to Mr Karibasappa K G, who helped in preparing the chapter on Finite State Machines. We are obliged to Mr Kiran Malagi, Ms Deepa Garag, Mr Praveen Itagi, and our colleagues in the college, for their invaluable assistance during preparation of the manuscript. Many other colleagues also supported our idea of writing this book; we wish to thank all of them. We are grateful to the students whom we taught over the years for their contribution to the content of this book.

We thank the publishing team of Universities Press (India), Pvt. Ltd, Hyderabad, especially Mr Vamsi Deepak, Mr Prabhakar Reddy, Ms Sreelatha Menon, Mr Ramesh M R and Mr Kallol, who were instrumental in bringing out this book within the stipulated time.

Last but not the least, we wish to express our gratitude to our respective wives and children for patiently putting up with the inconveniences during the preparation of the manuscript of this book. We could not have brought this book to fruition without their whole-hearted support.

Basavaraj S Anami
Venkanna S Madalli

About the Authors

Basavaraj S Anami is Principal, KLE Institute of Technology, Hubballi, and is one among the few pioneers who laid the groundwork for propagating computer science education in North Karnataka. Earlier, he was Professor and Head, Department of Computer Science, Basaveshwar Engineering College Bagalkot. He has been teaching computer science at the undergraduate and postgraduate levels for the past 33 years and has 13 years of research experience. Currently guiding five students to their Ph.D. degrees, he has authored two books in the subject and has to his credit about 100 papers published in journals and conference proceedings.

Dr Anami received his M.Tech. degree in computer science and engineering from Indian Institute of Technology (IIT) Madras in 1985 and earned his Ph.D. in computer science from University of Mysore, Mysore. His research interests include image processing, natural language processing and knowledge based systems development.

Venkanna S Madalli is Professor, Department of MCA, KLE Institute of Technology, Hubballi. He received his postgraduate degree in mathematics from Karnatak University, Dharwad (India) in 1987 and was awarded a doctorate degree in applied mathematics by the same university in 1996. He has also contributed a good number of publications in international and national journals of repute.

Dr Madalli has 22 years of teaching experience. Besides computer science, he has also taught mathematics and electronics at the postgraduate level in Karnatak University, where he served as faculty for ten years. Later, he moved to B V Bhoomaraddi College of Engineering and Technology, Hubballi, as faculty in the department of mathematics. He is now a visiting faculty at P C Jabin Science College, Hubballi, where he teaches physics to M. Sc. students and at Karnatak University, where he handles courses in MCA and M.Sc. (Mathematics). His areas of academic interest include fluid mechanics, numerical techniques and applied mathematics.

1 Overview of Disciplines in Mathematics

Mathematics can broadly be classified into two categories: analytical methods and numerical methods. These two categories are discussed in the following sections.

1.1 ANALYTICAL METHODS

Solutions to certain problems are obtained by using only analytical or exact methods. These solutions are often useful and provide excellent means for studying the behaviour of physical systems. However, analytical solutions are not always possible for all types of problems. They have limited practical importance, as a majority of physical problems are complex and non-linear in nature. In science and technology for example, the solutions for a majority of problems are in the form of ordinary differential and partial differential equations. Differential equations are simple and important, and this domain has grown in parallel with the development of mathematics, since the period of Newton and Leibniz. Differential equations are central to a wide variety of applications. As a case in point, let us apply ordinary differential equations on Newton's second law of motion, which is used to determine the terminal velocity of a free-falling body near Earth's surface.

$$F = ma \quad \text{and} \quad a = \frac{F}{m} \tag{1.1}$$

where

a = rate of change of velocity with respect to time = $\dfrac{dv}{dt}$

F = the net force on a free-falling body and is given as $F = F_D + F_U$

That is,
F_D is the downward force and is represented as 'mg'.
F_U is the upward force and is represented as '$-cv$'.
By substituting F and a in (1.1),

$$m\frac{dv}{dt} = mg - cv \quad \text{or} \quad \frac{dv}{dt} = \frac{mg - cv}{m} \tag{1.2}$$

Hence, the simple-looking differential equations are capable of describing complex and chaotic behaviour of physical systems.

1.1.1 Partial Differential Equations

In general, physical phenomena in the domains of science and technology such as fluid dynamics, electricity, magnetism, mechanics, optics and heat flow are described by means of partial differential equations (PDE). In fact, most engineering problems are modelled using PDE. In contrast to ordinary differential equations in which the unknown function depends on only one variable, in PDEs the unknown functions depend on several variables. For example, a temperature function, $\vec{T}(x,t)$, depends on both variables, i.e., location '*x*' and time '*t*'.

1.1.1.1 *Why partial differential equations are useful*

Most of the natural laws of physics, such as Maxwell's equations, Newton's law of cooling, the Navier–Stoke's equations, Newton's equations of motion and Schrodinger's equation in quantum mechanics are stated in terms of PDEs. PDEs are also used extensively in several engineering problems in the areas of electricity and magnetism, electrostatic field theory, fluid dynamics, aerodynamics and elasticity.

1.2 TRANSFORMS

Transforms play a vital role in solving engineering problems. A mathematical transformation is used to simplify the solution of a problem. For example, the logarithm is used to simplify multiplication, division and problems involving the *n*th power of a variable *x*. To multiply or divide two numbers, the numbers are initially transformed into the logarithmic domain, wherein the process of addition and subtraction is performed. Then, the inverse transformation (antilogarithm) is performed to obtain the product or quotient of the original numbers. The purpose of using a transformation is to create a new domain in which it is easier to handle the problem being investigated. Once results have been obtained in the new domain, they can be inverse-transformed to give the desired results in the original domain. The Laplace transform is an example of integral transforms. Another integral transform widely used by engineers, is the Fourier transform.

1.2.1 Laplace Transform

Laplace transforms are helpful in transforming differential equations in the time (*t*) domain into algebraic equations in the frequency (*s*) domain. Solving the algebraic equations for the desired unknowns, their values as functions of time may be found by taking inverse transforms. Laplace transforms finds particular application in the fields of signals and linear systems analysis. Laplace transforms leads to a more unified approach and provides engineers with a greater insight into the behaviour of systems. It has distinct advantages over the classical approach. Engineers are interested not only in system analysis, but also in system synthesis or design. Their objective is to study a system's response to specific inputs, with a view to learn more about the systems and to improve or control them. Laplace transforms is an effective tool that helps engineers achieve this.

1.2.2 Fourier Transforms

Fourier transform is used for decomposing signals into their constituent frequencies and oscillatory functions. It represents signals in frequency domain and transforms one complex value function of a real variable into another. They are used in solving PDEs as well as in the analysis of signals and systems, communication, statistics, quantum physics, optics and antennae. Discrete Fourier transforms are used in discrete signal analysis in the field of communication theory, speech processing and image processing.

1.3 NUMERICAL METHODS

In practice, an engineer would finally obtain results in numerical form. For example, from a set of data derived from an experiment, inferences may have to be drawn, or a system of linear algebraic equations may have to be solved. The main purpose of numerical methods is to provide efficient methods for obtaining numerical answers to such problems. Further, numerical methods are important for the following reasons.

1. Numerical methods are extensively powerful problem-solving tools. They are capable of handling large systems of equations, nonlinearities, and complicated geometries that are not uncommon in engineering problems and are often impossible or difficult to solve analytically.
2. One has to use commercially available software packages to adopt numerical methods. The intelligent use of these programs is often predicted on the knowledge of the basic theory underlying the methods.
3. One can design programs to solve the problems without buying expensive software packages only when one is conversant with numerical methods and good at computer programming.
4. Numerical methods are efficient vehicles for learning to use computers. They are especially well-suited to illustrate the power as well as the limitations of computers.
5. Numerical methods provide vehicles for reinforcing the understanding of mathematics. The methods reduce higher mathematics to basic arithmetic operations.

1.4 DISCRETE MATHEMATICS

Discrete mathematics is the study of mathematical structures that are fundamentally discrete rather than continuous. Computer functions that deal with 1s and 0s are examples of functions that are discrete in nature. In contrast to real numbers, the objects studied in discrete mathematics, such as integers, graphs, set theory, relations, mathematical induction, coding theory and logic, do not vary smoothly but have distinct, separated values. Therefore, discrete mathematics excludes topics in 'continuous mathematics' such as calculus and integration. Discrete objects can often be enumerated as integers. It has been characterised, more formally, as the branch of mathematics dealing with countable sets. However, there is no exact, universally agreed, definition of the term 'discrete mathematics'.

1.4.1 Computer Science Analogy

Concepts and notations from discrete mathematics are useful in studying and describing objects and problems in branches of computer science, such as computer algorithms, programming languages, cryptography, automated theorem proving, software engineering, computer networks. It is an emerging field which is being used to deal with newer applications. The input given to an algorithm is recognised as an input set and the corresponding output form as an output set, while the program is visualised as a function. These relations are used in database management systems. The logic finds application in knowledge representation and the development of intelligent systems. Coding theory is used in computer security. Usage of inductive proofs is required when analytical proofs either become too involved or are not possible. The syntax of computer languages is governed by formal grammar and formal languages. Therefore, the knowledge of discrete mathematics is essential for computer science students.

SUMMARY

The concept of discrete mathematics is not new to us; we find its direct application in computer science. It is a foundation course for computer science students. Fundamental knowledge of discrete mathematic also helps in computer science research. Without discrete mathematics, practical computer science courses cannot be appreciated.

2 Fundamentals of Mathematical Logic

In Chapter 1, we have discussed about the different disciplines in mathematics. Mathematical Logic belongs to the area of discrete mathematics that finds applications in other areas of mathematics. In this chapter, we discuss symbolic logic and predicate logic. Applications of logic is analysing mathematical arguments is presented.

2.1 INTRODUCTION

'Mathematical Logic' is one of the disciplines of mathematics. It is also known as 'Mathematics of Reasoning'. It was first explained by the Greek philosopher Aristotle (384–322 BC). In his treatise on this topic, Aristotle presented the principles for deductive reasoning. His works provide a foundation for the study of all branches of knowledge. Later, several scientists have contributed to the field of logic, namely German mathematician Gottfried Whithelm Leibniz, George Boole, Augustus De Morgan, Sanders Peirce and others. Mathematical reasoning (logic) is an art of reasoning that has emerged from a study on the use of language in argument and persuasion. Logic is very helpful in judging the correctness of a sequence of reasoning. It is used in the verification of algorithms, set definitions, knowledge representations, design of expert systems and other areas. The language called 'PROLOG' is built on the framework of logic. The word PROLOG is an abbreviation for 'programming in logic', and it was first conceived by Alain Colmerauer in the early 1970s. PROLOG is used in 'artificial intelligence' (AI). AI is the sub-field of computer science concerned with the concepts and methods of symbolic inference by computer. It is also associated with the use of symbolic knowledge representation by a computer to make inferences and solve problems faster than humans. Logic is used in natural language processing, theorem proving, game theory, automatic answering systems, ontologies, control systems and graphical user interfaces. Nowadays, logic is applied in the area of computer security. The area of logic is also divided into two categories, namely symbolic logic and first-order predicate logic.

2.2 SYMBOLIC LOGIC

Symbolic logic is also called 'sentential logic or statement calculus'. In mathematical discourse and other day-to-day activities, one encounters the declarative English language statements or sentences.

The meaning of these sentences can be modified by using words such as 'NOT', 'AND', 'OR', 'IF-THEN-ELSE', 'IF-AND-ONLY-IF' and 'EXCLUSIVE-OR'. These words are called statement connectives. These statements are further divided into elementary and composite or complex sentences. Elementary statements are simple declarative statements. Two or more elementary statements are joined with connectives to form a complex or compound statement. A statement that is modified by the connective NOT is called the negative or the negation of the given statement. For example, negative of the statement 'Rama goes to school' is 'NOT Rama goes to school'. In the real world, as the usage would be 'Rama does not go to school'. Consider the following statements:

1. 2 is not a prime.
2. 6 is a composite number.

Some of the compound statements are as follows:

3. 2 is prime AND 6 is a composite number.
4. IF 2 is not a prime THEN 6 is a composite number.
5. 2 is not a prime IF-AND-ONLY-IF 6 is a composite number.
6. 2 is prime OR 6 is a composite number.

AND, OR connectives are also known as conjunction and disjunction operation, respectively. The symbols used for connectives are as follows:

AND ($\wedge$), OR ($\vee$), IF-THEN ($\rightarrow$), IF-AND-ONLY-IF ($\leftrightarrow$), EXCLUSIVE-OR (Θ) and NOT($\neg$). These symbols are used for connectives in logical expressions. Consider the following statements.

7. I go to school.
8. The sun is shining.
9. Peter is six feet tall.
10. Mary dances well.
11. It is raining today.
12. It is snowing today.

The above statements are called 'declarative sentences' or 'assertive sentences' in the English language. Assertive sentences are either TRUE or FALSE but not both, are called propositions in statement or sentential calculus. All propositions are assertions, but the reverse may not always be true. Sentences such as 'Please give me the book', 'When are you coming home?' and 'What a beautiful picture!' are imperative, interrogative, and exclamatory sentences respectively, and hence are not propositions. Consider the sentence: 'I tell a lie'. This sentence can have truth value TRUE but not FALSE. If the sentence is TRUE, then one can take the opposite of what is being said. But when the sentence is assigned FALSE value, it leads to contradiction. If this statement is FALSE, then it is implied that the statement is true, but it is already stated to be a lie. This contradiction is popularly known as the 'Liar Paradox' in computer science. Hence, this sentence cannot take FALSE value and is not a proposition. Consider the following sentences:

13. The pot is hot or cold.
14. I pass or fail in the examination.
15. I go to Bangalore by train or by road.

Sentences (13) and (14) are compound sentences wherein the word OR is used in exclusive sense, since the pot can be in any one of the states at a given time. Similarly, one can be in the state of either pass or fail but not both at a time. In the sentence 'I read or listen to music', OR is used in an inclusive sense, as one can do both at the same time. The connective OR is used in two ways, but it is used mostly in the inclusive sense. Consider the following expression: $x + 2 = 4$. It is not a proposition because the truth value of the statement depends upon the value of the variable x. If $x = 2$ then $x + 2 = 4$ is true. Otherwise, it is false.

Based on the sentences from (7) to (12), the following statements are defined using the different connectives.

16. I go to school or it is raining today.
17. If Mary dances well, then Peter is six feet tall.
18. Mary dances well if and only if Peter is six feet tall.
19. I do not go to school.
20. It is raining or it is snowing today.
21. Mary dances well implies that Peter is six feet tall.

Sentences from (16) to (21) are called compound sentences. The sentences can be expressed in a more concise manner using symbols for connectives and variables for propositions. In sentential calculus, capital letters are used to denote the propositions. Let the sentences (7) to (12) be represented by the symbols in parentheses as in the following.

7. I go to school (*T*)
8. The sun is shining (*S*)
9. Peter is six-feet tall (*P*)
10. Mary dances well (*M*)
11. It is raining today (*R*)
12. It is snowing today (*I*)

Sentences from (16) to (21) are expressed in the following symbolic form:

(i) $(T \vee I)$
(ii) $(M \rightarrow P)$
(iii) $(M \leftrightarrow P)$
(iv) $\neg T$
(v) $(R \vee I)$
(vi) $(M \rightarrow P)$

Truth values of compound sentences are determined by the truth values of associated individual elementary sentences and the rules that are followed for the evaluation of the corresponding connective or logical operation. These rules for determining the truth value of a compound sentence for a given connective are expressed in the form of a truth table. Truth tables from 2.1 to 2.6 give the evaluation of truth values for different connectives. P and Q are the propositional variables. The expression $(P \vee Q)$ is true when P is true or Q is true or both are true.

Table 2.1 Truth table for OR ($\vee$) connective

P	*Q*	$(P \vee Q)$
F	F	F
T	F	T
F	T	T
T	T	T

Table 2.2 Truth table for AND (∧) connective

P	**Q**	**(P ∧ Q)**
F	F	F
T	F	F
F	T	F
T	T	T

The AND operation results in the value TRUE when its operands, individual elementary sentences, are TRUE. Otherwise, the expression is FALSE.

Table 2.3 Truth table for IMPLICATION (→) connective

P	**Q**	**(P → Q)**
F	F	T
T	F	F
F	T	T
T	T	T

The connective IF-THEN is also called an IMPLICATION operation or conditional operation. It is read as IF *P* THEN Q. The proposition *P* is called an 'antecedent or premise or hypothesis', and the proposition Q is called the 'consequence or conclusion'. The IMPLICATION is true when the combination of *P* and Q are (FF), (FT) and (TT). It is false when *P* is true and Q is false. The premise value does not dictate the truth value of IMPLICATION. There are many ways of interpreting IMPLICATION.

1. *P* implies Q
2. *P* only if Q
3. *P* is a sufficient condition for Q
4. Q provided that *P*
5. Q if *P*
6. Q is a necessary condition for *P*

It is interesting to see that IMPLICATION is true for (FT). Consider the English language statement, 'If I win the lottery then I will buy a car'. Let *P* be the statement 'I win the lottery' and Q be the statement 'I will buy a car'. The third row of Table 2.3, where *P* is false and Q is true, corresponds to a case wherein I have not won the lottery but I have bought the car. The statement is true because a car could be bought by some other means, such as a loan. If I do not buy a car in spite of winning the lottery, then I am at false. In order to corroborate the argument, let us take another example statement. 'If I am in Hubballi then I am in Karnataka'. Corresponding to the same case of row three, I am not in Hubballi does not necessarily mean that I am not in Karnataka. I might be in another place in Karnataka. This is an important interpretation used in inferences and proving many theorems in set theory.

Table 2.4 Truth table for DOUBLE IMPLICATION (↔) connective

P	**Q**	**(P ↔ Q)**
F	F	T
T	F	F

Continued

Continued

P	**Q**	**(P ↔ Q)**
F	T	F
T	T	T

The double IMPLICATION is also called bi-directional IMPLICATION. Table 2.4 almost resembles the AND truth table except in the first row. The propositions *P* and Q are both necessary and sufficient conditions. Given the case that *P* is true, one can infer that Q is going to be true. Consider the statement 'I will buy the car iff I win the lottery'. If you see him in a car then you can infer that he has won the lottery and *vice versa.*

Table 2.5 Truth table for Exclusive-OR (Θ) connective

P	**Q**	**(P Θ Q)**
F	F	F
T	F	T
F	T	T
T	T	F

Exclusive-OR connective, given in Table 2.5, is similar to OR connective except the fact that the last condition of both propositions being true is excluded and hence the name. The negation of any proposition is illustrated in Table 2.6. If *P* is true then ¬*P* is false and *vice versa.*

Table 2.6 Truth table for NOT (¬) connective

P	**¬P**
F	T
T	F

Example 2.1: Consider the sentence 'If it is foggy tonight then either John must stay at home or he must take a taxi' which is conditional. The elementary sentences are as follows: it is foggy tonight (*F*), John must stay at home (*H*) and John must take a taxi (*T*). The conclusion is an exclusive OR of *H* and *T*. The sentence is of the form (*A*→*B*), where the premise *A* and conclusion *B* are further rewritten as follows:

$$F \rightarrow (H \Theta T)$$

Example 2.2: Consider the sentence 'If either the Pirates or the Cubs lose and the Giants win, then the Dodgers will be out of the first place and, moreover, I will lose the bet'. It is an implication. This compound sentence has the elementary sentences, and the propositional variables are given in parentheses. The Pirates lose (*P*), The Clubs lose (*C*), The Giant's win (*G*), Dodger's will be out of first place (*D*) and I will lose the bet (*B*). The expression in symbolic form is of the form (*A* → *B*), where the premise *A* and conclusion *B* are further rewritten as follows:

$$((P \vee C) \wedge G) \rightarrow (D \wedge G)$$

Example 2.3: Consider the sentence 'If labour or management is stubborn, then the strike will be settled iff the government obtains an injunction, but troops are not sent into the mills', which is conditional.

The elementary sentences are as follows: labour is stubborn (L), management is stubborn (M), strike will be settled (S), government obtains an injunction (G), and troops are sent into the mills (R). The logical expression is as follows:

$$(L \vee M) \rightarrow (S \leftrightarrow (G \wedge (\neg R)))$$

When the sentences involve many elementary sentences and connectives, it is imperative to use the truth table approach to determine the truth value of the entire expression.

Example 2.4: Consider the sentence 'If I wear a pink shirt, then I will pass the examination. I have worn a pink shirt. Therefore, I will pass the examination'. The sentence is true by observation. The Truth Table 2.7 is used to arrive at the validity of the sentence. The sentences are as follows: I wear a pink shirt (P) and I pass the examination (E). The expression can be shown in the following form ($A \rightarrow B$), where the premise A and conclusion B are further rewritten as follows:

$$((P \rightarrow E) \wedge P) \rightarrow E$$

The columns in Table 2.7 are numbered from left to right to avoid entering sub-expressions in the columns. The column numbers are used in the expressions. From Column 5, all entries are true, and hence, the sentence is valid.

Table 2.7 Truth table for $((P \rightarrow E) \wedge P) \rightarrow E$

P	E	$(P \rightarrow E)$	$(3) \wedge (1)$	$(4) \rightarrow (2)$
F	F	T	F	T
F	T	T	F	T
T	F	F	F	T
T	T	T	T	T

Example 2.5: Consider the expression $P \rightarrow (P \vee Q)$ and find its validity. The truth table corresponding to the expression is shown in Table 2.8. Further, the expression $(P \wedge (\neg P \wedge Q))$ is false for all the truth values of the propositional variables and is shown in Table 2.9.

Table 2.8 Truth table for $((P \rightarrow (P \vee Q))$

P	Q	$(P \vee Q)$	$(1) \rightarrow (3)$
F	F	F	T
F	T	T	T
T	F	T	T
T	T	T	T

Table 2.9 Truth table for $(P \wedge (\neg P \wedge Q))$

P	Q	$\neg P$	$(3) \wedge (2)$	$(1) \wedge (4)$
F	F	T	F	F
F	T	T	T	F
T	F	F	F	F
T	T	F	F	F

Column 5 of Table 2.9 contains all the entries as false and it shows that the given expression is always false.

On the basis of logic, an expression that is always true, irrespective of the truth values of the propositional variables, is called a 'tautology'. Likewise, an expression that is always false irrespective of the truth values of the propositional variables, is called a 'contradiction'. If we have a mix of truth values for a given expression, it is called a 'contingency'. The expression shown in Tables 2.7 and 2.8 are tautologies, and the expression in Table 2.9 is a contradiction. Consider the expression $Q \wedge (\neg R \rightarrow P)$ given in Table 2.10, involving the variables P, Q and R.

Table 2.10 Truth table for $Q \wedge (\neg R \rightarrow P)$

P	*Q*	*R*	$\neg R$	(4) → (1)	(2) ∧ (5)
F	F	F	T	F	F
F	F	T	F	T	F
F	T	F	T	F	F
F	T	T	F	T	T
T	F	F	T	T	F
T	F	T	F	T	F
T	T	F	T	T	T
T	T	T	F	T	T

The expression illustrated in Table 2.10 is an example of contingency. These expressions are governed by the laws of logic.

2.3 LAWS OF LOGIC

In logic, it is ascertained that the given two logical expressions are equivalent or the same. Here, the support of laws similar to other areas in mathematics is considered. The common laws include the following: law of double negation, DeMorgan's laws, commutative laws, associative laws, distributive laws, idempotent laws, identity laws, inverse laws, domination laws and absorption laws.

Example 2.6: Consider the two expressions $(\neg P \vee Q)$ and $(P \rightarrow Q)$. Now, the question is, are they equivalent? To find out the similarity, the truth table is given in Table 2.11. From Table 2.11, it is evident that the columns corresponding to the two expressions, namely $(\neg P \vee Q)$ and $(P \rightarrow Q)$ are alike. Since the values are alike, we call them logically equivalent to each other, and one can be replaced by the other in expressions.

Table 2.11 Truth table for $(\neg P \vee Q)$ and $(P \rightarrow Q)$

P	*Q*	$\neg P$	(3) ∨ (2)	(1) → (2)
F	F	T	T	T
F	T	T	T	T
T	F	F	F	F
T	T	F	T	T

Let *P*, Q and *R* be the propositional variables. *T* and *F* represent the tautology and contradiction, respectively.

1.	$\neg\neg P \leftrightarrow P$	Law of double negation
2.	$\neg(P \vee Q) \leftrightarrow \neg P \wedge \neg Q$ $\neg(P \wedge Q) \leftrightarrow \neg P \vee \neg Q$	DeMorgan's laws
3.	$(P \vee Q) \leftrightarrow (Q \vee P)$ $(P \wedge Q) \leftrightarrow (Q \wedge P)$	Commutative laws
4.	$P \wedge (Q \wedge R) \leftrightarrow (P \wedge Q) \wedge R$ $P \vee (Q \vee R) \leftrightarrow (P \vee Q) \vee R$	Associative laws
5.	$P \wedge (Q \vee R) \leftrightarrow (P \wedge Q) \vee (P \wedge R)$ $P \vee (Q \wedge R) \leftrightarrow (P \vee Q) \wedge (P \vee R)$	Distributive laws
6.	$(P \wedge P) \leftrightarrow P$ $(P \vee P) \leftrightarrow P$	Idempotent laws
7.	$(P \vee F) \leftrightarrow P$ $(P \wedge T) \leftrightarrow P$	Identity laws
8.	$(P \vee \neg P) \leftrightarrow T$ $(P \wedge \neg P) \leftrightarrow F$	Inverse laws
9.	$(P \vee T) \leftrightarrow T$ $(P \wedge F) \leftrightarrow F$	Domination laws
10.	$P \vee (P \wedge Q) \leftrightarrow P$ $P \wedge (P \vee Q) \leftrightarrow P$	Absorption laws

Example 2.7: Prove the distributive laws using the $P \wedge (Q \vee R) \leftrightarrow (P \wedge Q) \vee (P \wedge R)$ and $P \vee (Q \wedge R) \leftrightarrow (P \vee Q) \wedge (P \vee R)$ truth table approach. The solution for $P \wedge (Q \vee R) \leftrightarrow (P \wedge Q) \vee (P \wedge R)$ is given in Table 2.12, in which Columns 5 and 8 are alike. Therefore, the expressions are logically equivalent. The proof for the second expression is left to the reader.

Table 2.12 Truth table for $P \wedge (Q \vee R) \leftrightarrow (P \wedge Q) \vee (P \wedge R)$

P	Q	R	(Q ∨ R)	(1) ∧ (4)	(P ∧ Q)	(P ∧ R)	(6) ∨ (7)
F	F	F	F	F	F	F	F
F	F	T	T	F	F	F	F
F	T	F	T	F	F	F	F
F	T	T	T	F	F	F	F
T	F	F	F	F	F	F	F
T	F	T	T	T	F	T	T
T	T	F	T	T	T	F	T
T	T	T	T	T	T	T	T

2.4 PRINCIPLE OF DUALITY

Consider a statement *S* containing only the connectives '∨' and '∧'. The dual of *S* denoted by S^d is the new statement obtained by interchanging '∨' with '∧' and '∧' with '∨' in the given statement. Each occurrence of *F* is changed to *T* and *vice versa*. Let *P* and Q be two elementary statements or sentences. The dual of *P* is *P*.

If the elementary sentence is negated ($\neg P$) then its dual is $\neg P$. The expressions ($P \wedge \neg Q$) and ($P \vee \neg Q$) are dual to each other. Consider the following expression:

$$P \wedge (Q \vee \neg R) \vee (Q \wedge F)$$

Its dual is $P \vee (Q \wedge \neg R) \wedge (Q \vee T)$, which is obtained by interchanging $\wedge$ and $\vee$ and the logical constant F with T. The principle of duality states that for s and t, two statements involving only connectives $\vee$ and $\wedge$, if $s \leftrightarrow t$ then $s^d \leftrightarrow t^d$. Consider the distributive laws and their duals.

$$P \wedge (Q \vee R) \leftrightarrow (P \wedge Q) \vee (P \wedge R)$$

Let s be $P \wedge (Q \vee R)$ and t be $(P \wedge Q) \vee (P \wedge R)$. The duals of s and t are s^d: $P \vee (Q \wedge R)$ and t^d: $(P \vee Q) \wedge (P \vee R)$. The validity can be examined using the truth table approach. Consider Table 2.13. The columns 5 and 8 are identical and hence $s^d \leftrightarrow t^d$.

Table 2.13 Truth table for $P \vee (Q \wedge R) \leftrightarrow (P \vee Q) \wedge (P \vee R)$

P	**Q**	**R**	**($Q \wedge R$)**	**(1) $\vee$ (4)**	**($P \vee Q$)**	**($P \vee R$)**	**(6) $\wedge$ (7)**
F	F	F	F	F	F	F	F
F	F	T	F	F	F	T	F
F	T	F	F	F	T	F	F
F	T	T	T	T	T	T	T
T	F	F	F	T	T	T	T
T	F	T	F	T	T	T	T
T	T	F	F	T	T	T	T
T	T	T	T	T	T	T	T

Example 2.8: Consider the two elementary statements P and Q where P: John is six feet tall and Q: Mary dances well. The implication $P \rightarrow Q$: If John is six feet tall, then Mary dances well. Suppose we want to negate this implication, that is, $\neg (P \rightarrow Q)$. This is read as 'It is not the case that if John is six-feet tall then Mary dances well'. We can use the laws of logic and rewrite the expression. We know that the two expressions ($P \rightarrow Q$) and ($\neg P \vee Q$) are logically equivalent. Therefore, we can write as follows:

$$(P \rightarrow Q) \leftrightarrow (\neg P \vee Q)$$
$$\neg (P \rightarrow Q) \leftrightarrow \neg (\neg P \vee Q)$$
$$\neg (P \rightarrow Q) \leftrightarrow (P \wedge \neg Q)$$

Therefore, $\neg (P \rightarrow Q)$ is read as 'John is six-feet tall and Mary does not dance well' or 'John is six-feet tall but Mary does not dance well'. Here, it can be observed that negation of implication does not start with the word 'if'. Consider the following expressions:

(i) $P \rightarrow Q$
(ii) $Q \rightarrow P$
(iii) $\neg P \rightarrow \neg Q$
(iv) $\neg Q \rightarrow \neg P$

It is possible to write truth tables for each of these expressions. Table 2.14 gives the truth values. From the table, it is evident that $(P \rightarrow Q) \leftrightarrow (\neg Q \rightarrow \neg P)$. Therefore, the expression ($\neg Q \rightarrow \neg P$) is called a contra-positive of the implication ($P \rightarrow Q$).

Table 2.14 Truth table for $(P \rightarrow Q)$, $(Q \rightarrow P)$, $(\neg P \rightarrow \neg Q)$ and $(\neg Q \rightarrow \neg P)$

P	$\neg P$	Q	$\neg Q$	$P \rightarrow Q$	$\neg P \rightarrow \neg Q$	$Q \rightarrow P$	$\neg Q \rightarrow \neg P$
F	T	F	T	T	T	T	T
F	T	T	F	T	F	F	T
T	F	F	T	F	T	T	F
T	F	T	F	T	T	T	T

Similarly, the expression $(\neg P \rightarrow \neg Q)$ is a contra-positive of $(Q \rightarrow P)$. The expression $(Q \rightarrow P)$ is called converse of $(P \rightarrow Q)$. The expression $(\neg P \rightarrow \neg Q)$ is called inverse of $(P \rightarrow Q)$. Consider the following statements:

1. If line A is parallel to B, then line B is parallel to A.
2. If the set A is equal to set B then the set B is equal to set A

In these cases, the converse of the statements is also true.

Example 2.9: Consider the month of April. The Thanksgiving day and Good Friday are consecutive days. Let the propositional variable P be 'Today is Thanksgiving' and Q be 'Tomorrow is Friday'.

1. $(P \rightarrow Q)$ means 'If today is Thanksgiving then tomorrow is Friday', implication is true.
2. $(\neg Q \rightarrow \neg P)$ means 'If tomorrow is not Friday then today in not Thanksgiving', contra-positive is also true.
3. $(Q \rightarrow P)$ means 'If tomorrow is Friday then today is Thanksgiving', converse is true.
4. $(\neg P \rightarrow \neg Q)$ means 'If today in not Thanksgiving then tomorrow is not Friday', the inverse is also true.

Consider the converse and inverse of an implication. For the other months, in addition to April, Thursday will be followed by Friday. On these days, the proposition P is false, but the statement Q is true. The expressions $(Q \rightarrow P)$ and $(\neg P \rightarrow \neg Q)$ are false. One has to be careful while dealing with converse and inverse of an implication. The laws of logic can be used in the simplification of given compound statements and simplified expressions can be tested for validity, as one is the equivalent of the other.

Example 2.10: Consider the expression $(P \vee Q) \wedge (\neg (\neg P \wedge Q))$ and simplify it. The laws are applied as under.

$(P \vee Q) \wedge (\neg (\neg P \wedge Q)) \leftrightarrow (P \vee Q) \wedge (\neg \neg P \vee \neg Q)$	Applying DeMorgan's law
$\leftrightarrow (P \vee Q) \wedge (P \vee \neg Q)$	Law of double negation
$\leftrightarrow [(P \vee Q) \wedge P] \vee [(P \vee Q) \wedge \neg Q]$	Distribution law
$\leftrightarrow [(P \wedge P) \vee (P \wedge Q)] \vee [(P \wedge \neg Q) \vee (Q \wedge \neg Q)]$	Expanded terms
$\leftrightarrow [P \vee (P \wedge Q)] \vee [(P \wedge \neg Q)]$	Absorption law
$\leftrightarrow P \vee (P \wedge \neg Q)$	Absorption law
$\leftrightarrow P$	

Therefore, $(P \vee Q) \wedge (\neg (\neg P \wedge Q)) \leftrightarrow P$

Example 2.11: Consider the expression $((P \vee Q) \wedge (P \vee \neg Q)) \vee Q$ and the reasons at each step in the simplifications of compound statement.

$((P \vee Q) \wedge (P \vee \neg Q)) \vee Q \leftrightarrow ((P \vee (Q \wedge \neg Q)) \vee Q$	Distribution law
$\leftrightarrow (P \vee F) \vee Q$	Inverse law
$\leftrightarrow (P \vee Q)$	Identity law

Example 2.12: Consider the expression $[(P \to Q) \wedge (\neg Q \wedge (R \vee \neg Q))]$ and the reasons at each step in the simplification of the compound statement.

$[(P \to Q) \wedge (\neg Q \wedge (R \vee \neg Q))] \leftrightarrow (P \to Q) \wedge \neg Q$	Absorption law
$\leftrightarrow (\neg P \vee Q) \wedge \neg Q$	Substitution for implication
$\leftrightarrow \neg Q \wedge (\neg P \vee Q)$	Commutative law
$\leftrightarrow (\neg Q \wedge \neg P) \vee (\neg Q \wedge Q)$	Distributive law
$\leftrightarrow (\neg Q \wedge \neg P) \vee F$	Inverse law
$\leftrightarrow \neg (Q \vee P)$	DeMorgan's law

2.5 RULE OF INFERENCE

The validity of a given compound statement of the form $P \to Q$ is tested with truth tables. If the compound statement has N propositional variables, then the entries in the table become 2^N. This method becomes more clumsy and prone to errors. The analysis of a compound statement depends upon the logical relationships among the elementary statements. Each elementary statement can be assigned a truth value. The compound statement is composed of these elementary statements joined by connectives and receives the truth value in accordance with the truth tables of different connectives. Therefore, one can develop a technique called 'rules of inference'. Using these techniques, one can effectively arrive at a conclusion taking into consideration only those premises which are true. It is necessary to logically arrive at the conclusion from the premises.

$P_1 \wedge P_2 \wedge P_3 \wedge P_4 \dots P_n \to Q$, where each P_i is a premise.

There are commonly used rules of inference, namely *Modus Ponens* (also called rule of detachment), *Modus Tollens*, law of the syllogism, resolution principle, rule of conjunction, rule of disjunctive syllogism, rule of contradiction, rule of conjunctive simplification, rule of disjunctive amplification, rule for proof by cases, rule of constructive dilemma and rule of destructive dilemma. They are discussed as follows.

2.5.1 Modus Ponens

Consider the statement 'If I wear a pink shirt then I will pass the examination. I have worn a pink shirt. Therefore, I will pass the examination'. Let P be the statement 'I wear a pink shirt', Q be the statement 'I will pass the examination'. The compound statement is put in symbolic form as $[(P \to Q) \wedge P] \to Q$, where the premise is $[(P \to Q) \wedge Q]$ and the conclusion is Q. The premise comprises of two sub-premises, namely $(P \to Q)$ and P. The argument can be written as follows:

$$\begin{array}{l} (P \to Q) \\ P \\ \hline \therefore\ Q \end{array}$$

The expression $[(P \to Q) \wedge P] \to Q$ is a tautology. Modus Ponens is also called the 'rule of detachment'. The word 'modus ponens' is in Latin and means 'the method of inference'. Consider another statement. 'If I am in Hubballi, then I am in Karnataka'. 'I am in Hubballi. Therefore, I am in Karnataka'. This is also a perfect example for Modus Ponens. Consider the Example 2.12.

Example 2.13: Show that $[((P \vee Q) \to T) \wedge (P \vee Q)] \to T$ is a tautology using a truth table. The solution is given in Table 2.15. The last column contains all True values.

Table 2.15 Truth table for $[((P \vee Q) \rightarrow T) \wedge (P \vee Q)] \rightarrow T$

P	Q	T	$(P \vee Q)$	$(P \vee Q) \rightarrow T$	$(3) \wedge (4)$	$(6) \rightarrow T$
F	F	F	F	T	F	T
F	F	T	F	T	F	T
F	T	F	T	F	F	T
F	T	T	T	T	T	T
T	F	F	T	F	F	T
T	F	T	T	T	T	T
T	T	F	T	F	F	T
T	T	T	T	T	T	T

2.5.2 Modus Tollens

Consider the expression $[(P \rightarrow Q) \wedge \neg Q] \rightarrow \neg P$. It is called Modus Tollens. The word 'modus tollens' means a 'method of denying'. Consider the statement 'If I wear a pink shirt (P) then I pass the examination (Q)'. I have not passed the examination ($\neg Q$). Therefore, I have not worn the pink shirt ($\neg P$).

$$\begin{array}{ll} & (P \rightarrow Q) \\ & \neg Q \\ \hline \therefore & \neg P \end{array}$$

We know that $(P \rightarrow Q)$ is logically equivalent to $(\neg Q \rightarrow \neg P)$. One can verify this using the truth table.

2.5.3 Law of Syllogism

Consider the expression $((P \rightarrow Q) \wedge (Q \rightarrow R)) \rightarrow (P \rightarrow R)$; this is known as 'law of syllogism'. Here, P, Q and R are the individual statements.

$$\begin{array}{ll} & (P \rightarrow Q) \\ & (Q \rightarrow R) \\ \therefore & (P \rightarrow R) \end{array}$$

The situation of syllogism is common in real world. Consider the statement 'If I study hard (P), then I pass the examination (Q). If I pass the examination (Q), then I get a good job (R)'. From this, we can infer that 'If I study hard then I get a good job'. This is also called 'chain rule'.

Example 2.14: Consider the following statements and argument based on these statements.

1. Kamala is preparing food (K).
2. If Kamala is preparing food then Kamala is not going to school ($\neg S$).
3. If Kamala is not going to school then her father does not make her take the examination ($\neg E$).
4. Therefore, Kamala's father does not make her take the examination.

The above argument is expressed using the following expression. We can use the law of syllogism.

$$(K \wedge (K \rightarrow \neg S) \wedge (\neg S \rightarrow \neg E)) \rightarrow \neg E$$

The sub-expression $(K\to \neg S) \wedge (\neg S \to \neg E)$ is logically equivalent to $(K \to \neg E)$. The above expression reduces to $(K \wedge (K \to \neg E))$. Using Modus Ponens, it is logically equivalent to $\neg E$. Modus Tollens, Modus Ponens and the law of syllogism is very useful in drawing inference.

$$K$$
$$(K\to \neg S)$$
$$(\neg S \to \neg E) \therefore \neg E$$

2.5.4 Resolution Principle

Resolution principle is also known as 'Rule of Disjunctive Syllogism'. Consider the following expression.

$$((P \vee Q) \wedge \neg P) \to Q$$

In the expression $(P \vee Q)$, P is not true. Therefore, Q is true. It is again useful in the inference process.

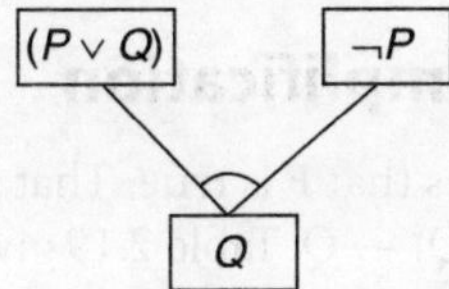

Truth values of this statement are shown in Table 2.16. The truth values in Column 6 are all True, and hence, it is a tautology.

Table 2.16 Truth table for $((P \vee Q) \wedge \neg P) \to Q$

P	**Q**	**¬P**	**(P ∨ Q)**	**(P ∨ Q)∧ ¬P**	**(5) → Q**
F	F	T	F	F	T
F	T	T	T	T	T
T	F	F	T	F	T
T	T	F	T	F	T

2.5.5 Rule of Conjunction

If the two statements P and Q are true, then the conjunction of P and Q is true, that is, $(P \wedge Q)$ is true. Table 2.17 gives the truth value of this statement. Column 4 contains all entries as true.

Table 2.17 Truth table for $(P \wedge Q) \to (P \wedge Q)$

P	**Q**	**(P ∧ Q)**	**(3) → (3)**
F	F	F	T
F	T	F	T
T	F	F	T
T	T	T	T

2.5.6 Rule of Contradiction

Consider the statement P and F as contradiction. The expression $(\neg P \rightarrow F) \rightarrow P$ is a tautology. The rule of contradiction is useful in the inference process. We use the proof by contradiction in testing the truth value of the given statement. Table 2.18 gives the truth values. Column 5 has all entries as True.

$$\therefore \quad \frac{\neg P \rightarrow F}{P}$$

Table 2.18 Truth table for $(\neg P \rightarrow F) \rightarrow P$

P	**¬P**	**F**	**¬P → F**	**(4) → P**
F	T	F	F	T
T	F	F	T	T

2.5.7 Rule of Conjunctive Simplification

The statements P and Q being true, implies that P is true. That is, $(P \wedge Q) \rightarrow P$. The statements P and Q being true, also implies that Q is true. $(P \wedge Q) \rightarrow Q$. Table 2.19 gives the truth values of this rule. Columns 4 and 5 are True values, and hence, both the expressions are tautologies.

$$\therefore \quad \frac{(P \wedge Q)}{P} \qquad \therefore \quad \frac{(P \wedge Q)}{Q}$$

Table 2.19 Truth table for $(P \wedge Q) \rightarrow P$

P	**Q**	**(P ∧ Q)**	**(3) → P**	**(3) → Q**
F	F	F	T	T
F	T	F	T	T
T	F	F	T	T
T	T	T	T	T

2.5.8 Rule of Disjunctive Amplification

The statement P be true, and it implies that $(P \vee Q)$ is true. This is given as follows and also supported by the Truth Table 2.20. The fourth column has all the values as True.

Table 2.20 Truth table for $P \rightarrow (P \vee Q)$

P	**Q**	**(P ∨ Q)**	**P → (P ∨ Q)**	**Q → (P ∨ Q)**
F	F	F	T	T
F	T	T	T	T
T	F	T	T	T
T	T	T	T	T

2.5.9 Rule for Proof by Cases

Let P, Q and R be the elementary statements. If $P \rightarrow R$ and $Q \rightarrow R$ be true expressions, then $(P \vee Q) \rightarrow R$ is True. Table 2.21 gives the truth values for this rule.

Table 2.21 Truth table for $((P \rightarrow R) \wedge (Q \rightarrow R)) \rightarrow (P \vee Q) \rightarrow R$

P	Q	$(P \vee Q)$	R	$P \rightarrow R$	$Q \rightarrow R$	(5) ∧ (6)	$(P \vee Q) \rightarrow R$	(7) → (8)
F	F	F	F	T	T	T	T	T
F	F	F	T	T	T	T	T	T
F	T	T	F	T	F	F	F	T
F	T	T	T	T	T	T	T	T
T	F	T	F	F	T	F	F	T
T	F	T	T	T	T	T	T	T
T	T	T	F	F	F	F	F	T
T	T	T	T	T	T	T	T	T

Columns 7 and 8 are identical, and hence, Column 8 has all true values. Therefore, the expression is a tautology.

$$
\begin{array}{l}
(P \rightarrow R) \\
(Q \rightarrow R) \\
\hline
\therefore \quad (P \vee Q) \rightarrow R
\end{array}
$$

2.5.10 Rule of Destructive Dilemma

Let P, Q, R and S be the elementary statements. If $P \rightarrow Q$ and $R \rightarrow S$ and $(\neg Q \vee \neg S)$ then $(\neg P \vee \neg R)$. This is illustrated by the Truth Table 2.22. Column 10 contains all entries True, and hence, the given expression is a tautology.

Table 2.22 Truth table for $((P \rightarrow Q) \wedge (R \rightarrow S) \wedge (\neg Q \vee \neg S)) \rightarrow (\neg P \vee \neg R)$

P	Q	R	S	$P \rightarrow Q$	$R \rightarrow S$	$\neg Q \vee \neg S$	(5) ∧ (6) ∧ (7)	$\neg P \vee \neg R$	(8) → (9)
F	F	F	F	T	T	T	T	T	T
F	F	F	T	T	T	T	T	T	T
F	F	T	F	T	F	T	F	T	T
F	F	T	T	T	T	T	T	T	T
F	T	F	F	T	T	T	T	T	T
F	T	T	T	T	T	F	F	T	T
F	T	T	F	T	F	T	F	T	T
F	T	T	T	T	T	F	F	T	T
T	F	F	F	F	T	T	F	T	T
T	F	F	T	F	T	T	F	T	T

Continued

Continued

P	**Q**	**R**	**S**	**P→Q**	**R→S**	**¬Q∨¬S**	**(5)∧(6)∧(7)**	**¬P∨¬R**	**(8)→(9)**
T	F	T	F	F	F	T	F	F	T
T	F	T	T	F	T	T	F	F	T
T	T	F	F	T	T	T	T	T	T
T	T	F	T	T	T	F	F	T.	T
T	T	T	F	T	F	T	F	F	T
T	T	T	T	T	T	F	F	F	T

Therefore, the following is the argument.

$$\begin{array}{ll} & P \to Q \\ & R \to S \\ & \neg Q \vee \neg S \\ \hline \therefore & \neg P \vee \neg R \end{array}$$

2.5.11 Rule of Constructive Dilemma

Consider the elementary statements P, Q, R and S. If $P \to Q$ and $R \to S$ and $P \vee R$, then $Q \vee S$. Table 2.23 gives the truth values. It is a tautology.

Table 2.23 Truth table for $((P \to Q) \wedge (R \to S) \wedge (P \vee R)) \to (Q \vee S)$

P	**Q**	**R**	**S**	**P→Q**	**R→S**	**P∨R**	**(5)∧(6)∧(7)**	**Q∨S**	**(8)→(9)**
F	F	F	F	T	T	F	F	F	T
F	F	F	T	T	T	F	F	T	T
F	F	T	F	T	F	T	F	F	T
F	F	T	T	T	T	T	T	T	T
F	T	F	F	T	T	F	F	T	T
F	T	T	T	T	T	T	T	T	T
F	T	T	F	T	F	T	F	T	T
F	T	T	T	T	T	T	T	T	T
T	F	F	F	F	T	T	F	F	T
T	F	F	T	F	T	T	F	T	T
T	F	T	F	F	F	T	F	F	T
T	F	T	T	F	T	T	F	T	T
T	T	F	F	T	T	T	T	T	T
T	T	F	T	T	T	T	T	T	T
T	T	T	F	T	F	T	F	T	T
T	T	T	T	T	T	T	T	T	T

Therefore, the following is the argument.

$$\begin{array}{c} P \to Q \\ R \to S \\ P \vee R \\ \hline Q \vee S \end{array}$$

$\therefore$

Rules for simplification of the given logical expression are given in Table 2.24 for reference, where *P*, *Q*, *R* and *S* are the elementary statements.

Table 2.24 Rules for simplification

Sl. No	Rule of simplification/Inference	Remarks
1	$[(P \to Q) \wedge P] \to Q$	Modus Ponens or rule of detachment
2	$[(P \to Q) \wedge \neg Q] \to \neg P$	Modus Tollens
3	$((P \to Q) \wedge (Q \to R)) \to (P \to R)$	Law of syllogism or chain rule
4	$((P \vee Q) \wedge \neg P) \to Q$	Resolution principle or rule of disjunctive syllogism
5	$(P \wedge Q) \to (P \wedge Q)$	Rule of conjunction
6	$(\neg P \to F) \to P$	Rule of contradiction
7	$(P \wedge Q) \to P.$	Rule of conjunctive simplification
8	$P \to (P \vee Q)$	Rule of disjunctive amplification
9	$((P \to R) \wedge (Q \to R)) \to (P \vee Q) \to R$	Rule of proof by cases
10	$((P \to Q) \wedge (R \to S) \wedge (\neg Q \vee \neg S)) \to (\neg P \vee \neg R)$	Rule of destructive dilemma
11	$((P \to Q) \wedge (R \to S) \wedge (P \vee R)) \to (Q \vee S)$	Rule of constructive dilemma

2.6 METHOD OF RESOLUTION

An approach called 'method of resolution' is used in determining the truth value of logical expression. In this step, the following steps are used.

1. Bring the logical expression in the form $P \to Q$.
2. Negate the expression, that is, $\neg(P \to Q) \leftrightarrow (P \wedge \neg Q)$.
3. Bring the expression to conjunctive normal form (CNF).
4. List the disjuncts.
5. Attempt the contradiction using the disjuncts.

Logical expression is said to be in CNF when it has the conjunction of disjunctions of propositional variables negated or non-negated. For example, P, $(P \wedge Q)$, $(P \vee Q) \wedge (R \vee S)$, $\neg P$, $(\neg P \wedge Q)$, etc. are in CNF. CNF is analogous to the product of sums form of Boolean Algebra. Similarly, the disjunctive normal form (DNF) has disjunctions of conjunctions of propositional variables negated or non-negated. For example, P, $(P \vee P)$, $(P \wedge = Q) \vee (R \wedge S)$, $\neg P$, $(\neg P \vee Q)$, etc. The CNF is used in the inference process.

Example 2.15: Consider the Modus Ponens and its truth value using the method of resolution, which uses the resolution principle.

$$((P \to Q) \wedge P) \to Q$$

Procedure:

Step 1: The expression is in the form $P \to Q$
Step 2: Negate the expression and we get the following expression.

$$((P \to Q) \wedge P) \wedge \neg Q$$

The implication symbol needs to be resolved, and thus, we get the following expression.

$$(\neg P \vee Q) \wedge P \wedge \neg Q$$

Step 3: The expression is in CNF.
Step 4: The following are the disjunctions.

$$(\neg P \vee Q), P \text{ and } \neg Q$$

Step 5: Attempting at contradiction

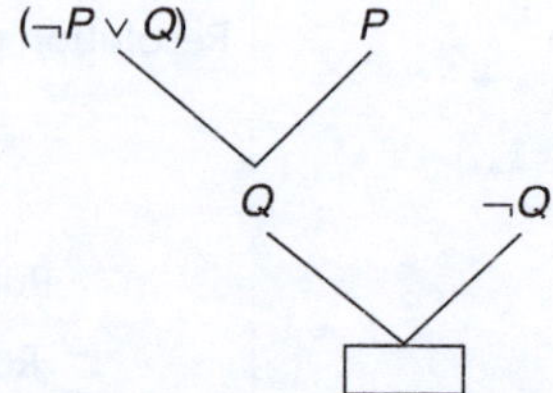

Consider one disjunction at a time to obtain the contradiction. Therefore, the expression is true. This is because of negation of the expression in the beginning of the method. If we do not get the contradiction, then the given expression is incorrect.

Example 2.16: Consider the following expression and find its truth value using the method of resolution, which uses the resolution principle.

$$((P \to Q) \wedge P) \to \neg Q$$

Step 1: The expression is in the form $P \to Q$
Step 2: Negate the expression and we get the following expression.

$$((P \to Q) \wedge P) \wedge Q$$

The implication symbol needs to be resolved, and thus, we get the following expression.

$$(\neg P \vee Q) \wedge P \wedge Q$$

Step 3: The expression is in CNF.
Step 4: The following are the disjuncts.

$$(\neg P \vee Q), P, \text{ and } Q$$

Step 5: Attempting contradiction

One disjunct is considered at a time. We have not got the contradiction, and hence, the given expression is incorrect.

2.7 PREDICATE LOGIC

Consider the statement 'All Americans own a car. John is an American and hence he owns a car'. In symbolising this sentence, the power of propositional logic becomes inadequate, as we need to use one propositional variable for each citizen of America. We need to use a large number of variables. This is impractical. Consider another sentence, 'There exists one wise student in the class'. We need to quantify the words like *All, There exist, Every,* and *Some,* which are used in our day-to-day conversation. Consider another statement '$x+2=4$', whose truth value depends upon the value of x and x is a variable. Consider the statement 'John owns a car', in which the sub-string 'owns a car' is called the predicate. Predicate resembles a function and returns either true or false. The statement is symbolised as follows:

Owns_car(John), Owns_car(x)

In the above expression, owns_car is a predicate and John is a constant. The x is a variable, which is assigned the value John. Consider some more statements and their equivalent expressions.

1. Every rational number is a real number. For every x, $Q(x) \rightarrow R(x)$ where $Q(x) - x$ is a rational number and $R(x) - x$ is a real number.
2. Some real numbers are rational.
   For some x, $Q(x) \wedge R(x)$
3. x is married to y.
   Married to (x, y) or $M(x, y)$
4. Human beings are mortal. For every x, $H(x) \rightarrow R(x)$ where $H(x) - x$ is human being and $R(x) - x$ is mortal.
5. All the students are wise. For every y, $W(y)$
6. There exists a good picnic place. For some x, $G(x)$
7. x is less than y, $L(x, y)$

2.7.1 Quantifiers

The words All, There exist, Every and Some quantify things in the world. In predicate logic, symbols are used to represent these quantifiers. $\forall x$ and $\exists x$ represent universal and existential quantifiers, respectively. $\forall x$ is read as 'For all x' and 'For every x'. $\exists x$ is read as 'For some x' and 'There exist x'. The statements (i) to (vii) are further rewritten in the following manner. The quantifiers are said to be bounded by the variable.

1. $\forall x\ Q(x) \rightarrow R(x)$
2. $\exists x\ Q(x) \wedge R(x)$
3. $M(x, y)$
4. $\forall x\ H(z) \rightarrow T(z)$
5. $\forall x\ W(y)$
6. $\exists x\ G(x)$
7. $L(x, y)$

The quantifiers' positions are important while converting the given statements to predicate logic. For example, consider the following two logical expressions and the quantifiers are read from left to right.

1. $\forall x\ \exists y\ M(x, y)$
2. $\exists y\ \forall x\ M(x, y)$

The expression (i) means, 'for all x, there exists y such that x is married to y'. If x is a member from a set of male candidates and y belongs to a set of female candidates, then the logical expression means that every male has a counterpart, which makes some sense. On the other hand, the expression (ii) means there is a female for all x such that x is married to y, which does not make any sense. This is further illustrated by another example.

Consider the arithmetic expression $x + y = 10$. Let $P(x, y)$ represent this arithmetic expression and is read as 'x plus y is equal to 10' Let x and y be drawn from an integer set.

1. $\forall x\ \exists y\ P(x, y)$ is read as 'for every integer x, there exists an integer y such that $x + y$ is equal to 10'. The statement makes sense as once x is selected, the integer y corresponds to $(10 - x)$. On the other hand:
2. $\exists y \forall x\ P(x, y)$ is read as 'there exists an integer y for every x such that $x + y$ is equal to 10'. The statement does not make sense and is considered false. This is because once an integer y is selected, the only value that x assumes is $(10 - y)$. If we assume that the statement is true, it means that all x are same and equal to $(10 - y)$.
3. From the aforesaid argument, it is evident that the expressions (1) and (2) are not equivalent. That is $\forall x\ \exists y\ P(x, y) \neq \exists y \forall x\ P(x, y)$.

2.7.2 Negation of Statements with Quantifiers

Consider the expression $\forall y\ W(y)$, where the predicate $W(y)$ represents the statement 'x is wise in the class'. The expression is read as 'everyone in the class is wise'. Consider the negation of the expression $\forall y\ W(y)$. The expression is written as under.

$$\neg\forall y\ W(y)$$

This is equivalent to reading as 'not everyone is wise'. This means that 'there exists one who is not wise'. That is, $\exists y\, W(y)$. Therefore, $\neg\forall y\ W(y) \leftrightarrow \exists y \neg W(y)$. Similarly, consider the following expression.

$$\neg\exists y\, W(y).$$

This is read as 'not there exists one who is wise'. This equivalent to saying 'everyone is not wise' and is written as $\forall y\ \neg W(y)$. Therefore, $\neg\exists y\, W(y). \leftrightarrow \forall y\ \neg W(y)$. Table 2.25 gives the rules for negating the statements with quantifiers.

Table 2.25 Negation of Quantifiers

$\neg\forall y\ W(y) \leftrightarrow \exists y \neg W(y).$
$\neg\exists y W(y). \leftrightarrow\ \forall y\ \neg W(y)$
$\neg[\forall y \neg\ W(y)] \leftrightarrow\ \exists y \neg\neg W(y) \leftrightarrow\ \exists y W(y)$
$\neg[\exists y \neg W(y)]. \leftrightarrow\ \forall y\ \neg\neg W(y) \leftrightarrow\ \forall y\ W(y)$
$. \leftrightarrow \neg W(y). \leftrightarrow \neg W(y)$

The logical equivalence and implications of statements and quantifiers in addition to other logical operators, namely, and $\vee$ are given as in Table 2.26.

Table 2.26 Logical Implications and Equivalences

$\exists y\,[W(y) \wedge P(y)] \rightarrow [\exists y\, W(y) \wedge \exists y\, P(y)]$
$\exists y\,[W(y) \vee P(y)] \leftrightarrow [\exists y\, W(y) \vee \exists y\, P(y)]$
$\forall y\,[W(y) \wedge P(y)] \leftrightarrow [\forall y\, W(y) \wedge \forall y\, P(y)]$
$\forall y\,[W(y) \vee P(y)] \rightarrow [\forall y\, W(y) \vee \forall y\, P(y)]$
$\wedge P(y)] \rightarrow \wedge P(y)] \vee P(y)] \leftrightarrow \vee P(y)] \wedge P(y)] \wedge P(y)] \vee P(y)]$

Example 2.17: Convert the associative law for addition of integers $x + (y + z) = (x + y) + z$ using predicate logic.

Using universal quantifiers, we can write the logical expression as $\forall x\ \forall y\ \ (\forall z\ (x + (y + z) = (x + y) + z)$. The same is written as $\forall y\ \forall x\ \forall z\ (x + (y + z) = (x + y) + z)$, $\forall z\ \forall y\ \forall x\ (x + (y + z) = (x + y) + z), x + (y + z) = (x + y) + z)$ and so on. There will be 3! ways for writing this expression and all are equivalent. In some books, it is written as follows, where $\forall y\ \forall x\ \forall z$ is written in short form as $\forall x, y, z\forall$.

$$\forall x, y, z\ (x + (y + z) = (x + y) + z)$$

Similarly, when multiple existential quantifiers are used, it is written as $\exists y, x, z$. These short forms help in representing the expression neatly.

Example 2.18: Consider the statement 'there exists integers x and y such that $x + y = 10$'. The symbolic form of this statement uses the predicate say plus (x, y) for the addition $x + y$ and equal $(z, 10)$ for z equals 10. The symbolic form of the given statement is as given under.
$\exists x, y\ (x + y = 10)$ or $\exists x, y$ equal (plus(x, y), 10)

Example 2.19: Consider the predicate $P(x)$: $x^2 = 2x$ and is drawn from the set of integers. Let us determine whether each of the following statements is true or false:

(i) $P(0)$	(ii) $P(1)$	(iii) $P(2)$
(iv) $P(-2)$	(v) $\exists x\ P(x)$	(vi) $\forall x\ P(x)$

The value of $P(0)$: $0^2 = 2 \times 0 = 0$ and hence, it is true. The value of $P(1)$: $1^2 = 2 \times 1$, which is false. The value of $P(2)$: $2^2 = 2 \times 2 = 4$ and hence it is true. The value of $P(-2)$: $(-2)^2 = 4 = 2 \times (-2) = -4$, which is false. The value of $\exists x\ P(x)$ is true because $P(0)$ and $P(2)$ are true. $\forall x\ P(x)$ is false because $P(1)$ is not true. Therefore, (i), (iii) and (v) are true and others are false.

Example 2.20: Consider the following statements and identify the bound variables and free variables in them:

(i) $\forall x\ \exists z\ [\cos(x + y) = \sin(z - x)]$
(ii) $\exists x, y\ [x^2 - y^2 = z]$

In (i), y and z are bound variables and x is free as it is not bound by any quantifier. In (ii), the variables x and y are bound; the variable x is free.

Example 2.21: Consider the following statements for negation and simplification:

(i) $\exists x\ (P(x) \vee Q(x))$

(ii) $\forall y\ (P(y) \wedge \neg Q(y))$

(iii) $\forall y\ (P(y) \to Q(y))$

(iv) $\exists x\ ((P(x) \vee Q(x)) \to R(x))$

The negation of (i) is written as $\neg\exists x\ (P(x) \vee Q(x))$ and is logically equivalent to $\forall x \neg(P(x) \vee Q(x))$. Further, it is written as $\forall x\ (\neg P(x) \wedge \neg Q(x))$. Therefore, $\neg\exists x\ (P(x) \vee Q(x)) \leftrightarrow \forall x\ (\neg P(x) \wedge \neg Q(x))$.

The negation of (ii) is written as $\neg\forall y\ (P(y) \wedge \neg Q(y))$ and is logically equivalent to $\exists x\ \neg(P(y) \wedge \neg Q(y))$. Further, it is simplified as $\exists x\ (\neg P(y) \vee Q(y))$. Therefore, $\neg\forall y\ (P(y) \wedge \neg Q(y)) \leftrightarrow \exists x\ (\neg P(y) \vee Q(y))$.

The negation of (iii) is written as $\neg\forall y\ (P(y) \to Q(y))$ and is logically equivalent to $\exists x\ \neg(P(y) \to Q(y))$. Further, it is simplified as $\exists x\ (P(y) \wedge \neg Q(y))$. Therefore, $\neg\forall y\ (P(y) \to Q(y)) \leftrightarrow \exists x\ (P(y) \wedge \neg Q(y))$. We have used the logical equivalence: $\neg\ (A \to B) \leftrightarrow \neg\ (\neg A \vee B) \leftrightarrow (A \vee \neg B)$.

The negation of (iv) is written as $\neg\exists x\ ((P(x) \vee Q(x)) \to R(x))$ and is logically equivalent to $\forall x \neg((P(x) \vee Q(x) \to R(x))$. Further, it is written as $\forall x\ ((P(x) \vee Q(x)) \wedge \neg R(x))$. Therefore, $\neg\exists x\ ((P(x) \vee Q(x)) \to R(x)) \leftrightarrow \forall x\ ((P(x) \vee Q(x)) \wedge \neg R(x))$.

2.7.3 Proving the Logical Expressions

The logical identities used for sentential or propositional logic hold good for predicate logic, except that the quantifiers need to be eliminated, and the predicates variables be substituted with constants. For example, consider $\forall x\ W(x)$, which means 'everyone is wise'. This being the case, one can infer W (John), which means 'John is wise'. The constant 'John' when substituted, the symbolic expression, it is similar to an expression in propositional logic. The laws discussed in the earlier sections hold good in the predicate logic. This is also called rule of universal specification in some books. If we say that $\exists x\ P(x)$ then P (John) is also true.

Example 2.22: Consider the statement 'All Americans own a car. John is an American, and therefore, he owns a car'. Let $A(x)$: x is an American and $C(x)$: x owns a car be the predicates. Let $A(J)$ and $C(J)$ represent the predicates 'John is an American' and 'John owns a car', respectively. The statement is symbolised as under:

$$\begin{array}{lr} & \forall x\ A(x)) \to C(x) \\ & A(J) \\ \hline \therefore & C(J) \end{array}$$

The complete symbolic expression is $\forall x\ (A(x)) \to C(x) \wedge A(J)) \to C(J)$. The validity of this statement can be made as given under.

1. $\forall x\ A(x)) \to C(x)$ is a premise
2. $A(J)$ is a premise
3. $A(J) \to C(J)$ is obtained by the rule of universal specification.
4. Therefore it is $C(J)$ by Modus Ponens.

Hence, the statement is true.

Example 2.23: Consider the statement 'No child is enrolled for work in hotels. Kumar is enrolled for work in a hotel. Therefore, Kumar is not a child'. Let $C(x)$: x is a child. $E(x)$: x is enrolled for work in a hotel. Using these predicates, the above statement is written in symbolic form as under:

$$\begin{array}{r} \forall x\, C(x)) \rightarrow \neg E(x) \\ E(k) \\ \hline \therefore \quad \neg C(k) \end{array}$$

Therefore, the complete expression is written as $\forall x\, (C(x)) \rightarrow \neg E(x) \wedge E(x)) \rightarrow \neg C(x)$. The validity of this statement can be made as given under.

1. $\forall x\, C(x) \rightarrow \neg E(x)$ is a premise
2. $E(k)$ is a premise
3. $C(k) \rightarrow \neg E(k)$ is obtained by the rule of universal specification
4. $E(k) \rightarrow \neg C(k)$ This is because $C(k) \rightarrow \neg E(k) \leftrightarrow E(k) \rightarrow \neg C(k)$
5. $\therefore \neg C(k)$ Hence, the statement is true

Example 2.24: Giving reasons establish the validity of the following arguments:

$$\begin{array}{c} \forall\, x\, P(x) \vee Q(x) \\ \forall x \neg P(x) \wedge Q(x) \rightarrow R(x) \\ \hline \end{array}$$

(i) $\therefore \forall x \neg R(x) \rightarrow P(x)$ $\forall x\, P(x) \vee Q(x)$	is a premise
(ii) $P(c) \vee Q(c)$	is obtained by rule of universal specification and c is constant
(iii) $\forall x \neg P(x) \wedge Q(x) \rightarrow R(x)$	is a premise
(iv) $\neg P(c) \wedge Q(c) \rightarrow R(c)$	is obtained by rule of universal specification
(v) $\neg R(c) \rightarrow \neg(\neg P(c) \wedge Q(c))$	This is because $(P \rightarrow Q) \leftrightarrow (\neg Q \rightarrow \neg P)$
(vi) $\neg R(c)$	Premise is assumed
(vii) $P(c) \vee \neg Q(c)$	Modus Ponens is used
(viii) $(P(c) \vee Q(c)) \wedge (P(c) \vee \neg Q(c))$	Rule of conjunction
(ix) $(P(c) \vee (Q(c)) \wedge \neg Q(c))$	Distributive law is used
(x) $P(c)$	$P(c) \vee \text{False} \leftrightarrow P(c)$
(xi) $\therefore \forall x \neg R(x) \rightarrow P(x)$	Generalisation

Hence, the statement is valid. The rules stated in the Section 2.3 are to be remembered and used in the simplification of expressions.

ADDITIONAL SOLVED EXAMPLES

Example W2.1: Show that S is a valid conclusion, from the following premises.

1) $(P \rightarrow \neg Q)$, 2) $(Q \vee R)$, 3) $(\neg S \rightarrow P)$ and 4) $\neg R$.

Solution:

Consider the premises (1) and (3)

$$(P \rightarrow \neg Q) \wedge (\neg S \rightarrow P).$$

Using the chain rule, we obtain the expression

$(\neg S \rightarrow \neg Q)$. Let us call it expression (5).

Using expressions (5) and (2),

$$(\neg S \rightarrow \neg Q) \wedge (Q \vee R)$$

By applying the resolution principle, we get $(S \vee R)$.

Let us call it expression (6).
Using (6) and (4), we obtain, by the resolution principle,

$$((S \vee R) \wedge \neg R) \rightarrow S.$$

Hence S is a valid conclusion.

Example W2.2: Show that t is a valid conclusion, from the following premises.

1) $(P \rightarrow Q)$, 2) $(Q \rightarrow R)$, 3) $(R \rightarrow S)$, 4) $(P \vee t)$ and 5) $\neg S$.

Solution:
$(P \rightarrow Q) \wedge (Q \rightarrow R)$. Using the chain rule, we obtain the expression

$(P \rightarrow R)$. Let us call it expression (6).

From (6) and (3) we have, $(P \rightarrow R) \wedge (R \rightarrow S)$.
Using the chain rule, we obtain the expression

$(P \rightarrow S)$. Let us call this (7).

From (7) and (4) we have, $(P \rightarrow S) \wedge (P \vee t)$
Using resolution principle, we obtain,

$(S \vee t)$, Let us call this (8)

From (8) and (4), $(S \vee t) \wedge \neg S$.
So, using the resolution principle, we obtain t.
Hence t is a valid conclusion.

Example W2.3: Determine whether each of the following inferences is valid or invalid, and furnish some evidence using inference rules.

(a) If today is David's birthday, then today is January 24.
Today is January 24.
Therefore, today is David's birthday.

(b) The days are becoming longer.
The nights are becoming shorter if the days are becoming longer.
Therefore, nights are becoming shorter.

Solution:

(a) Let the propositions be denoted as follows:
D: Today is David's birthday.
J: Today is January 24.
The logical expression corresponding to the argument is obtained by the premises,

$$((D \rightarrow J) \wedge J) \rightarrow D$$

Since it is Modus Ponens, the argument is valid.

(b) Let the propositions be denoted as follows:
L: The days are becoming longer.
S: The days are becoming shorter.
The logical expression corresponding to the argument is obtained by the premises,

$$((L \rightarrow S) \wedge L) \rightarrow S$$

Since it is Modus Ponens, the argument is valid.

Example W2.4: Determine the validity of the following argument.
If I study well then I will pass.
If I do not go to a movie, then I will study.
I failed.
Therefore, I went to a movie.

Solution:
Let the following propositional variables denote different statements.
s: I study well, *m*: I go to a movie and *p*: I will pass
The argument is expressed as

1. $s \rightarrow p$
2. $\neg m \rightarrow s$
3. $\neg p$ Therefore, *m*.

From 1) and 2), using the chain rule, we get $\neg m \rightarrow p$. Let us call this expression 4) $\neg m \rightarrow p$. Using 4) and 3) and Modus Tollens, we get $\neg(\neg m)$, which is *m*.

Example W2.5: Give the steps and the reasons needed to establish the validity of the argument

(i) $\forall x\,(p(x) \vee q(x))$
(ii) $\forall x\,((\neg p(x) \wedge q(x)) \rightarrow r(x))$
(iii) Therefore, $\forall x\,(\neg r(x) \rightarrow p(x))$

Solution:

1. $\forall x\,(p(x) \vee q(x))$	It is a premise.
2. $p(c) \vee q(c)$	Rule of universal specification.
3. $\forall x\,((\neg p(x) \wedge q(x)) \rightarrow r(x))$	It is premise.
4. $\forall x\,((\neg p(c) \wedge q(c)) \rightarrow r(c))$	Rule of universal specification.
5. $\neg r(c) \rightarrow \neg(\neg p(c) \wedge q(c))$	
6. $\neg r(c) \rightarrow (p(c) \vee \neg q(c))$	Double negation.
7. $\neg r(c)$	Premise is assumed.
8. $p(c) \vee \neg q(c)$	6) and 7) with Modus Ponens.
9. $(p(c) \vee q(c)) \wedge (p(c) \vee \neg q(c))$	Resolution process.
10. $p(c) \vee ((q(c) \vee \neg q(c))$	Distributive law.
11. $p(c)$	

Therefore $\forall x\,(\neg r(x) \rightarrow p(x))$

SUMMARY

In this chapter, we have tried to give a fair insight of logic to the reader. The limitations of propositional logic and the advantages of first-order predicate logic in computer science are covered. Logic forms the backbone of any knowledge-based system or in other words, any intelligent systems. There are other kinds of logic in practice, namely modal logic, informal logic, temporal logic, philosophical logic and the like. The topic on logic is a foundation course for many fields in mathematics and also courses in computer science. Many different aspects of software engineering, from database management, through programming language design to artificial intelligence have benefitted from the discoveries of modern logic. It is also used in automated theorem proving.

EXERCISES

1. Determine which of the following statements are propositions:
   (a) The movie is thrilling.
   (b) Rama and Sita went to the forest.
   (c) Can you help me to solve this problem?
   (d) You please do it.
   (e) $X+5=10$
   (f) $X+Y=10$
   (g) I tell lie.
   (h) I am passing the buck to others.
2. Convert the following sentences to symbolic form:
   (a) I and my friend will attend the birthday party today at 8.00 PM.
   (b) Neither Karnataka nor Maharashtra won the inter-state cricket match.
   (c) If I am in Chennai, then I am in Tamil Nadu.
   (d) If I wear a pink shirt then I will pass in the examination. I have worn a pink shirt. Therefore, I will pass in the examination.
   (e) If Kumar studies then he will pass in discrete mathematics. Kumar studies if he does not play. Kumar has passed discrete mathematics. Therefore, Kumar has studied.
   (f) I go to shopping if and only if I get salary for the month.
3. Convert the following symbolic expressions to equivalent English sentences:
   (a) $(\neg P \vee \neg Q) \wedge \neg R$ (b) $(Q \wedge T) \rightarrow R$ (c) $(Q \rightarrow P) \wedge R$ (d) $(P \vee Q \vee T) \wedge \neg R$
   (e) $(P \leftrightarrow Q) \rightarrow \neg R$ (f) $(\neg P \wedge Q \wedge T \wedge \neg R)$
   where *P*: It is raining today. *Q*: It is cloudy today. *R*: I wear rain coat. *T*: It is thundering today.
4. Simplify the following symbolic expressions:
   (a) $((P \leftrightarrow Q) \rightarrow \neg R) \wedge ((\neg P \vee \neg Q) \wedge \neg R)$
   (b) $(\neg P \wedge Q \wedge T \wedge \neg R) \rightarrow ((P \leftrightarrow Q) \rightarrow \neg R)$
   (c) $((Q \rightarrow P) \wedge R) \leftrightarrow ((P \vee Q \vee T) \wedge \neg R)$
   (d) $((P \leftrightarrow Q) \rightarrow \neg R) \rightarrow ((Q \wedge T) \rightarrow R)$
5. Draw the truth table for the expressions corresponding to 2(d) and 2(e).
6. Prove that the statements given in 2(d) and 2(e) are valid using the resolution principle.
7. From the symbolic formulae given below, identify those (a) that are well-formed (b) that are tautologies and (c) that are contradictions.
   (a) $(P \rightarrow (P \vee Q))$
   (b) $((P \rightarrow (\neg P)) \rightarrow \neg P)$
   (c) $((\neg Q \wedge P) \wedge Q)$
   (d) $((P \wedge Q) \leftrightarrow P)$
8. Write the formulae given in Exercise (7) in conjunctive normal form.
9. If there is a chance of rain or Kamala loses her umbrella then Kamala will not sweep the house. If the temperature is more than 30°C, then it rarely rains. Today, the temperature is 30°C and Kamala uses her umbrella. Therefore, Kamala sweeps the house. Convert this argument into symbolic form and test its validity.
10. Either I shall go home or stay and have a drink. I shall not go home. Therefore, I shall stay and have a drink. (use *H*, *S*).
11. If the races are fixed or the gambling houses are crooked, then the tourist trade will decline and the town will suffer. If the tourist trade decreases, then the police force will be happy. The police force is never happy. Therefore, the races are not fixed. (use *R*, *H*, *D*, *S*, *P*).

12. Represent the following statements in predicate logic:
    (a) Every apple is red.
    (b) All human beings are mortal.
    (c) Some men are clever.
    (d) Some real numbers are rational.
    (e) All human beings are mortal. John is a human being. Therefore, John is mortal.
    (f) x is the father of the mother of y.
    (g) Some women lawyers are politicians.
    (h) Every city in India is clean.
13. Express the following symbolic statements in equivalent English sentences:
    (a) $\forall x\,(P(x) \to Q(x))$
    (b) $\forall x\,(S(x) \to T(x))$
    (c) $\forall x\,(P(x) \to \neg Q(x))$
    (d) $\exists x\, S(x) \wedge T(x)$ Assume the predicates as per your convenience.
14. Show that $P(x) \wedge (\forall x\, Q(x)) \to \exists x\, P(x) \wedge Q(x)$
15. Show that $\forall x\, P(x) \vee \forall x\, Q(x) \to \forall x\,(P(x) \vee Q(x))$
16. Show that $\neg\, \forall x \forall y\, P(x, y) \leftrightarrow \exists x \exists y\, P(x, Y)$
17. If $P(x, y)$, $Q(x, y)$ and $R(x, y)$ represent three statements, simplify the following expression.

$$\neg\, \forall x\, \exists y\, ((P(x, y) \wedge Q(x, y)) \to \exists x\, R(x, y))$$

18. Name the bound and free variables in the following statements:
    (a) $\forall y\, \exists x\, ((\cos\,(x + y) = \sin\,(z - x))$
    (b) $\exists x \exists y\, (x^2 + y^2 = z^2)$
    (c) $\forall x \forall y\, \forall z\, (x^2 + y^2 = z^2)$
    (d) $\forall x\, (\sin^2 x + \cos^2 x = 1)$
19. Infer from the following premises:
    (a) All law abiders pay the taxes. Ashok is a law abider. Therefore ….
    (b) All integers are rational numbers. The real number π is not a rational number. Therefore ….
    (c) Some students in the class are irregular. Therefore …
    (d) Everyone is wise in the class. Therefore …
20. Prove that the following logical expression is valid.

$$\begin{array}{ll} & \forall x\,(P(x) \to (Q(x) \wedge R(x))) \\ & \forall x\,(P(x) \wedge S(x)) \\ \hline \therefore & \forall x\,(Q(x) \wedge R(x)) \end{array}$$

21. Prove the validity of the statement 'No human beings are quadrupeds. All women are human beings. Therefore, no women are quadrupeds'.
22. Prove the validity of the statement 'Everyone who buys a ticket receives a prize. Therefore, if there are no prizes then nobody buys a ticket'.
23. Prove the validity of the statement 'Every rational number is a real number. There is a rational number. Therefore, there is a real number'.
24. Prove the validity of the statement 'Some persons admire Ashok. Some persons like no one who admires Ashok. Therefore, some persons are not liked by all the persons'.
25. Prove the validity of the statement 'Every member of the committee is wealthy and Republican. Some committee members are old. Therefore, there are some old Republicans'.

3

Mathematical Induction

In Chapter 2, we discussed the importance of logic in computer science. Logic is used in many topics in discrete mathematics. In this chapter, we deal with the topic of mathematical induction (MI), and here too, logic comes into play.

3.1 INTRODUCTION

Mathematical induction is a technique used to prove results over a set of natural numbers, for example Z^+, which is the set of positive integers. It is a process of inferring the truth for a general statement from particular cases. A statement may be true with reference to more than a hundred cases, yet we cannot conclude it to be true in general. MI is an important technique that can be used to make assertions, prove results about the complexity of algorithms, provide corrections to certain types of computer programs, establish theorems about graphs and trees, and so on. It is to be noted that MI is not a tool for deriving formulae or theorems.

In general, MI is used to prove a statement $S(n)$ as true over all positive integers n, where $S(n)$ is a predicate. A proof by MI has three parts, namely, the basis clause, an induction clause and an external clause.

(i) In the basis clause, we show that $S(n)$ is true for some initial values taken by 'n'.

(ii) In the inductive clause, we assume that the statement is true for some arbitrary k, $k \le n$, that is, $S(k)$ is true, and we try to prove that $S(k+1)$ is also true. We then generalise by induction that $S(n)$ is true for any 'n'.

(iii) In the external clause, we make the statement that the statement is true if and only if it is proved for the basis clause and the induction clause.

In some books, people use the word 'step' for clause, and hence we can read the above as the basis step, induction step and external step. The external step is just a statement, and we omit this step during the proof. We have two types of MI, namely, weak induction and strong induction.

In the case of strong induction, we prove the statement $S(n)$ for all the values of 'n' and then generalise. But in the case of weak induction, we prove for the basis and induction steps and then generalise. An analogy can be drawn with respect to temperature measurement in Bangalore city. We measure the temperature in each locality of Bangalore city and then take the average of all the measurements as the temperature of Bangalore city. This is strong induction and, often, not a viable method.

Instead, we make measurements at a few specified localities, where weather stations are placed, and the average of the measurements at these places is given as the temperature of Bangalore city. This is what is called weak induction. In proofs, we use weak induction more often than strong induction.

We prove S(1) is true, where '*n*' takes initial value 1.

We assume S(*k*) is true and prove S(*k* + 1) is true for $1 \le k \le n$

$\therefore$ S(*n*) is true $\forall n \in Z^+$

Example 3.1: Prove that the sum of the first '*n*' positive integers is $\frac{n(n+1)}{2} \forall n \in Z^+$

Proof: Let the statement be,

$$S(n) = 1+2+3+\cdots+n = \frac{n(n+1)}{2} \forall n \in Z^+.$$

(i) Basis step: For $n = 1$, we have

$$S(n) = \frac{n(n+1)}{2}.$$

$$S(1) = \frac{1(1+1)}{2} = 1.$$

(ii) Induction step.

We assume that the statement S(*n*) is true for some arbitrary positive integer *k*, where $1 \le k \le n$. We shall prove it is true for S(*k* + 1).

$$S(k) = 1+2+3+\cdots+k = \frac{k(k+1)}{2}$$

$$S(k+1) = (1+2+3+\cdots+k)+(k+1)$$

$$= \frac{k(k+1)}{2}+(k+1) = \frac{k(k+1)+2(k+1)}{2}$$

$$= \frac{(k+1)(k+2)}{2}$$

Therefore, if S(*k*) is true, S(*k* + 1) is true, and hence by the principle of MI, the result is true $\forall n \in Z^+$.

Example 3.2: Prove that,

$$S(n) = 1^3+2^3+3^3+\cdots+n^3 = \frac{n^2(n+1)^2}{4}$$

Proof:

(i) Basis step: For $n = 1$, we have,

$$S(1) = 1^3 = 1 = \frac{1^2(1+1)^2}{4}$$

(ii) Induction step:

We assume that the statement S(*n*) is true for some arbitrary positive integer *k*, $1 \le k \le n$. We shall prove it is true for (*k* + 1).

$$= \frac{(k+1)^2((k+1)+1)^2}{4}$$

$$S(k)=1^3+2^3+3^3+\cdots+k^3=\frac{k^2(k+1)^2}{4}$$

$$S(k+1)=1^3+2^3+3^3+\cdots+k^3+(k+1)^3$$
$$=\frac{k^2(k+1)^2}{4}+(k+1)^3$$
$$=\frac{k^2(k+1)^2+4(k+1)^3}{4}$$
$$=\frac{(k+1)^2(k^2+4(k+1))}{4}$$
$$=\frac{(k+1)^2(k+2)^2}{4}$$

Hence, by the principle of MI, the statement S(*n*) is true $\forall n \in Z^+$.

Example 3.3: Show that, $2^n > n^2$, whenever *n* is a positive integer and is greater than 4.

Proof: Let $S(n): 2^n > n^2$. We have to show that the statement S(*n*) is true for all integers $n \geq 5$.

(i) Basis step: For $n = 5$, $S(5): 2^5 > 5^2$, which is true as 32 > 25, and hence, the statement is true for $n = 5$.
(ii) Induction step: We assume that the statement S(*n*) is true for some arbitrary positive integer *k*, $5 \leq k \leq n$. We shall prove it is true for $(k + 1)$.

$$S(k): 2^k > k^2, \text{ where } 5 \leq k \leq n.$$
$$S(k+1): 2^{k+1} = 2.2^k > 2k^2$$
$$\text{and } 2k^2 = k^2 + k^2 > k^2 + 4k$$
$$> k^2 + 2k + 1 = (k+1)^2$$

Hence $S(k+1)$ is true.

Using mathematical induction, we have proved that S(*n*): $2^n \geq n^2$ is true for $n > 4$.

Example 3.4: Suppose that there are *n* people in a room, ($n \geq 1$) and all of them greet one another. Prove that $n\frac{(n-1)}{2}$ greetings will have been exchanged.

Proof:

(i) Basis step: For $n = 1$, we have $1\frac{(1-1)}{2} = 0$.
For a single person, there is no one to greet.
(ii) Induction step: We assume that the number of greetings for some arbitrary *k* number of people is $\frac{k(k-1)}{2}$, $1 \leq k \leq$ n. We shall prove the formula holds good for $(k + 1)$ people.
We can write $S(k): \frac{k(k-1)}{2}$.
When the $(k + 1)$th person enters, he or she greets *k* people, so for $n = k + 1$

$$S(k+1) = k\frac{(k-1)}{2} + k = \frac{k^2 - k + 2k}{2}$$

$$S(k+1) = \frac{k^2 + k}{2} = k\frac{(k+1)}{2}$$

$$S(k+1) = \frac{[(k+1)-1][k+1]}{2}$$

Hence, by the principle of MI, the result is true for $n = k + 1$. The result can also be proved by the combination theory.

$$nC_r = \frac{n!}{(n-r)!}$$

$1C_2 = 0$, a single person cannot greet himself.
$2C_2 = 1$, Two persons can greet each other only once.
Similarly, $3C_2 = 3$, $4C_2 = 6$ and so on.

Example 3.5: Given $n^4 < 10^n$ for any positive integer $n \geq 2$ then prove that $(n+1)^4 < 10^{n+1}$.

Proof: Let $n^4 < 10^n$

(i) Basis step: The initial value n can take is 2. For $n = 2$, we have, $S(2): 2^4 < 10^2$, which is true, as $16 < 100$.

(ii) Induction step: We assume that the statement S(n) is true for some arbitrary positive integer $k + 1$, where k is an integer greater than 2.

i.e.
$$k^4 < 10^k \ (k+1)^4 < 10^{k+1} \qquad (1)$$

$$(k+1)^4 < 10^{k+1}$$

Now
$$(k+1)^4 = k^4 + 4k^3 + 6k^2 + 4k^4 + 1$$

$$10(k+1)^4 = 10k^4 + 40k^3 + 60k^2 + 40k + 10 \qquad (2)$$

$$(k+2)^4 = k^4 + 8k^3 + 24k^2 + 32k + 16 \qquad (3)$$

Subtracting (3) from (2) gives

$$10(k+1)^4 - (k+2)^2 = 9k^4 + 32k^3 + 36k^2 + 8k - 6$$

When k is greater than 2, the right hand side of the above equation is positive.

$$10(k+1)^4 - (k+2)^2 = 9k^4 + 32k^3 + 36k^2 + 8k - 6 > 0$$

$$10(k+1)^4 > (k+2)^2 \text{ or } (k+2)^2 < 10(k+1)^4$$

But from (1), $(k+1)^4 < 10^{k+1}$

$\therefore$
$$(k+1)^4 < 10.10^{k+1} = 10^{k+2}$$

$\therefore$
$$(k+2)^2 < 10^{k+2}$$

This implies S(k + 2) is true.
S(2) is true and if S(k + 1) is true, then S(k + 2) is true.
Therefore, S (n + 1) is true $\forall n \geq 2$.

Example 3.6: Use the principle of MI to prove that $S(n) = nC_0 + nC_1 + nC_2 + \cdots + nC_n = 2^n, \forall n \in Z^+$.

Proof:

(i) Basis step:
For $n = 1$, we have,

$$S(1) = 1C_0 + 1C_1 = 1 + 1 = 2^1.$$

Hence, it is true for $n = 1$.

Induction step: We assume that the statement S(n) is true for some arbitrary positive integer k, where k is an integer greater than or equal to 1.

Let $$S(k) = kC_0 + kC_1 + kC_2 + \cdots + kC_k = 2^k \quad (1)$$

Hence, we have to show that the result is true for $n = k + 1$, i.e., we have to prove that,

$$S(k+1) = (k+1)C_0 + (k+1)C_1 + (k+1)C_2 + \cdots + (k+1)C_{(k+1)} = 2^{k+1} \quad (2)$$

We know that,

$(k+1)C_0 = 1 = kC_0; (k+1)C_{k+1} = 1 = kC_k$ and $(k+1)C_r = kC_r + kC_{r-1}$ For $1 \leq r \leq k$, the left hand side of equation (2) can be written as

$kC_0 + (kC_0 + kC_1) + (kC_1 + kC_2) + \ldots + (kC_{k-1} + kC_k) + kC_k = 2(kC_0 + kC_1 + kC_2 + \cdots + kC_k = 2.2^k = 2^{k+1}$

Therefore if S(k) is true, then S(k + 1) is true.
Hence, by MI the result is true $\forall n \geq 1$.

Example 3.7: Show by MI that $1 + 2 + 2^2 + \cdots + 2^n = 2^{n+1} - 1, \forall n \geq 0$.

Proof:

Let $p(n) = 1 + 2 + 2^2 + \cdots + 2^n$

We have to prove that the statement is true in the domain of all non-negative integers.
Basis step: when $n = 0$, $2^0 = 1 = 2^1 - 1 = 1$. This shows that p(0) *is true.*

(i) Induction step: For the inductive hypothesis, we assume that p(k) is true for some arbitrary positive integer k.

$$p(k) = 1 + 2 + 2^2 + \cdots + 2^k = 2^{k+1} - 1$$

To prove the result is true for (k + 1), we have to prove

$$p(k+1) = 1 + 2 + 2^2 + \cdots + 2^k + 2^{k+1} = 2^{k+2} - 1$$

Let us evaluate the left hand side of the above expression.

$$= (1 + 2 + 2^2 + \cdots + 2^k) + 2^{k+1}$$

$$=(2^{k+1}-1)+2^{k+1}$$
$$=2.2^{k+1}-1$$
$$=2^{k+2}-1$$

This implies that p($k+1$) is true.
Therefore, if p(k) is true, then p($k+1$) is true.
That is, $p(n)=1+2+2^2+\cdots+2^n=2^{n+1}-1, \forall n \geq 0.$

Example 3.8: Use MI to prove the formula for the sum of a finite number of terms of a geometric progression.

$$s(n)=a+ar+ar^2+ar^3+\cdots+ar^n=\frac{ar^{n+1}-1}{r-1}, \tag{1}$$

where $r \neq 1$, and n is a non-negative integer.

Proof:

(i) Basis step: For ($n=0$),

$$s(0)=a=\frac{ar^{0+1}-a}{r-1}=a$$

Hence, it is true for the basis.

(ii) Induction step: For the inductive hypothesis, we assume that s(k) is true for some arbitrary positive integer k

$$s(k)=a+ar+ar^2+ar^3+\cdots+ar^k=\frac{ar^{k+1}-1}{r-1} \tag{2}$$

For completeness, if p(k) is true, then we have to prove p($k+1$) is also true.

Add ar^{k+1} to both sides of equation (2).

$$a+ar+ar^2+ar^3+\cdots+ar^k+ar^{k+1}=\frac{ar^{k+1}-1}{r-1}+ar^{k+1} \tag{3}$$

Now, solving the right hand side of equation (3),

$$\frac{ar^{k+1}-a}{r-1}+ar^{k+1}=\frac{ar^{k+1}-a+ar^{k+2}-ar^{k+1}}{r-1}=\frac{ar^{k+2}-a}{r-1} \tag{4}$$

Combining (3) and (4) we get,

$$a+ar+ar^2+ar^3+\cdots+ar^k+ar^{k+1}=\frac{ar^{k+2}-a}{r-1}$$

Hence, s($k+1$) is true.
Therefore, by MI the formula for the sum of n terms of a geometric series is correct.

Example 3.9: Prove that for any positive integer n, 7 divides $3^{2n+1}+2^{n+2}$.

Proof:

(i) Let $p(n): 7 \mid 3^{2n+1}+2^{n+2}$, for all n., Basis step: Let $n=1$, then $3^{2.1+1}+2^{1+2}=35$, which is divisible by 7.
Hence, p(n) is true for $n=1$.

(ii) Induction step: We assume that p(k) is true for some arbitrary positive integer k, $k \geq 0$. We prove it is true for p(k + 1).

$$p(k): 7 \mid 3^{2k+1} + 2^{k+2}$$

For completeness, if p(k) is true, then we prove that p(k + 1) is also true.
Consider p(k + 1).

Let
$$u = 3^{2(k+1)+1} + 2^{(k+1)+2}$$
$$= 3^2.3^{2k+1} + 2^{(k+2)}.2$$
$$= 9.3^{2k+1} + 2.2^{(k+2)}$$
$$= (7+2)3^{2k+1} + 2.2^{(k+2)}$$
$$= 7.3^{2k+1} + 2(3^{2k+1} + 2^{(k+2)})$$

Now, $7 \mid 3^{2k+1} + 2^{k+2}$ by our assumption and $7 \mid 7.3^{2k+1}$ clearly.
Hence, if p(k) is true, then p(k + 1) is true.
Therefore by MI, the statement is true for all n.

$$3^{2k+1} + 2^{k+2} = 7m, \text{where } m \in I.$$

Example 3.10: Prove that for any positive integer n, 5 divides $n^5 - n$.

Proof:
Let p(n) be the given statement , that is, $p(n): 5 \mid (n^5 - n)$.

(i) Basis step: Let $n = 1$, then 5 divides $1^5 - 1 = 0$.
The statement is true for $n = 0$.

(ii) Induction step: We assume that p(n) is true for $n = k$, where $1 \leq k \leq n$. $k \geq 1$.
Hence $p(k) = 5 \mid (k^5 - k)$ is divisible by 5.
Hence, $k^5 - k = 5m$, where $m \in I$.
For completeness, if p(k) is true, then we prove p(k + 1) is true.

$$p(k+1) = (k+1)^5 - (k+1) = (k^5 + 5k^4 + 10k^3 + 10k^2 + 5k + 1) - (k+1)$$
$$= (k^5 - k) + 5(k^4 + 2k^3 + 2k^2 + k)$$
$$= (5m + 5(k^4 + 2k^3 + 2k^2 + k)$$
$$= 5(m + k^4 + 2k^3 + 2k^2 + k)$$

This shows that $(k+1)^5 - (k+1)$ is a multiple of 5, that is, 5 divides $(k+1)^5 - (k+1)$.

Thus, the statement p(k + 1) is true whenever the statement p(k) is true for $k \geq 1$. Hence by MI, the statement p(n) is true for every positive integer 'n'.

Example 3.11: Show that $n^3 - 7n+3$ is divisible by 3, for all positive integers *n*.

Proof:
Let $p(n)=3|(n^3-7n+3), \forall n, p(n)$ is in the domain of all positive integers.

1. Basis step: Let $n=1$. $p(1)=1^3-7.1+3=-3$, which is divisible by 3.
   Hence p(*n*) holds for $n=1$.
2. Induction Step: Let *k* be a positive integer such that p(*k*) holds.

$$p(k):3|(k^3-7k+3) \text{ or } (k^3-7k+3)=3\text{m for } m \in I$$

If p(*k*) is true , then we prove that p(*k* + 1) is also true For n = k+1,

$$p(k+1)=(k+1)^3-7(k+1)+3$$
$$=(k^3+3k^2+3k+1)-7(k+1)+3$$
$$=(k^3-7k+3)+3(k^2+k-2)$$

By induction hypothesis $3|(k^3-7k+3)$, that is $(k^3-7k+3)=3m$

$$p(k+1)=3m+3p(k+1)=3m+3(k^2+k-2)$$

Now 3 divides (k^3-7k+3) and 3 divides $3(k^2+k-2)$

Hence, we find that 3 divides $(k^3-7k+3)+3(k^2+k-2)$.
This implies that p(*k* + 1) is true.
If p(*k*) is true, then p(*k* + 1) is true.
Therefore, by MI, the given expression is true for all *n*.

Example 3.12: Consider the Fibonacci series $f_1, f_2, f_3, \ldots$, where $f_1 = 1, f_2 = 1$ and the *n*th Fibonacci number is given by the following expression.

$$f_n = f_{n-1} + f_{n-2}$$
$$f_3 = f_2 + f_1 = 1+1=2$$
$$f_4 = f_3 + f_2 = 2+1=3$$

We use here the recursive definition of Fibonacci numbers to prove many properties of these numbers.

- Show by the principles of MI that

$$f_n \geq u^{n-2} \text{ for } n \geq 3, \text{ where } u = \frac{1+\sqrt{5}}{2}$$

Proof:

Let $p(n): f_n \geq u^{n-2}$, where $u = \dfrac{1+\sqrt{5}}{2}$

We have to prove that for all $n \geq 3$ p(*n*) is true.

1. Basis step: we have to prove p(n) is true for all $n \geq 3$.
   Hence, we verify that p(3) is true.

   For $n = 3$, $u = \dfrac{1+\sqrt{5}}{2} < 2$ and $f_3 = 2$

   $f_3 = 2 > u = \dfrac{1+\sqrt{5}}{2} < 2$. Hence, p(3) is true.

2. Induction step: Assume that p(k) is true for all integers k, such that $3 \leq k \leq n$.

$$p(k): f_k \geq u^{k-2}$$

   Now let us show that p(k + 1) is true.

   Let $u = \dfrac{1+\sqrt{5}}{2}$, u is a solution of the equation $x^2 - x - 1 = 0$.

   This also implies that

$$u^2 = \left(\frac{1+\sqrt{5}}{2}\right)^2 = \frac{1}{4}\left(1+5+2\sqrt{5}\right) = \frac{1}{2}\left(3+\sqrt{5}\right) = \frac{1}{2}\left(2+1+\sqrt{5}\right) = u+1$$

   Now, $u^{k-1} = u^2 u^{k-3} = (u+1)u^{k-3} = u^{k-2} + u^{k-3} \leq f_k + f_{k-1}$

   Hence, $f_{k+1} = f_k + f_{k-1}$ for $k \geq 3$.

   Thus $f_{k+1} = f_k + f_{k-1} \geq u^{k-1}$ and p(k + 1) is true.

   From MI, it follows that $f_n \geq u^{n-2}\ \forall n \geq 3$.

Example 3.13: The factorial of a non-negative integer n is defined by

$$n! = 1; \text{ if } n = 0 \text{ and}$$

$$n! = n*(n-1)!; \text{ if } n = 0$$

Consider the following algorithm that determines the factorial of a non-negative integer. The algorithm is given by,
Algorithm: factorial (n)
Input: positive integer 'n'
Output: factorial of 'n'

```
Begin
i := 0;
fact := 1;
while( i < n ) do ;
begin
i := i + 1 ;
fact : = fact * i ;
return (fact);
End
```

Let the predicate p(n) be $fact_n = fact_{n-1} * i_n$. This is the loop invariant in this algorithm. We will prove by MI that $p(n)$ is true $\forall n > 0$.

Proof:

(i) Basis step: The $fact_0$ and i_0 correspond to the values of these variables before entering the loop. $i_0 = 0$, and $fact_0 = 1$. Hence, p(0) is true.

(ii) Induction step: Assume that p(*k*) is true for some *k*, $0 \le k \le n$.

$$fact_k = fact_{k-1} * i_k.$$

Given that p(*k*) is true we will prove it is true for p(*k* + 1). After a pass through the loop, i is increased by 1 and the fact value is multiplied by the immediate previous iteration value of fact.

Thus, the predicate p(*k* + 1): $fact_{k+1} = fact_{k-1} * i_k * i_{k+1}$

$$= fact_k * i_{k+1}$$
$$= fact_k * i_{k+1}.$$

Therefore if p(*k*) is true, then p(*k* + 1) is true.
Hence, by MI the predicate is true for any '*n*'.
An alternative way of proving the same predicate is as follows. The predicate for this loop is given by,

$$p(k): fact(k) = k!\ 0 \le k \le n.$$

1. Basis step: Before the loop is executed, i = 0 and fact (0) = 0! = 1.
   Hence, p(0) is true.
   Thus fact (*k*) = *k*! i(*k*) = *k*, and $k < n$.
   For completeness, if p(*k*) is true, then we must prove p(*k* + 1) is also true. One more pass through the loop.

$$i(k+1) = i(k) + i = k + 1$$

$$fact(k+1) = fact(k) * i(k+1)$$
$$= k!\ (k+1)$$
$$= (k+1)\ k!$$
$$= (k+1)!$$

Thus, p(*k* + 1) is true. When the loop terminates, p(*k*) is true and *i* < *n* is false, as i = *n* is true. Therefore, when the loop terminates, p(*n*) = p(*i*): fact = *i*! = *n*!.

Example 3.14: Consider the following code which yields the square of the given positive integer.

```
Function square (x)
begin:
1 a = 0
2 b = 0
3 while ( b < > x ) do
4 a = a + x
5 b = b + 1
6 return (a)
end
```

From lines 4 and 5, we know that every time the loop is executed a is added to *x* and b is incremented by 1. Consider the predicate P(n) representing $a_n = x * b_n\ \forall n \ge 0$ and assume P(*k*) is true for some $1 \le k \le n$. We shall prove it for P(*k* + 1).

Proof:

(i) Basis step: When the loop is not entered, the values of a and b are zeroes. That is, P(0) is true.

$$a_0 = x * b_0$$

(ii) Induction step: Let us assume that the predicate is true for some positive integer k and $0 \le k \le n$.

$$P(k): a_k = x * b_k$$

We shall prove it is true for $P(k+1)$: $a_{k+1} = x * b_{k+1}$. When we pass through the loop onceafter k times, $a_{k+1} = a_k + x$ and $b_k = b_k + 1$.

Consider $P(k+1)$: $a_{k+1} = a_k + x$

$$= x * b_k + x$$

$$= x * (b_k + 1)$$

$$a_{k+1} = x * b_{k+1}$$

We have proved that if $P(k)$ is true, then $P(k+1)$ is true.
Hence, by MI, it follows that the predicate is true for any positive integer n.

Example 3.15: Consider the sets $A_1, A_2, A_3, \ldots, A_n$ and prove by MI the following expression, which is an extended version of De Morgan's laws. Let the predicate $P(n)$ represent the following expression.

$$\overline{\left(\bigcup_{i=1}^{n} Aj\right)} = \bigcap_{i=1}^{n} \overline{Aj}$$

Proof:

(i) Basis step: For $P(1) = \overline{A1} = \overline{A1}$, which is true.
(ii) Induction step: Let the expression $P(k)$ be true for some positive integer k; we will show it is true for $P(k+1)$.

$$\overline{\left(\bigcup_{i=1}^{k+1} Aj\right)} = \overline{A_1 \cup A_2 \cup A_3 \cdots \cup A_k \cup A_{k+1}}$$

$$= \overline{(A1 \cup A2 \cup A3 \ldots \cup Ak) \cup Ak+1}$$

$$= \overline{A1 \cup A2 \cup A3 \ldots \cup Ak} \cap \overline{Ak+1}$$

$$= \bigcap_{i=1}^{k} \overline{Aj} \cap \overline{Ak+1}$$

$$= \bigcap_{i=1}^{k+1} \overline{Aj}$$

Thus, if the predicate is true for $P(k)$, we have proved it is true for $P(k+1)$, and hence, by MI, $P(n)$ is true for all n.

Example 3.16: Let $\{a_n\}$ be a sequence of natural numbers such that $a_1 = 5, a_2 = 13$. Prove that $a_n = 2^n + 3^n$, for all natural numbers (n).

Proof: Let the statement $P(n): a_n = 2^n + 3^n$

(i) Basis step: For $n = 1$ and $n = 2$

$a_1 = 2^1 + 3^1 = 5$ and $a_2 = 2^2 + 3^2 = 13$. Hence, P(1) and P(2) are true.

(ii) Induction step: Let the statement P(k) be true for some positive integer k; now, we will show it is true for P(k + 1).

Let $P(k): a_k = 2^k + 3^k$ be true for some natural number k.

$$P(k+1): a_{k+1} = 2^{k+1} + 3^{k+1}$$

Then

$$\begin{aligned} a_{k+2} &= 5\,a_{k+1} - 6\,a_k \\ &= 5\left[2^{k+1} + 3^{k+1}\right] - 6\left[2^k + 3^k\right] \\ &= 5.2^k.2 + 5.3^k.3 - 6.2^k - 6.3^k \\ &= 10.2^k + 15.3^k - 6.2^k - 6.3^k \\ &= 4.2^k + 9.3^k \\ &= 2^{k+2} + 3^{k+2} \end{aligned}$$

Hence, the predicate p(k + 1) is true for $n = k + 1$ and also holds for $n = k + 2$.
Therefore, $a_n = 2^n + 3^n$ is valid for all natural numbers.

Example 3.17: Prove that $(3+\sqrt{5})^n + (3-\sqrt{5})^n$ is an even integer for all natural numbers n.

Proof: Let the function be $f(n) = á^n + â^n$, where $á = 3 + \sqrt{5}$ and $â = 3 - \sqrt{5}$.

(i) Basis step: For $n = 1$ and $n = 2$:
$f(1) = 6, f(2) = 28$; both are even integers, and hence the statement is true.

(ii) Induction step: Suppose $f(k)$ and $f(k + 1)$ are both even integers for some positive integer k. Consider the case for $n = k + 2$. Note that α and β are roots of the equation $x^2 - 6x + 4 = 0$

$$x = \frac{6 \pm \sqrt{36-16}}{2} = \frac{6 \pm \sqrt{20}}{2} = 3 \pm \sqrt{5}$$

$$\begin{aligned} f(k+2) &= á^{k+2} + â^{k+2} \\ &= á^k(6á - 4) + â^k(6á - 4) \\ &= 6(á^{k+1} + â^{k+1}) - 4(á^k + â^k) \\ &= 6f(k+1) - 4f(k) \end{aligned}$$

It follows that $f(k + 2)$ is also an even integer. By MI, we conclude that $f(n)$ is an even integer.

Example 3.18: For all positive integers '*n*', prove that

$$f(n): \sin è + \sin 2è + \sin 3è + \cdots + \sin nè = \sin\frac{(n+1)è}{2}\sin\frac{nè}{2}\csc\frac{è}{2}$$

Proof:

(i) Basis step: For $n = 1$, f(1) is given by

$\sin è \sin\frac{è}{2}\csc\frac{è}{2} = \sin è$ $(\because \sin\theta\csc\theta = 1)$. So the formula holds for $n = 1$.

(ii) Induction step: Suppose the formula holds for $(n = k)$, that is,

$f(k): \sin è + \sin 2è + \sin 3è + \cdots + \sin kè = \sin\frac{(k+1)è}{2}\sin\frac{kè}{2}\csc\frac{è}{2}$ Consider the case for $f(k+1)$.

$$f(k+1): \sin è + \sin 2è + \sin 3è + \cdots + \sin kè + \sin(k+1)è$$

$$= \sin\frac{(k+1)è}{2}\sin\frac{kè}{2}\csc\frac{è}{2} + \sin(k+1)è$$

$$= \sin\frac{(k+1)è}{2}\sin\frac{kè}{2}\csc\frac{è}{2} + 2\sin\frac{(k+1)è}{2}\cos\frac{(k+1)è}{2} \qquad \because \sin è = 2\sin\frac{è}{2}\cos\frac{è}{2}$$

$$= \sin\frac{(k+1)è}{2}\csc\frac{è}{2}\left[\sin\frac{kè}{2} + 2\sin\frac{è}{2}\cos\frac{(k+1)è}{2}\right]$$

$$= \sin\frac{(k+1)è}{2}\csc\frac{è}{2}\left[\sin\frac{kè}{2} + \left[\sin\left(\frac{è}{2}+\frac{(k+1)è}{2}\right) + \sin\left(\frac{è}{2}-\frac{(k+1)è}{2}\right)\right]\right]$$

using $2\sin A\cos B = \sin(A+B) + \sin(A-B)$

$$= \sin\frac{(k+1)è}{2}\csc\frac{è}{2}\left[\sin\frac{kè}{2} + \sin\frac{(k+2)è}{2} - \sin\frac{kè}{2}\right]$$

$$= \sin\frac{(k+1)è}{2}\csc\frac{è}{2}\sin\frac{(k+2)è}{2}$$

$$= \sin\frac{(k+1+1)è}{2}\sin\frac{(k+1)è}{2}\csc\frac{è}{2}$$

Therefore if f(*k*) is true, then f(*k* + 1) is true.
By the principle of mathematical induction, the formula holds for all positive '*n*'.

Example 3.19: Prove by mathematical induction that the arithmetic mean (A.M) is always greater than or equal to the geometric mean (G.M). Or prove that for all positive integers $(a_1, a_2, a_3 \ldots a_n)$

$$\left(\frac{a_1 + a_2 + a_3 + \cdots + a_n}{n}\right) \geq \sqrt[n]{a_1 \cdot a_2 \cdot a_3 \ldots a_n}$$

Proof:

(i) Basis step: Consider $(\sqrt{a_1} - \sqrt{a_2})^2 \geq 0$; we obtain

$\frac{(a_1 + a_2)}{2} \geq \sqrt[2]{a_1 . a_2}$; the inequality holds good for $(n = 2)$.

Since the arithmetic mean can be obtained only if there are at least two numbers, we started with $(n = 2)$.

(ii) Induction Step: Suppose the inequality holds good for $n = 2k$ for some positive integer k. Consider the case ($n = 2k$); using the case (n = 2) and the induction hypothesis, we have,

$$\frac{a_1 + a_2 + - - + a_{2k}}{2k} = \frac{1}{k}\left[\frac{a_1 + a_2}{2} + \frac{a_3 + a_4}{2} + - - + \frac{a_{2k-1} + a_{2k}}{2}\right]$$

$$\geq \frac{\sqrt{a_1 a_2} + \sqrt{a_3 a_4} + - - + \sqrt{a_{2k-1} a_{2k}}}{k}$$

$$\geq \sqrt[k]{\sqrt{a_1 a_2}\,\sqrt{a_3 a_4} \ldots \sqrt{a_{2k-1} a_{2k}}}$$

$$= \sqrt[2k]{a_1 a_2 \ldots\ldots\ldots a_{2k}}$$

The inequality also holds good for ($n = 2k$). Hence, the inequality holds for all positive powers of 2.

Example 3.20: Let $(x > -1)$ be a real number. Prove that $(1+x)^n \geq 1 + nx$ for all natural numbers n.

Proof:

(i) Basis step: for $n = 1$,
$(1 + x)^1 = 1 + 1(x)$ $(1+x)' = 1 + 1(x)$. The inequality holds good for $n = 1$.

(ii) Induction Step: Assume the inequality holds good for some positive k.
Therefore $(1+x)^k \geq 1 + kx$. For the case $n = k + 1$, we have,

$$(1+x)^{k+1} = (1+x)^k (1+x) \geq (1+kx)(1+x)$$

$$= 1 + (k+1)x + kx^2$$

$$= 1 + \left[k(1+x) + 1\right]x$$

$$= 1 + (k'+1)x;\ \text{taking}\ k(1+x) = k'$$

Hence, if the inequality holds for the case $n = k$, it also holds for $n = k + 1$.
It follows by MI that $(1+x)^n \geq 1 + nx$ for all natural numbers 'n'.

3.2 JOKES AND PARADOXES

Let the predicate P(n) represent the statement that one can drink any amount (n units) of water when feeling thirsty. We will prove the statement 'when one feels thirsty, one will be able to drink n units of water' using mathematical induction.

Basis step: Clearly, the statement holds for $n = 1$ because one certainly wants to drink some water when feeling thirsty.

Induction step: Suppose the statement holds for $n = k$, that is, when one feels thirsty, one is able to drink k drops of water. Consider the case $n = k + 1$. By assumption, when one feels thirsty one is able to drink k drops of water. Being thirsty, one certainly is able to drink one more drop. So the statement holds for $n = k + 1$, as well.

By the principle of mathematical induction, the statement holds for all natural numbers n. In other words, when one feels thirsty, one can drink any amount of water. **Thus, one is able to swallow up all the water in the oceans! Is it not a joke?**

Consider an example from the book on Calculus by. M. Spivak. Seventeen professors in the Department of Mathematics of a certain University meet every week for a discussion on research. They have a rule that *if any professor finds mistakes in his own papers, he must resign.*

For a long time no professor has resigned. This does not mean that no professor has ever made any mistakes in his papers. Rather, everyone has made mistakes, and every other professor has discovered it. In other words, everyone knows that every other professor has made mistakes, but does not know his own mistakes.

During the last meeting of a year, a research assistant says he wants to tell everyone something. He points out that every professor has made mistakes in his papers, and that some other professor has always discovered the mistake. All the professors are made to resign at the seventeenth meeting of the subsequent year.

One interesting question is that the research assistant has pointed out something every professor knows. Why did nobody resign before the research assistant exposed them? Why did every professor resign after this 'well- known' remark? The mathematics professors were busy in finding the mistakes of other fellow-professors and did not resign until the research assistant pointed out mistakes in the papers written by each professor.

ADDITIONAL SOLVED EXAMPLES

Example W3.1: The *n*-th term of an Arithmetic progression is given by

$$T_n = a + (n - 1)d,$$

where a = first term, d = common difference.

Show by the principle of mathematical induction, that the sum of the first *n*-terms of an AP is

$$S_n = \frac{n}{2}[2a+(n-1)d]$$

Solution:

Let $S_n = a+(a+d)+(a+2d)+\text{-----}+(a+(n-1)d = \frac{n}{2}[2a+(n-1)d]$ (I)

1. Basis of Induction:
   Let $n = 1$, then LHS $= a$, and RHS $= \frac{1}{2}[2a+(1-1)d] = a$
   $\therefore$ S(n) is true for $n = 1$.
2. Induction step:
   Assume that S(n) is true for $n = k$.

$$S_n = a+(a+d)+(a+2d)+\text{-----}+(a+(k-1)d = \frac{k}{2}[2a+(k-1)d] \quad \text{(II)}$$

Let $n = k + 1$

The LHS of Equation (I) then becomes

$$= a+(a+d)+(a+2d)+\text{-----}+(a+(k+1-1)d.$$

$$= a+(a+d)+(a+2d)+\text{-----}+(a+(k-1)d)+(a+kd).$$

Using the above result in equation (II)

$$= \frac{k}{2}[2a+(k-1)d]+[a+kd]$$

$$= ak + \frac{k(k-1)}{2}d + a + kd$$

$$= a(k+1) + \frac{k(k-1)}{2}d + kd$$

$$= a(k+1) + \left[\frac{k^2 - k + 2k}{2}\right]d$$

$$= a(k+1) + \frac{k(k+1)}{2}d \quad \text{(III)}$$

Put $n = k + 1$ in RHS of (I).

$$= \frac{(k+1)}{2}[2a + (k+1-1)d]$$

$$= \frac{(k+1)}{2}[2a + kd]$$

$$= a(k+1) + \frac{k(k+1)d}{2} \quad \text{(IV)}$$

From (III) and (IV), LHS = RHS for $n = 1$, $n = k$ and $n = k + 1$. Hence, by mathematical induction we can say that S(n) is true for every positive integer n.

Example W3.2: Let $A = \begin{bmatrix} 1 & \alpha \\ 0 & 1 \end{bmatrix}$ be a matrix, then prove by mathematical induction that

$$A^n = \begin{bmatrix} 1 & \alpha n \\ 0 & 1 \end{bmatrix}$$

Solution:

1. Basis of Induction:
   For $n = 2, 3$ let us do the actual multiplication

$$A^2 = A\,A = \begin{bmatrix} 1 & \alpha \\ 0 & 1 \end{bmatrix}\begin{bmatrix} 1 & \alpha \\ 0 & 1 \end{bmatrix} = \begin{bmatrix} 1+0 & \alpha+\alpha \\ 0 & 1 \end{bmatrix} = \begin{bmatrix} 1 & 2\alpha \\ 0 & 1 \end{bmatrix}$$

$$A^3 = A^2\,A = \begin{bmatrix} 1 & 2\alpha \\ 0 & 1 \end{bmatrix}\begin{bmatrix} 1 & \alpha \\ 0 & 1 \end{bmatrix} = \begin{bmatrix} 1 & 3\alpha \\ 0 & 1 \end{bmatrix}$$

   Hence the result is true for $n = 2, 3$

2. Induction step:
   Assume that A^n is true for $n = k$

$$A^k = \begin{bmatrix} 1 & k\alpha \\ 0 & 1 \end{bmatrix}$$

   Let $n = k + 1$

$$A^{k+1} = A^k\,A = \begin{bmatrix} 1 & k\alpha \\ 0 & 1 \end{bmatrix}\begin{bmatrix} 1 & \alpha \\ 0 & 1 \end{bmatrix}$$

$$= \begin{bmatrix} 1 & \alpha + k\alpha \\ 0 & 1 \end{bmatrix} = \begin{bmatrix} 1 & \alpha(k+1) \\ 0 & 1 \end{bmatrix}$$

Which shows that the result is true for $n = k + 1$.
Hence A^n is true for $n = 2, 3$; $n = k$ and $n = k + 1$. By mathematical induction we infer that A^n is true for every positive integer n.

Example W3.3: Prove by mathematical induction that

$$S(n) = \frac{1}{2} + \frac{1}{2^2} + \frac{1}{2^3} + --- + \frac{1}{2^n} = 1 - \frac{1}{2^n} \qquad \text{(I)}$$

Solution:

1. Basis of induction:
   Let $n = 1$. Then L H S $= \frac{1}{2}$ and RHS $= 1 - \frac{1}{2} = \frac{1}{2}$
   Hence LHS = RHS for $n = 1$.

   Induction step:
   Let us assume that S(n) is true for $n = k$.

$$S(k) = \frac{1}{2} + \frac{1}{2^2} + \frac{1}{2^3} + --- + \frac{1}{2^k} = 1 - \frac{1}{2^k} \qquad \text{(II)}$$

   Let $n = k + 1$
   Then from (I), LHS is

$$= \frac{1}{2} + \frac{1}{2^2} + \frac{1}{2^3} + -- + \frac{1}{2^k} - + \frac{1}{2^{k+1}}$$

   Using equation (II) for $k + 1$,

$$= 1 - \frac{1}{2^k} + \frac{1}{2^{k+1}} = 1 - \left[\frac{1}{2^k} - \frac{1}{2^{k+1}}\right] = 1 - \frac{1}{2^{k+1}}$$

   This shows that S(n) is true for $n = 1$, $n = k$, and $n = k + 1$. Hence by mathematical induction we can infer that S(n) is true for every positive integer n.

Example W3.4: Prove by mathematical induction that

$$S(n) = \frac{1}{1.2} + \frac{1}{2.3} + \frac{1}{3.4} + --- + \frac{1}{n(n+1)} = \frac{1}{n+1} \qquad \text{(I)}$$

Solution:

1. Basis of Induction:

   Let $n = 1$, then LHS $= \frac{1}{1(1+1)} = \frac{1}{2}$

   RHS $= \frac{1}{1+1} = \frac{1}{2}$

   Hence LHS = RHS for $n = 1$.
2. Induction step:
   Assume that n is true for $n = k$

$$S(k) = \frac{1}{1.2} + \frac{1}{2.3} + \frac{1}{3.4} + --- + \frac{1}{k(k+1)} = \frac{1}{k+1} \qquad \text{(II)}$$

Let $n = k + 1$
Then LHS of (I) becomes

$$= \frac{1}{1.2} + \frac{1}{2.3} + \frac{1}{3.4} + --- + \frac{1}{k(k+1)} + \frac{1}{(k+1)(k+2)}$$

Using equation (II) we get

$$\frac{1}{(k+1)} + \frac{1}{(k+1)(k+2)} = \frac{k^2+2k+1}{(k+1)(k+2)} = \frac{(k+1)(k+1)}{(k+1)(k+2)}$$

$$= \frac{k+1}{k+2} = \frac{k+1}{(k+1)+1}$$

$$\text{RHS} = \frac{1}{(k+1)+1}$$

This shows that S(n) is true for $n = 1$, $n = k$ and $n = k + 1$. Hence by mathematical induction we can infer that S(n) is true for every positive integer n.

Example W3.5: Prove by mathematical induction that

$$S(n) = 1.3 + 2.4 + 3.5 + ----- + n(n+2) = \frac{n}{6}(n+1)(2n+7) \qquad \text{(I)}$$

Solution:

1. Basis of Induction:
   Let us check for $n = 1$
   For $n = 1$, LHS = $1(1 + 2) = 3$
   RHS = $\frac{1}{6}(1+1)(2+7) = 3$
   LHS = RHS for $n = 1$.
2. Induction step:
   Assume that S(n) is true for $n = k$
   From equation (I)

$$S(k) = 1.3 + 2.4 + 3.5 + ----- + k(k+2) = \frac{k}{6}(k+1)(2k+7)$$

Let us evaluate for $n = k + 1$
From equation (I), LHS

$$= 1.3 + 2.4 + 3.5 + ----- + k(k+2) + (k+1)(k+1+2)$$

Using equation (II) we get

$$= \frac{k}{6}(k+1)(2k+7) + (k+1)(k+3)$$

$$= (k+1)\left[\frac{k}{6}(2k+7) + (k+3)\right]$$

$$= (k+1)\left[\frac{2k^2+13k+18}{6}\right]$$

$$= \frac{(k+1)(k+2)(2k+9)}{6}$$

Let $n = k+1$ in RHS of equation (I) $= \frac{(k+1)}{6}(k+1+1)(2(k+1)+7)$

$$= \frac{(k+1)(k+2)(2k+9)}{6}$$

This shows that S(n) is true for $n = 1$, $n = k$, and $n = k+1$. Then by mathematical induction it is proved that S(n) is true for every positive integer n.

SUMMARY

MI is a useful method of proving mathematical statements. The method can be extended to prove statements about more general well-founded structures such as trees. This generalisation is also known as structural induction. There are variations of MI, namely, complete induction and transfinite induction. MI is useful in computer science for verifying algorithms, wherein we write the loop invariants. It is also used in proving theorems in graph theory with much more ease compared to analytical proofs.

EXERCISES

1. Prove that $1^2+2^2+3^2+\cdots+(2n+1)^2 = \frac{(n+1)(2n+1)(2n+2)}{3}$

   where n is a nonnegative integer.
2. Prove that for every positive integer n,

$$1.2.3+2.3.4+\cdots+n\,(n+1)\,(n+2) = \frac{n\,(n+1)(n+2)(n+3)}{4}$$

3. Prove that $\frac{1}{1.3}+\frac{1}{3.5}+\frac{1}{5.7}+\cdots+\frac{1}{(2n-1)(2n+1)} = \frac{n}{2n+1}$
4. Prove that for every positive integer n,

$$1.2+2.3+\cdots+n\,(n+1) = \frac{n\,(n+1)(n+2)}{3}$$

5. Use induction to prove that,

$$1^2-2^2+3^2\cdots+(-1)^{n-1}n^2 = (-1)^{n-1}\frac{n(n+1)}{2} \text{ for all positive integers } n.$$

6. For all $n \in z : n \geq 0$ prove that 2^{2n+1} is divisible by 3.
7. Show that $11^n - 4^n$ is divisible by 7, $\forall n \in z^+$
8. Prove by MI that,

$$7^n + 5 \text{ is divisible by 3, } \forall n \in z^+, n \in z^+.$$

9. Use MI to prove that $4|(6.7^n + 2.3^n)$ for all non-negative integers n.

10. Prove that $5^n + 3$ is divisible by 4 for all non-negative integers.
11. Show that the following algorithm correctly determines the largest element in a list of n elements. What is the loop invariant?
    Algorithm: largest element in a list of n elements.
    Input: List of n elements.
    Output: Largest element in the list.

```
Begin
i := 1;
x := L[1];
while i < n do
begin
i := i + 1;
if x < L[i] then
x : = L[i];
end
return (x);
End
```

12. Prove using MI the Euclidean algorithm of finding the GCD of the given two positive integers.
13. Prove that $1 + 2 + 3 + \cdots + n = (2n + 1)^2/8$
14. Prove by MI that if a set A has n elements then P(A) has 2^n elements.
15. Prove that p(n) representing $1 + 5 + 9 + \cdots + (4n - 3) = (2n + 1)(n - 1)$ is true.
16. $1^2 \times 2 + 2^2 \times 3 + \cdots + n^2 \times (n+1) = \dfrac{n(n+1)(n+2)(3n+1)}{12}$
17. $\cos\theta + \cos 2\theta + \cdots + \cos n\theta = \sin\dfrac{(n+1)\theta}{2}\cos\dfrac{n\theta}{2}\csc\dfrac{\theta}{2} - 1.$
18. The Lucas Sequence is defined by 1, 3, 4, 7, 11, 18, 29, $\cdots$. where $a_1 = 1$, $a_2 = 3$, and $a_n = a_{n-1} + a_{n-2}$ for $n \geq 3$. Prove by MI that $a_n < (1.75)^n$ for all positive integers n.
19. Prove that $(4007)^n - 1$ is divisible by 4005.
20. Prove by mathematical induction that

$$1 \times 1! + 2 \times 2! + \cdots + n \times n! = (n + 1)! - 1.$$

4 Introduction to Set Theory

The theory of sets was introduced by German mathematician Georg Cantor (1845–1918). Set theory is considered a prerequisite for many areas in mathematics such as abstract algebra, combinatorics, relations, topology, graphs, rings, vector spaces, discrete mathematics, fuzzy sets and the like. Set theory is considered the foundation for all other areas in mathematics. Many books on mathematics have an introductory course on set theory. The layman calls a set a 'collection' or a 'class', and hence these words are synonymous. In computer science, a set is used as a data structure. The computer language Pascal supports the concept of sets and operation on sets. If we look at a program, its input and output, these form the sets. When we test the software, we design the test cases, which again form sets. One can view a program to be a collection of functions in C language. In other words, we can say that it is a set of functions. Sets are of two kinds, finite and infinite. If all the elements or members of the set, when enumerated, are finite in number, the set is called finite; otherwise, it is called infinite. There are different ways of defining sets.

4.1 SETS AND SET DEFINITIONS

A set is defined as a collection of well-defined objects. The objects or things are called members or elements of the set. For example, a set of computer science books, a set of metal chairs, a set of programs, a set of electronic gadgets, a set of real numbers, a set of integers and the like. A set is said to be well defined if we are able to say whether the given object is a member of the given set or not. The membership of an object to a set is denoted by $\in$. We formally denote the sets by capital letters and also by meaningful names. For example, A, B, Z, integer-set, colours, etc. The members or elements of the set are enclosed within the set symbols '{' and '}', also called curly brackets. Consider the set $S = \{1, 2, 3, 4, 5\}$. 'The object 1 is a member of S' is concisely represented as $1 \in S$. Since 6 is not present in S, it is denoted by $6 \notin S$. The two sets, $S_1 = \{1, 2, 3, 4\}$ and $S_2 = \{2, 1, 4, 3\}$ are equal because they have the same number and types of elements. The order in which the elements or members are listed is unimportant. Sets are allowed to have repetition of members.

The sets $S_3 = \{1, 1, 2, 3\}$ and $S_4 = \{1, 2, 3\}$ are the same as they have the same members, irrespective of the repetition. Usually, the set elements are listed only once. Consider the set $\mathbb{Q} = \{a/b \mid a, b \in \mathbb{Z}$ and $b \neq 0\}$, which represents the set of rational numbers and is read as '$\mathbb{Q}$ is a set that contains elements that are expressed as a/b such that a and b belong to the set of integers $\mathbb{Z}$ and b is not equal to zero'.

The symbol '|' is read as 'such as'. The common sets and their descriptions are given in Table 4.1. The sets are defined over the universe of discourse represented as 'U'. It should not be confused with the operator symbols used in set theory.

Table 4.1 Common set notations and description

(i)	$\mathbb{Z}$ = the set of integers	$\{..., -3, -2, -1, 0, 1, 2, 3, ...\}$
(ii)	$\mathbb{N}$ = set of non-negative integers	$\{1, 1, 2, 3, 4, ...\}$
(iii)	$\mathbb{Z}^+$ = set of positive integers	$\{1, 2, 3, 4, ...\}$
(iv)	$\mathbb{Q}$ = set of rational numbers	$\{a/b \mid a, b \in \mathbb{Z} \text{ and } b \neq 0\}$
(v)	$\mathbb{Q}^+$ = set of positive rational numbers	$\{r \in \mathbb{Q} \mid r > 0\}$
(vi)	$\mathbb{Q}^*$ = set of non-zero rational numbers	
(vii)	$\mathbb{R}$ = set of real numbers	
(viii)	$\mathbb{R}^+$ = set of positive real numbers	
(ix)	$\mathbb{R}^*$ = set of non-zero real numbers	
(x)	$\mathbb{C}$ = set of complex numbers	$\{x + iy \mid x, y \in \mathbb{R} \text{ and } i^2 = -1\}$
(xi)	$\mathbb{C}^*$ = set of non-zero complex numbers	
(xii)	$\mathbb{Z}n$ = first n positive integers	$n \in \mathbb{Z}^+$ and $\mathbb{Z}n = \{0, 1, 2, 3,... (n-1)\}$

Consider the set S = {1, 2, 3, 4, 5}. This is represented precisely as $S = \{x \mid 1 \leq x \leq 5\}$. This further reads as 'the set S contains the elements of the type "x" such that 1 is less than or equal to x and x is less than or equal to 5'.

Example 4.1: Let $\mathbb{Z}^+$ be the universal set. Define the following sets.

(i) Set of positive even integers
(ii) Set of positive odd integers
(iii) Set of squares of positive integers less than 50
(iv) Set of divisors of 45

Solution:

(i) $\{x \mid x = 2i \text{ and } i \in \mathbb{Z}^+\}$
(ii) $\{x \mid x = 2i + 1 \text{ and } i \in \mathbb{Z}^+\}$
(iii) {1, 4, 9, 16, 25, 36, 49}
(iv) {1, 3, 5, 9, 15, 45}

The number of elements or members of a given set is called the cardinality or size of the set. The cardinality of the set in Example 4.1 (iii) is 7, and that of Example 4.1 (iv) is 6. The cardinality of a given set A is denoted as |A| or #(A). We talk about cardinality of finite sets. The sets in Example 4.1 (iii) and (iv) are finite or countable, as the elements are listed. Examples 4.1 (i) and (ii) are examples of infinite sets. There are sets that can be mapped onto the set of non-negative integers, and are called countably infinite sets. There should be one-to-one correspondence between the member of a set and integers or natural numbers, one at a time. The count may never end. It can be stated in other words as 'a set is called countably infinite iff it has the same cardinality as the set of positive integers $\mathbb{Z}^{+}$. A set is countable iff it is finite or countably infinite. A set that is not countable is called uncountable or non-denumerable.

Example 4.2: We know that the set $\mathbb{Z}$ of integers is countably infinite. Let us define $f: N \to Z$ by

$$f(n) = (n/2) \quad \text{if } n \text{ is even;}$$
$$= -(n-1)/2 \text{ if } n \text{ is odd.}$$

Then f is a bijection from $\mathbb{N}$ to $\mathbb{Z}$. If there is no bijection between $\mathbb{N}$ and a set A, then A is called uncountable.

A set that does not contain any elements is called an empty set and is denoted by $\varnothing$. An alternative representation is $\varnothing = \{\ \}$. An empty set is also called a null set. The cardinality of an empty set, i.e. $|\varnothing| = 0$. A few examples of an empty set include a set of immortal persons, a set of real numbers whose squares are negative, a set of bicycles with no wheels, a set of months with 32 days, a set of computers that are in the 'on' and 'off' state, and so on. Sets can be defined in different ways.

4.1.1 Set Definitions

The methods of defining both finite and infinite sets are enumeration, logic and mathematical induction.

(i) **Logic:** Knowledge of predicate logic as discussed in Chapter 2, is useful in defining infinite sets. Let $P(x)$ be some predicate. A set of elements that satisfy this predicate qualify to be members of the set, and the set is represented as $S = \{x \mid P(x)\}$. This is read as 'S is a set of elements of the type x such that each x satisfies the predicate $P(x)$'. Consider the following definitions of certain sets.

(a) $A = \{x \mid P(x)\}$, where $P(x)$ represents the statement 'x is a positive even integer'.
(b) $B = \{y \mid Q(y)\}$, where $Q(y)$ represents the statement 'y is a basic colour'.
(c) $C = \{z \mid R(z)\}$, where $R(z)$ represents the statement 'z is a vowel'.

The definition given under (a) represents the set of even integers $A = \{2, 4, 6, \ldots\}$, which is an infinite set. Similarly, (b) and (c) represent finite sets $B = \{\text{red,blue,green}\}$ and $C = \{a, e, i, o, u\}$, respectively.

(ii) **Enumeration:** It is used for defining finite sets, wherein the members are listed within the set symbols. The following are a few examples of enumeration.

(a) $S = \{1, 2, 3, 4, 5\}$, which is defined over the universal set $\mathbb{Z}^{+}$.
(b) $A = \{2, 4, 6, \ldots\}$, which is defined over the universal set $\mathbb{Z}^{+}$.
(c) $C = \{a, b, \ldots, z\}$, which are letters of the English alphabet.

(iii) **Mathematical Induction:** In this method, we use basis, induction hypothesis and external clause as discussed in Chapter 3. In the basis step, we assume the basic element in the set, and it acts like a building block. In the induction step, we develop the rules to generate the subsequent elements in the set. In the external clause, we write a declaration that no element belongs to the set unless it is generated by the basis or the induction step. Consider the following examples.

(a) $A = \{2, 4, 6, \ldots\}$ set of even integers.
Basis step: Initially, $A = \{\ \}$. Let $2 \in A$. Now $A = \{2\}$.
Induction step: If $x \in A$, then $x + 2 \in A$. Thus, we generate subsequent elements.
When $x = 2$, the element 4 also belongs to A. When 4 belongs to A, the element $(4 + 2)$ also belongs to A, and so on.. Hence we get $A = \{2,4,6,\ldots\}$.
External step: It is declared that no element is a valid member of the set unless it is generated by basis or induction.

(b) S = {(), () (), (()), (() ()), ...} set of well-formed parentheses.

Basis step: Let () ∈ S and the set S = {()}.

Induction step: (i) If $x \in S$, then $(x) \in S$.

(ii) If $x \in S$ and $y \in S$, then $xy \in S$, where xy is the concatenation of x and y.

External step: It is declared that no element is a valid member of the set unless it is generated by basis or induction.

4.2 RUSSELL'S PARADOX

Bertrand Russell discovered the Russell's paradox applying the elements of logic that are used for the definition of sets. A mere collection of objects does not form a set. If every collection forms a set, then it leads to a paradox. Consider the statement 'S be a set of all subsets that are not members of themselves'. The question is 'Is $S \in S$?' There are two possibilities: (i) $S \in S$ and (ii) $S \notin S$.

(i) Let $S \in S$: From the definition of the given set, S should not belong to S. This means that if $S \in S$, then $S \notin S$.

$$S \in S \to S \notin S \tag{1}$$

(ii) Let $S \notin S$: Again, from the definition of the set, S should belong to S. This means that if $S \notin S$, then $S \in S$.

$$S \notin S \to S \in S \tag{2}$$

From (1) and (2), we can say that 'S is a member of S iff S is not a member of S', i.e. $S \in S \leftrightarrow S \notin S$, which is a paradox. Many paradoxes exist in mathematics. The Liar paradox and the Barber paradox in logic are very commonly discussed. Drinking the whole ocean is another paradox in mathematical induction.

4.3 SUBSET

A set A is called a proper subset of B, denoted as $A \subset B$ if and only if every element of A is also an element of B and, in addition, B contains at least one element more than A. B is also called a superset of A, and is denoted by $B \supset A$. When set B contains all the elements of A and also the same number of elements, the sets are said to be equal, and this is denoted by $A \subseteq B$. Consider the three sets A, B and C subsets of $\mathbb{Z}^+$.

$A = \{1, 2, 4, 5, 6\}$, $B = \{1, 2, 4\}$ and $C = \{1, 2, 4, 5, 6\}$. The sets are related as $B \subset A$ and $B \subset C$. The sets A and C are equal and are written as $A \subseteq C$. We can define the concept of subset using the following logic: $B \subset A \leftrightarrow \forall x\ (x \in B \to x \in A)$

The two sets A and C are said to be equal if and only if all the elements of A are in C and vice versa, and this is denoted by $A = C$. It is expressed using logic as follows:

$$A = C \leftrightarrow A \subset C \text{ and } C \subset A$$

$$\leftrightarrow \forall x\ (x \in A \to x \in C) \text{ and } \forall x\ (x \in C \to x \in A)$$

Example 4.3: Let $\mathbf{U} = \{a, b, c, d, e, f, x, y, z, \{a, b\}, \{b, c, d\}, \{b, c, d, a\}\}$ be the universal set. Determine whether the following statements are true or false for the sets defined over **U** as $A = \{a, b, c\}$, $B = \{a, b, c, d\}$, $C = \{\{b, c, d\}\}$, $D = \{a, b, c\}$, $E = \{b, c, d\}$.

(i) $A \subset B$

(ii) $A = D$

(iii) $C = E$

(iv) $B \supset C$

Solution:

(i) $A \subset B$ is true because every element of A is in B, and therefore A is a subset of B.

(ii) $A = D$ is true because every element of A and D are the same.

(iii) $C = E$ is not true because the set E appears as an element in C. There are three elements in E but only one element in C. The cardinalities are different.

(iv) $B \supset C$ is not true because B has four elements, but C has one element, which is a set of three elements.

Two sets are said to be disjoint when they do not have any elements in common. In Example 4.3, the sets C, D and E are disjoint. The sets A, B and E are not disjoint. An intersection operation on disjoint sets always results in an empty set or a null set.

Example 4.4: Prove that an empty set is a subset of every set.

Proof: Consider any set A defined over the given universal set. $\varnothing$ is an empty set. It is true that $\varnothing \subset A$.

$\varnothing \subset A \leftrightarrow \forall x\ (x \in \varnothing \rightarrow x \in A)$	definition of subset
$\leftrightarrow \forall x\ (\text{False} \rightarrow x \in A)$	$x \in \varnothing$ is false as no elements
$\leftrightarrow \forall x\ (\text{True})$	from truth table for implication

Therefore, it is true for all sets.

Example 4.5: Prove that every set is a subset of itself.

Proof: Consider any set A defined over the given universal set. We can write the following.

$A \subset A \leftrightarrow \forall x\ (x \in A \rightarrow x \in A)$	definition of subset
$\leftrightarrow \forall x\ (\text{True})$	from truth table for implication

Therefore, it is true for all sets.

Example 4.6: Prove that the empty set is unique.

Proof: Let us assume that there are two empty sets, $\varnothing_1$ and $\varnothing_2$. Now consider $\varnothing_1$ is an empty set and $\varnothing_2$ is any other set. From the definition of an empty set, we can write the following.

$$\varnothing_1 \subset \varnothing_2 \quad (1)$$

Now consider $\varnothing_2$ as an empty set and $\varnothing_1$ as any other set. From the definition of an empty set, we can say:

$$\varnothing_2 \subset \varnothing_1 \quad (2)$$

From expressions (1) and (2), we can say that $\varnothing_1 \subset \varnothing_2$ and $\varnothing_2 \subset \varnothing_1$, which means $\varnothing_1 = \varnothing_2$. Hence proved.

4.4 POWER SET

The set of all the subsets of a given set A is called the power set of A and is denoted by $P(A)$. For example, If $A = \varnothing$, $P(A)$ is written as follows:

If $A = \varnothing$, then $P(A) = P(\varnothing) = \{\varnothing\} = \{\{\ \}\}$, and the cardinality is 1. It is 2^0.

If $A = \{1\}$, then $P(A) = \{\varnothing, \{1\}\}$, and the cardinality of $P(A)$ is 2. It is 2^1.

If $A = \{1, 2\}$, then $P(A) = \{\varnothing, \{1\}, \{2\}, \{1, 2\}\}$, and the cardinality is 4. It is 2^2.

If A = {1, 2, 3}, then $P(A) = \{\varnothing, \{1\}, \{2\}, \{3\}, \{1, 2\}, \{1, 3\}, \{2, 3\}, \{1, 2, 3\}\}$, and the cardinality is 8. It is 2^3. From the above examples, we can generalise that if the cardinality of |A| is n, then the cardinality of $|P(A)|$ is 2^n.

4.5 SET OPERATIONS

The operations on sets include (i) union, (ii) intersection, (iii) difference, (iv) symmetric difference, (v) complement and (vi) Cartesian product. Each of these operations results in a new set. Let U = {1, 2, 3, 4, 5, 6, 7, 8, 9}.

(i) **Union:** Let A and B be two sets defined over the universal set. The set A union B is denoted by ($A \cup B$) and defined as under.

$$A \cup B = \{x \mid x \in A \text{ or } x \in B\}$$

Consider the set A = {1 ,2, 3} and B = {2, 3, 4, 5}.

$$A \cup B = \{1, 2, 3, 4, 5\}.$$

(ii) **Intersection:** The intersection of two sets, say A and B, is denoted by $A \cap B$ and is defined as follows:

$$A \cap B = \{x \mid x \in A \text{ and } x \in B\}$$

Considering the sets used for illustrating the union operation, $A \cap B = \{2, 3\}$.

(iii) **Difference:** The difference of two sets, say A and B, is denoted by A – B and is defined as under.

$$A - B = \{x \mid x \in A \text{ and } x \notin B\}$$

Considering the sets used for illustrating the union operation, A – B = {1}.

(iv) **Symmetric difference:** The symmetric difference of two sets, say A and B, is denoted by $A \oplus B$ and is defined as under.

$$A \oplus B = \{x \mid (x \in A \text{ or } x \in B) \text{ and } (x \notin A \text{ or } x \notin B)\}$$
$$A \oplus B = \{x \mid (x \in A \cup B) \text{ and } (x \notin A \cap B)\}$$
$$A \oplus B = (A - B) \cup (B - A)$$

Considering the sets used for illustrating the union operation, $A \cup B = \{1, 2, 3, 4, 5\}$, $A \cap B = \{2, 3\}$, and the set $x \notin (A \cap B)$ means $x \in \{1, 4, 5, 6, 7, 8, 9\}$.
Thus, $A \oplus B = \{1, 4, 5\}$.
Alternatively, we can obtain A – B = {1}, B – A = {4, 5} and $A \oplus B = \{1, 4, 5\}$, which is equal to $\{1\} \cup \{4, 5\}$.

(v) **Complement:** Complement of a set, say A, with respect to the universal set is denoted by $\bar{A}$ and is defined as follows:

$\bar{A} = \{x \mid x \notin A\}$. The element x belongs to **U**.

Consider A = {1, 2, 3}. The complement of A = $\bar{A}$ = {4, 5, 6, 7, 8, 9}.

(vi) **Cartesian product:** The Cartesian product of two sets gives a new set that has elements that are pairs. If three sets are involved, then the elements are triplets and so on. The Cartesian product of two sets is denoted by $A \times B$ and defined as under.
Let A = {1, 2, 3} and B = {a, b}.

$$A \times B = \{(a, b) \mid a \in A \text{ and } b \in B\}$$
$$A \times B = \{(1, a), (1, b), (2, a), (2, b), (3, a), (3, b)\}$$

There are totally six pairs. $|A \times B| = mn$, where $|A| = m$ and $|B| = n$.

Example 4.7: Consider $\mathbf{U} = \{1, 2, 3, 4, 5, 6, 7, 8, 9\}$, $A = \{1, 2, 3, 4, 5\}$, $B = \{3, 4, 5, 6, 7\}$, $C = \{7, 8, 9\}$ and $D = \{2\}$. We can obtain the following.

(i) $A \cup B = \{1, 2, 3, 4, 5, 6, 7\}$

(ii) $A \cap B = \{3, 4, 5\}$

(iii) $B \cap C = \{7\}$

(iv) $A \cap C = \varnothing$

(v) $A \cup C = \{1, 2, 3, 4, 5, 7, 8, 9\}$

(vi) $A \oplus B = \{1, 2, 6, 7\}$

(vii) $A \oplus C = \{1, 2, 3, 4, 5, 7, 8, 9\}$

(viii) $A - B = \{1, 2\}$

(ix) $\overline{A} = \{6, 7, 8, 9\}$

(x) $D \times B = \{(2, 3), (2, 4), (2, 5), (2, 6), (2, 7)\}$

(xi) $B \times D = \{(3, 2), (4, 2), (5, 2), (6, 2), (7, 2)\}$

The Cartesian product is not commutative. From (x) and (xi), it is clear that $B \times D \neq D \times B$.

Example 4.8: Consider the sets $A = \{1, 2\}$, $B = \{3, 4\}$ and $C = \{5, 6\}$. Obtain (i) $A \times B$, (ii) $(B \times A)$ and (iii) $(A \times B \times C)$.

(i) $A \times B = \{(1, 3), (1, 4), (2, 3), (2, 4)\}$

(ii) $B \times A = \{(3, 1), (3, 2), (4, 1), (4, 2)\}$

(iii) $A \times B \times C = \{(1, 3, 5), (1, 3, 6), (1, 4, 5), (1, 4, 6), (2, 3, 5), (2, 3, 6), (2, 4, 5), (2, 4, 6)\}$

Relative Complement: Let A and $B \in \mathbf{U}$. The relative complement of A in B, denoted by $B - A$, is given by $\{x \mid x \in B \text{ and } x \notin A\}$. Consider the sets from Example 4.7. The relative complements are A in B, B in A, C in A, A in C and A in A. A in B is denoted by $B - A = \{6, 7\}$, B in A is denoted by $A - B = \{1, 2\}$, C in A is denoted by $A - C = A$, A in C is denoted by $C - A = C$ and A in A is denoted by $A - A = \varnothing$.

4.6 LAWS OF SET THEORY

The various laws of set theory are listed in Table 4.2.

Table 4.2 Laws of set theory

(i)	$\overline{\overline{A}} = A$	Law of double complement
(ii)	$\overline{A \cup B} = \overline{A} \cap \overline{B}$ $\overline{A \cap B} = \overline{A} \cup \overline{B}$	DeMorgan's laws
(iii)	$A \cup B = B \cup A$ $A \cap B = B \cap A$	Commutative laws
(iv)	$A \cap (B \cap C) = (A \cap B) \cap C$ $A \cup (B \cup C) = (A \cup B) \cup C$	Associative laws
(v)	$A \cap (B \cup C) = (A \cap B) \cup (A \cap C)$ $A \cup (B \cap C) = (A \cup B) \cap (A \cup C)$	Distributive laws
(vi)	$A \cup A = A$ $A \cap A = A$	Idempotent laws
(vii)	$A \cap \overline{A} = \varnothing$ $A \cup \overline{A} = \mathbf{U}$	Inverse laws
(viii)	$A \cup \varnothing = A$ $A \cap \mathbf{U} = A$	Identity laws

Continued

Continued

(ix)	$A \cup \mathbf{U} = \mathbf{U}$ $A \cap \varnothing = \varnothing$	Domination laws
(x)	$A \cup (A \cap B) = A$ $A \cap (A \cup B) = A$	Absorption laws

4.7 PRINCIPLE OF DUALITY

It is interesting to note that the laws given in Table 4.2 are dual with respect to the operations $\cup$ and $\cap$. They deals with equality of set expressions. The two expressions involving sets and only the operations $\cup$ and $\cap$ are said to be duals if we are able to get one expression from the other by replacing $\cup$ by $\cap$ and vice versa, and $\varnothing$ by **U** and vice versa. The sets can occur complemented in the expression.

Example 4.9: Consider the laws of absorption. Let the expression $s = A \cup (A \cap B)$ and its dual $s^d = A \cap (A \cup B)$ be obtained by replacing $\cup$ by $\cap$ and vice versa. Both the expressions are equal. It is a generic case.

Example 4.10: Consider *s*: $A \cap \overline{A} = \varnothing$ and its dual is obtained replacing $\cup$ by $\cap$ and vice versa as s^d: $A \cup A = \mathbf{U}$. Here, $\varnothing$ is replaced by **U** and vice versa.

The laws listed in Table 4.2 motivate us to state the duality principle as 'if s is a law on sets involving $\cup$ and $\cap$, then its dual is also a law on sets'.

Example 4.11: Prove that $\overline{A \cup B} = \overline{A} \cap \overline{B}$, using the definitions of set operations.

Proof:

$\overline{A \cup B} = \{x \mid x \notin (A \cup B)\}$
$= \{x \mid x \notin A \text{ and } x \notin B\}$
$= \{x \mid x \in \overline{A} \text{ and } x \in \overline{B}\}$
$= \{x \mid x \in \overline{A} \cap \overline{B}\}$
$= \overline{A} \cap \overline{B}$. Here, we have used the concept of set definition.

Alternatively, we can consider an element belonging to the set. The proof is modified as follows:

Proof:

Let $x \in \overline{A \cup B} = x \notin (A \cup B)$
$= x \notin A \text{ and } x \notin B$
$= x \in \overline{A} \text{ and } x \in \overline{B}$
$= x \in \overline{A} \cap \overline{B}$

Example 4.12: Prove that $A \cap (B \cup C) = (A \cap B) \cup (A \cap C)$

Proof:

$A \cap (B \cup C) = \{x \mid x \in A \cap (B \cup C)\}$
$= \{x \mid x \in A \wedge x \in (B \cup C)\}$
$= \{x \mid x \in A \wedge (x \in B \vee x \in C)\}$, Symbols of logical and/or used
$= \{x \mid (x \in A \wedge x \in B) \vee (x \in A \wedge x \in C)\}$, $\wedge$ is distributed over v
$= \{x \mid x \in (A \cap B) \vee x \in (A \cap B)\}$
$= \{x \mid x \in (A \cap B) \cup (A \cap B)\}$
$= (A \cap B) \cup (A \cap B)$, hence proved.

Example 4.13: Prove that $(A - B) = A \cap \overline{B}$

Proof:

$$\begin{aligned}(A - B) &= \{x \mid x \in A \wedge x \notin B\}\\ &= \{x \mid x \in A \wedge x \in \overline{B}\}\\ &= \{x \mid x \in (A \cap \overline{B})\}\\ &= (A \cap \overline{B}), \text{ hence proved.}\end{aligned}$$

Example 4.14: Prove that $A - (A - B) = (A \cap B)$

Proof:

$$\begin{aligned}A - (A - B) &= A - (A \cap \overline{B})\\ &= A \cap \overline{(A \cap \overline{B})}\\ &= A \cap (\overline{A} \cup B)\\ &= (A \cap (\overline{A}) \cup (A \cap B)\\ &= \varnothing \cup (A \cap B)\\ &= (A \cap B)\end{aligned}$$

Example 4.15: Show that $A - (A \cap B) = A - B$

Solution:

$$\begin{aligned}A - (A \cap B) &= A \cap \overline{(A \cap B)}\\ &= A \cap (\overline{A} \cup \overline{B})\\ &= (A \cap \overline{A}) \cup (A \cap \overline{B})\\ &= \varnothing \cup (A \cap \overline{B})\\ &= (A \cap \overline{B})\\ &= A - B\end{aligned}$$

Example 4.16: Show that $A - (B \cup C) = (A - B) \cap (A - C)$

Solution:

$$\begin{aligned}A - (B \cup C) &= A \cap \overline{(B \cup C)}\\ &= A \cap (\overline{B} \cap \overline{C})\\ &= (A \cap \overline{B}) \cap (A \cap \overline{C})\\ &= (A - B) \cap (A - C)\end{aligned}$$

Example 4.17: Show that $A - (B - C) = (A - B) - C$ if and only if $A \cap C = \varnothing$.

Solution:

$$\begin{aligned}A - (B - C) &= A - A \cap (B - C)\\ &= A - A \cap (B - B \cap C)\\ &= A - (A \cap B) + (A \cap B \cap C) \qquad (1)\end{aligned}$$

$$\begin{aligned}(A - B) - C &= (A - B) - (A - B) \cap C\\ &= A - (A \cap B) - A - (A \cap B) \cap C\\ &= A - (A \cap B) - A \cap C + (A \cap B) \cap C \qquad (2)\end{aligned}$$

When $A \cap C = \varnothing$, expressions (1) and (2) are the same.

Example 4.18: Show that $A \cap (B \cup C) = (A \cap B) \cup (A \cap C)$, using the membership table.

Solution:

Similar to logic tables, we can write a membership table for the set expressions and show that the corresponding columns are the same. This is shown in Table 4.3. This is an alternative way of proving the law.

Table 4.3 Membership table for distributive law

A	B	C	$B \cup C$	$A \cap (B \cup C)$	$A \cap B$	$A \cap C$	$(A \cap B) \cup (A \cap C)$
0	0	0	0	0	0	0	0
0	0	1	1	0	0	0	0
0	1	0	1	0	0	0	0
0	1	1	1	0	0	0	0
1	0	0	0	0	0	0	0
1	0	1	1	1	0	1	1
1	1	0	1	1	1	0	1
1	1	1	1	1	1	1	1

Example 4.19: Show that $\overline{A \cup (B \cap C)} = (\overline{C} \cup \overline{B}) \cap \overline{A}$

Solution:

$\overline{A \cup (B \cap C)} = \overline{A} \cap (\overline{B \cap C})$

$= \overline{A} \cap (\overline{B} \cup \overline{C})$

$= \overline{A} \cap (\overline{C} \cup \overline{B})$ Commutative law for $\cup$

$= (\overline{C} \cup \overline{B}) \cap \overline{A}$ Commutative law for $\cap$

4.8 COMPUTER REPRESENTATION OF SETS

There are different ways of representing sets on computers. One can represent sets as arrays without ordering. In this method, the set operations become difficult as they involve a large amount of searching. Another method is to represent the sets as graphs. The union operation becomes simple pointer manipulation. However, there is a method of representing the sets according to arbitrary ordering of the universal set. This method of storing sets makes operations on sets easy.

Consider a finite universal set. Arbitrarily order the elements of a universal set. Let $a_1, a_2, \ldots a_n$. Define the sets on this universal set, and represent it as a bit string of length n wherein the ith bit is set to 1 if the element a_i belongs to the given subset; otherwise, the corresponding bit is set to zero.

Example 4.20: Consider the universal set **U** = {1, 2, 3, 4, 5, 6, 7, 8, 9, 10} and an increasing order imposed on the set **U**. What bit strings represent the subset of all odd integers in **U**, the subset of all even integers in **U** and the subset of integers not exceeding 5 in **U**?

Solution:

The bit string representing the subset of all odd integers {1, 3, 5, 7, 9} in **U** is given by

1010101010

The first, third, fifth, seventh and ninth bit positions contain 1, and other positions contain zeros. Similarly, the bit string corresponding to the subset of even integers {2, 4, 6, 8, 10} is given by

0101010101

The bit string corresponding to all integers that do not exceed 5, namely {1, 2, 3, 4, 5}, is given by

1111100000

Example 4.21: Obtain the complements of the sets given in Example 4.19, namely, A = {1, 3, 5, 7, 9}, B = {2, 4, 6, 8, 10} and C = {1, 2, 3, 4, 5} given with respect to **U**, as follows: $\bar{A}$ = {2, 4, 6, 8, 10} = B, and the bit string is 0101010101, which is obtained by replacing the 0s with 1s and vice versa, – in other words, taking 1's complement of the bit string corresponding to A. Similarly, the complement of C is 0000011111, which corresponds to all integers in **U** that exceed 5.

Example 4.22: Obtain the union and intersection of A and B defined in Example 4.21 and also by their string representations.

$$A \cup B = \{1, 2, 3, 4, 5, 6, 7, 8, 9, 10\}$$
$$A \cap B = \{\} = \emptyset$$

The bit strings corresponding to A and B are 1010101010 and 0101010101, respectively. The bit string corresponding to A ∪ B is obtained by logical OR of the bit strings and is 1111111111. This is the result of (1010101010) ∨ (0101010101), and the set it represents is {1, 2, 3, 4, 5, 6, 7, 8, 9, 10}.

The intersection of A and B is obtained by its bit strings using logical AND operation. The resulting string is (0000000000), which is equal to (1010101010) ∧ (0101010101). The corresponding set is { }.

The intersection of the sets A and C is the set {1, 3, 5}. This is obtained by AND operation on the bit strings, namely, (1010101010) ∧ (1111100000). The resulting string is 1010100000, which corresponds to the set {1, 3, 5}.

4.9 VENN DIAGRAMS

In Venn diagrams, we represent the universal set by a rectangle and the sets by circles. The circles are shown intersecting if they are not disjoint. The set B, which is a subset of the set A, is shown in Fig. 4.1.

The union, intersection and difference of two sets A and B are shown in Figs. 4.2 – 4.6 respectively. The hatched region in the diagram indicates the members of the new set. **U** is the universal set. The elements in the intersected region or regions indicate the elements common to the sets (Figs. 4.7 and 4.8). The disjoint sets do not have elements in common, and hence there is no intersection of sets (Fig. 4.9).

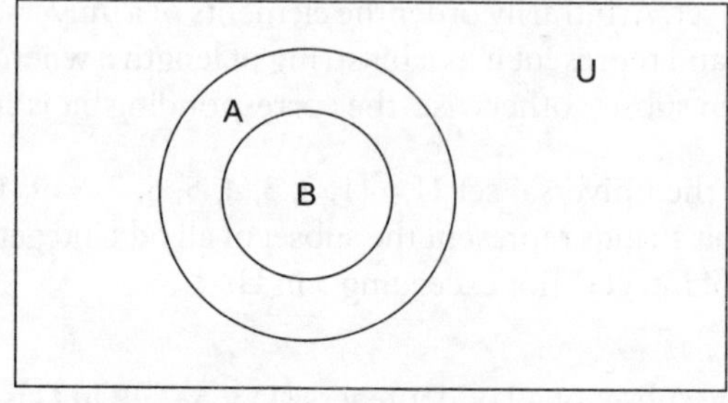

Figure 4.1 Set B is subset of set A

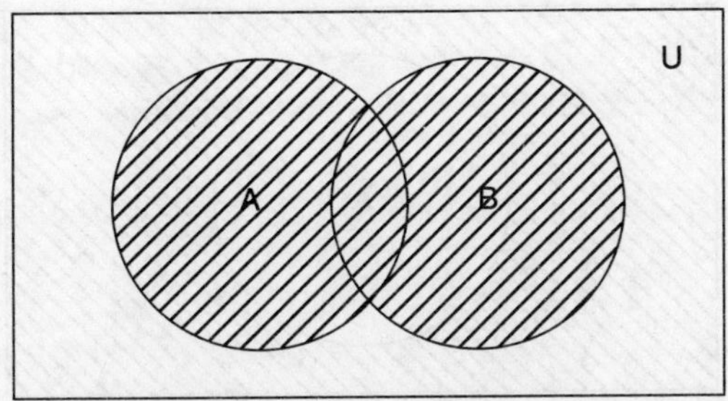

Figure 4.2 Union operation (A ∪ B)

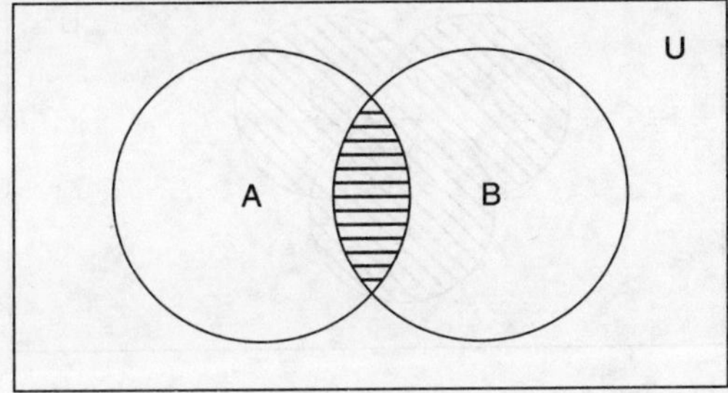

Figure 4.3 Intersection operation (A ∩ B)

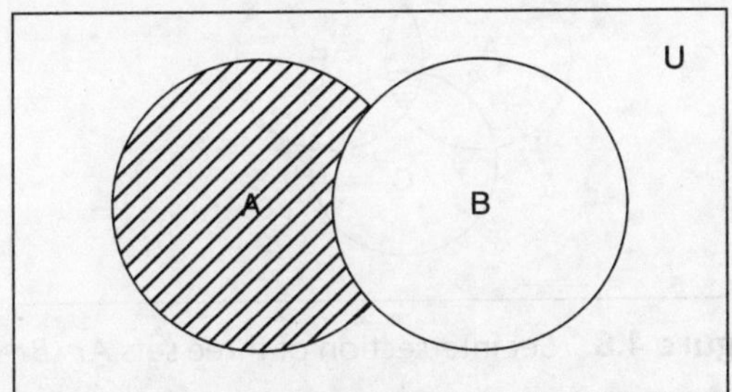

Figure 4.4 Set difference operation (A – B)

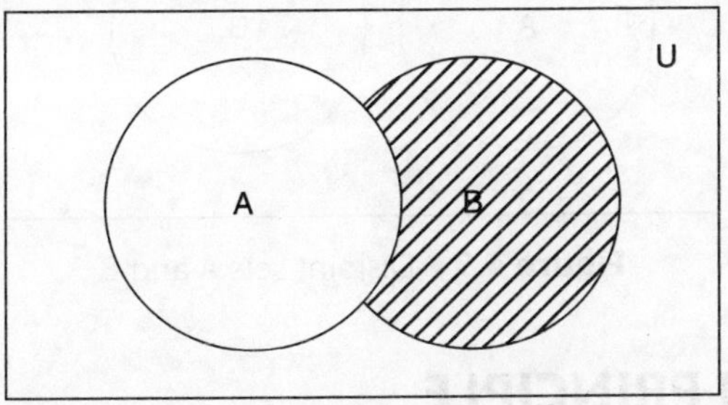

Figure 4.5 Set difference operation (B – A)

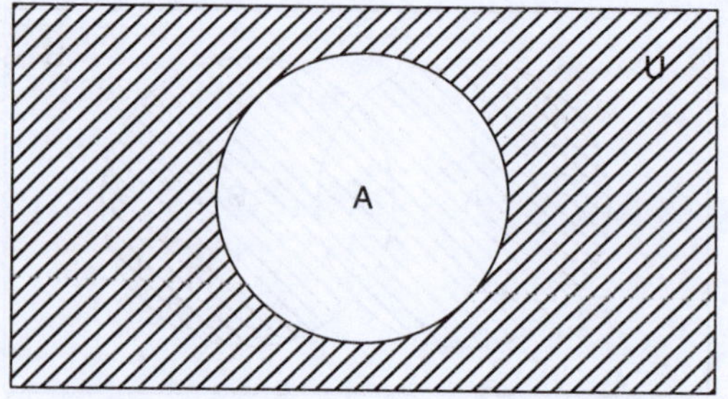

Figure 4.6 Set complement $\overline{A}$

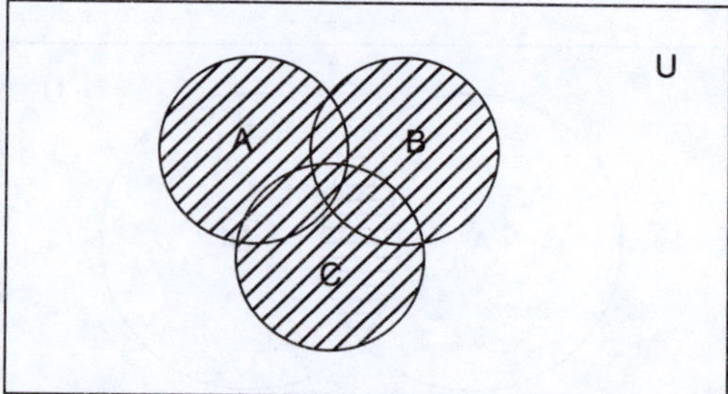

Figure 4.7 Set union of three sets $A \cup B \cup C$

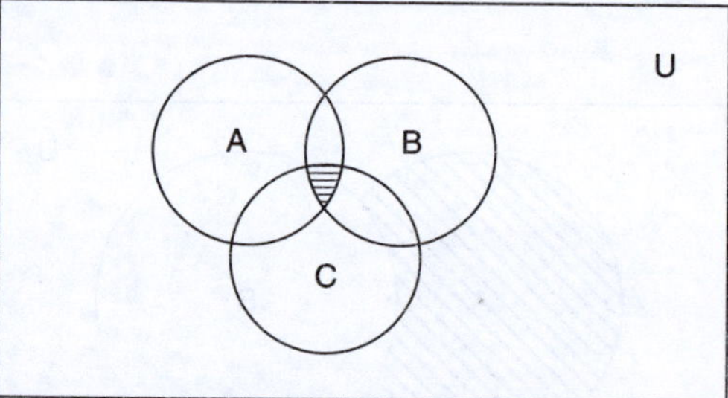

Figure 4.8 Set intersection of three sets $A \cap B \cap C$

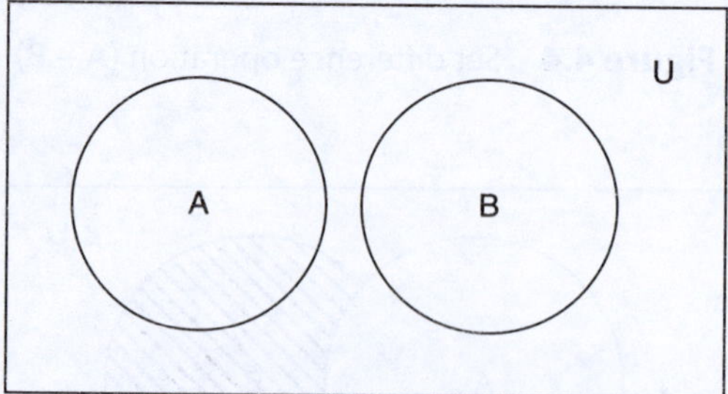

Figure 4.9 Disjoint sets A and B

4.10 ADDITION PRINCIPLE

In the case of sets, the union of two sets involves the common elements to be counted only once. When we work with cardinalities of individual sets, it is useful to consider the common elements to find the cardinality of |A ∪ B|. The cardinality of |A ∪ B| is equal to the arithmetic sum of cardinalities of the individual sets when the sets are disjoint. |A ∪ B| = |A| + |B|. If the sets are not disjoint, it is written as follows:

$$|A \cup B| = |A| + |B| - |A \cap B| \quad (1)$$

The elements of A ∩ B are counted twice, once in A and again in B, and hence it is deducted once. In the case of disjoint sets, |A ∩ B| = ∅. Equation (1) is called the Addition principle or the Principle of Inclusion and Exclusion. It can be extended to more than two sets.

Example 4.23: Consider A = {1, 2, 3, 4, 5} and B = {3, 5, 6, 8, 11, 13}. The addition principle can be verified as follows:

$$|A| = 5, |B| = 6, |A \cup B| = 9 \text{ and } |A \cap B| = 2.$$

The sets are A ∪ B = {1, 2, 3, 4, 5, 6, 8, 11, 13} and A ∩ B = {3, 5}. From the addition principle,

$$\begin{aligned}|A \cup B| &= |A| + |B| - |A \cap B| \\ &= 5 + 6 - 2 \\ &= 9, \text{ hence the principle.}\end{aligned}$$

The addition principle extended for three sets; |A ∪ B ∪ C|, is derived as follows:
We will consider D = B ∪ C and rewrite equation (1) as

$$|A \cup D| = |A| + |D| - |A \cap D|,$$

where the cardinality of set |D| = |B ∪ D| = |B| + |C| − |B∩C |, and |A ∪ B ∪ C| is written as

$$\begin{aligned}|A \cup B \cup C| &= |A| + |B| + |C| - |B \cap C| - |A \cap (B \cup C)| \\ &= |A| + |B| + |C| - |B \cap C| - |(A \cap B) \cup (A \cap C)|,\end{aligned}$$

where |(A ∩ B) ∪ (A ∩ C) | = | A ∩ B| + | A ∩ C| − |A ∩ B ∩ C | and |A ∪ B ∪ C| is written as

$$|A \cup B \cup C| = |A| + |B| + |C| - |A \cap B| - |A \cap C| - |B \cap C| + |A \cap B \cap C|$$

Thus, the formula is for three sets. Further, it can be generalised.

Example 4.24: A class of 30 students includes 13 who play the piano, 16 who play the guitar and 5 who play both the piano and the guitar. How many students in the class play the piano or the guitar? How many play neither instrument? The Venn diagram is given in Fig. 4.10.

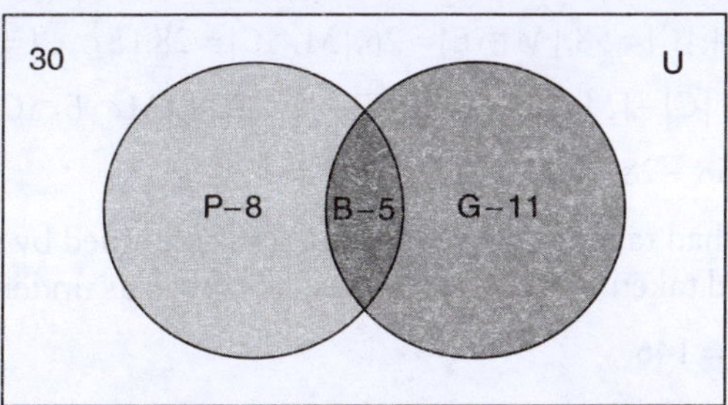

Figure 4.10 Venn diagram for example 4.24

Solution:
Students playing the piano or the guitar are 13 + 16 − 5 = 24, and students who play neither of these instruments are 30 − 24 = 6. Let P represent piano, G represent guitar and B represent both the instruments.

The Venn diagram of three arbitrary non-disjoint sets A, B and C divides the universal set U into $2^3 = 8$ regions. Note: A Venn diagram of four sets A1, A2, A3 and A4 will contain $2^4 = 16$ regions. The eight regions contain the elements listed here.

1. In A, B and C
2. In A and B, but not C
3. In A and C, but not in B
4. In B and C, but not in A
5. Only in A
6. Only in B
7. Only in C
8. In none of A, B, C

Example 4.25: In a group of 300 students, 64 had taken a mathematics course, 94 had taken an English course, 58 had taken a computer course, 28 had taken both a mathematics and a computer course, 26 had taken both an English and a mathematics course, 22 had taken both an English and a computer course and 14 had taken all the three courses. How many students had taken none of the three courses?

Solution:

Let M, E and C denote the sets of students taking mathematics, English and computer courses, respectively. The Venn diagram for Example 4.25 is given in Fig. 4.11.

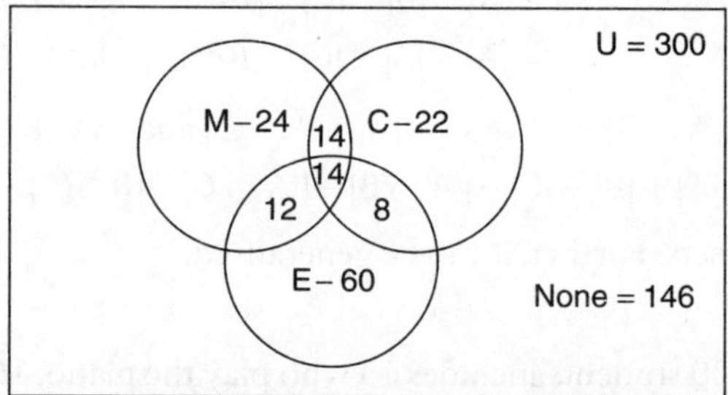

Figure 4.11 Venn diagram of Example 4.25

$$|M| = 64, |E| = 94, |C| = 58, |M \cap E| = 26, |M \cap C| = 28, |E \cap C| = 22, |M \cap E \cap C| = 14$$

$$|M \cup E \cup C| = |M| + |E| + |C| - |M \cap C| - |M \cap E| - |E \cap C| + |M \cap E \cap C|$$

$$= 64 + 94 + 58 - 28 - 26 - 22 + 14 = 154.$$

The number of students who had taken none of the courses is obtained by subtracting from the group the number of people who had taken one of the courses. It is given as under

$$= 300 - 154 = 146$$

Alternative Method:

Consider the Venn diagram given in Fig. 4.12. The regions in the Venn diagram are marked by integers from 1 to 8. Let R1 to R8 represent the cardinalities of these regions. The cardinality of $|M \cup E \cup C|$ is given by

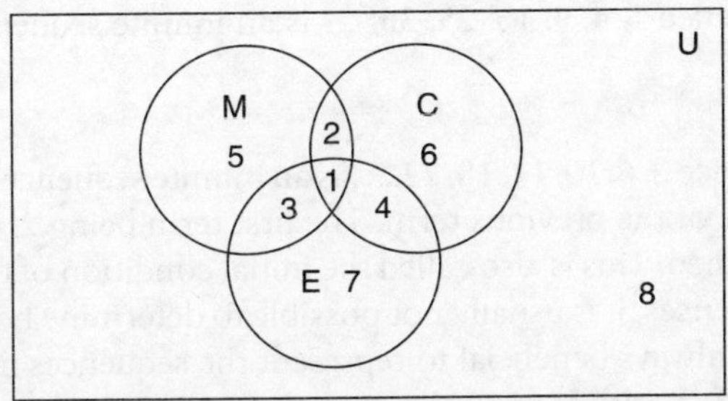

Figure 4.12 Alternative method for three sets problem

$$|M \cup E \cup C| = R1 + R2 + R3 + R4 + R5 + R6 + R7$$
$$|M| = R1 + R2 + R3 + R5 = 64 \quad |C| = R1 + R2 + R4 + R6 = 58 \quad |E| = R1 + R3 + R4 + R7 = 94$$
$$|M \cap C| = R1 + R2 = 28, \quad |M \cap E| = R1 + R3 = 26 \quad |E \cap C| = R1 + R4 = 22$$
$$|M \cap E \cap C| = R1 = 14$$

Thus, we get R1 = 14, R2 = 14, R3 = 12, R4 = 8, R5 = 24, R6 = 22, R7 = 60

$$\begin{aligned} |M \cup E \cup C| &= R1 + R2 + R3 + R4 + R5 + R6 + R7 \\ &= 14 + 14 + 12 + 8 + 24 + 22 + 60 \\ &= 154 \end{aligned}$$

The number of students who had taken none of the courses is R8 and is equal to 300 – 154 = 146. This approach is very useful when we are interested in knowing how many students took only mathematics, which is equal to 24. Other such questions are easily answered through this approach.

4.11 SEQUENCES

Consider the list of elements A = {1, 4, 3, 5, 2, 6} and B = {1, 2, 3, 4, 5, 6}. These two are sets and are equal as they have the same elements. Consider the same elements in different representations as C = (1, 4, 3, 5, 2, 6) and D = (1, 2, 3, 4, 5, 6). The two lists C and D are different as they represent two different sequences. In C, 4 is the second element, whereas in D it is the fourth element. Similarly, 2 is the second element in D, whereas it is the fifth element in C. Likewise, in the sequence, the elements are ordered. We talk about the first element, second element, third element and so on. The two sequences are said to be equal when their elements and the order of the elements are the same. Similar to sets, we also talk of finite and infinite sequences. In the case of sets the ordering is not essential. Thus, we can say that the ordered set is a sequence and the parentheses are used as sequence symbols.

Example 4.26: (A, B, C, D, E, …, Z) is a sequence of the uppercase alphabet in the English language. Here A is the first character, and Z is the twenty-sixth character.

Example 4.27: The binary representation of the decimal digit 5 is 0101, which is a string of 1's and 0's.

Example 4.28: The sequence 10010100111 is a finite sequence of 1's and 0's. The digit zero occurs in the second, third, fifth, seventh and eighth positions.

Example 4.29: The sequence 1, 4, 9, 16, 25, 36, … is an infinite sequence representing squares of positive integers.

Example 4.30: The sequence 2, 6, 10, 14, 18, 22, … is an infinite sequence where the value of the next term depends upon the value of the previous term. The first term being 2, the next terms are defined. The terms before 2 are undefined. This is also called the initial condition of the sequence.

In the case of infinite sequences, it is usually not possible to determine how the sequence progresses from a few initial terms. It is always beneficial to represent the sequences in a more compact manner. For example, in Example 4.29, the *n*th term in the sequence is given by $b_n = n^2$, where $(1 \leq n \leq \infty)$. The term depends upon its position, and hence it is called an explicit representation of the terms in the sequence.

On the other hand, consider Example 4.30. The successive terms differ from their preceding terms by an integer value 4. Thus, the next terms are obtained by adding 4 to the previous terms. The positions of the terms in the sequence are subscripted, and the sequence is defined as under.

$$b_1 = 2 \quad \text{and} \quad b_n = b_{(n-1)} + 4$$

This is called the recursive definition of the sequence. Every recursive formula must include the initial condition.

Example 4.31: Let $b_1 = 3$ and $b_n = b_{(n-1)} + 5$ represent the sequence 3, 8, 13, 18, 23, …

Example 4.32: Let $b_1 = 3$ and $b_n = 2 \times b_{(n-1)}$ represent the sequence 3, 6, 12, 24, 48, …

Example 4.33: Let $b_1 = 0$ and $b_2 = 1$. $b_n = b_{(n-1)} + b_{(n-2)}$ represent the sequence 0, 1, 1, 2, 3, 5, …, which is called the Fibonacci sequence.

Example 4.34: Let $b_n = 100 - b_n$ be the explicit definition and represent the sequence 99, 98, 97, …

An array represents a sequence. The contents of a stack represent a sequence. English words represent a sequence of characters. This sequence of characters is called a string. In computer languages, provisions are given to define strings and manipulate strings using built-in functions. In the C language environment, the 'string.h' header file provides all these facilities.

4.12 STRINGS

We define strings over an alphabet. An alphabet is a finite non-empty set of symbols. For example, the English language alphabet comprises 26 letters, punctuation marks etc. Every computer language also has an alphabet, from which we start when studying the language. We become acquainted with these symbols. We use the symbol Σ to denote the alphabet set; for example let $\Sigma = \{a, b, c\}$.

Word or String or Sentence: It is a finite sequence of symbols of the alphabet. For example, *a*, *b*, *c*, *ab*, *abc*, *abba*, are the sequences representing the strings or sentences.

Length of word or sentence: Let *w* be a word defined over the Σ alphabet. The length of *w*, denoted as $|w|$, represents the number of symbols in the word. For example, $|abc| = 3$ and $|a| = 1$. The two words $a_1 \ldots a_r$ and $b_1 \ldots b_s$ are equal if $(r = s)$ and $a_i = b_i$ each i, $1 \leq i \leq r$ or $1 \leq i \leq s$. We perform two operations on the strings called concatenation and Kleene's closure. Let $S_1 = \{a, b, c\}$ and $S_2 = \{d\}$ be two sets of strings.

$S_1 \bullet S_2 = \{ad, bd, cd\}$, where • is a concatenation operator. Every symbol of S_1 is joined with a symbol of S_2. Consider $S_1 = \{a, b\}$ and $S_2 = \{c, d\}$. The concatenation of these is given by $S_1 \bullet S_2 = \{ac, ad, bc, bd\}$. The number of elements in $S_1 \bullet S_2$ is equal to $|S_1| \times |S_2|$, which is $2 \times 2 = 4$. There are four elements in $S_1 \bullet S_2$. Further, the set of all words or sentences defined over an alphabet is called Kleene's closure. For example, $\Sigma = \{d\}$. Kleene's closure is denoted as $\Sigma^* = \{\lambda, d, dd, ddd, ddd, \ldots\}$. It is an infinite set. The symbol λ denotes a null string or word, whose length is zero. The words of length 0, 1, 2 and so on are present in the set Σ^*.

$\Sigma^+ = \Sigma^* - \{\lambda\} = \{d, dd, ddd, \ldots\}$. It is called the positive closure that excludes λ.

$\Sigma^* = \Sigma^+ \cup \{\lambda\}$. One important aspect to note here is that concatenation operation is not commutative, $a \bullet b \neq b \bullet a$, i.e. $ab \neq ba$. The strings are useful in regular expressions.

4.13 REGULAR EXPRESSIONS

The Regular sets, Regular expressions, Regular languages and Regular grammar are alternative representations. A regular expression defines the regular language for which it is possible to write regular grammar. The regular languages are recognised by finite state automata. Hence they are all equivalent. Regular expressions are used in the Unix operating system to denote commands and other utilities. Regular expressions are used in lex generators, text editors etc.

The given regular expression precisely defines a language. The regular expression is defined recursively as follows:

1. $\in$ is a regular expression, and ϕ is a regular expression.
2. Every symbol in the alphabet Σ is a regular expression.
3. Let R_1 and R_2 be two regular expressions. Then $R_1 \cup R_2$ is also a regular expression.
4. The concatenation of R_1 and R_2 is also a regular expression $(R_1 \bullet R_2)$.
5. The closure of R_1 or R_2 denoted by $(R_1)^*$ or $(R_2)^*$ is also a regular expression.
6. The notations (R_1) and (R_2) are also regular expressions.

Further, the language represented by the null string ($\in$) is denoted by $L(\in) = \{\in\}$, and the language generated by an empty set (ϕ) is denoted by $L(\phi) = \phi$. For '*a*' belonging to the alphabet Σ, $a \in \Sigma$, $L(a) = \{a\}$ is the language. Let R_1 and R_2 be regular expressions. In certain books $R_1 \cup R_2$ is also denoted by $R_1 + R_2$. The language represented by $R_1 + R_2$ is denoted by $L(R_1 + R_2) = L(R_1) \cup L(R_2)$. It is the union of language represented by each expression. The language represented by the concatenation of R_1 and R_2, denoted by $(R_1 \bullet R_2)$, is the concatenation of the languages represented by R_1 and R_2 and is denoted by $L(R_1) \bullet L(R_2)$. If R is a regular expression, R^* is also a regular expression (RE). Kleene's closure and language represented by R are related as $L(R^*) = (L(R))^*$. The operators used in RE are +, • and *. The priorities associated with these operators are as follows:

(i) * has the highest priority, (ii) • has the second-highest priority and (iii) + has the least priority. Consider the following examples illustrating the use of regular expressions.

Example 4.35: Let $\Sigma = \{0, 1\}$ be the alphabet. As per the recursive definition of RE, the symbols of the alphabet, namely 0 and 1, are REs. The null string $\in$ is an RE, and ϕ is an RE. Hence, ϕ, $\in$, 0 and 1 are the four REs possible over the Σ alphabet.

Example 4.36: Let $\mathbf{ab}^*$ be a regular expression. The language generated by this is $\{a, ab, abb, \ldots\}$. All the strings comprise the starting symbol **a** followed by zero or more **b**'s.

Example 4.37: (i) Let $\Sigma = \{a, b\}$. The language over Σ comprises all strings formed by the symbols of Σ and ending with **aab**. Give the regular expression. All the strings defined over the alphabet Σ are given by Σ^*, i.e. $\{a, b\}^* = \{\in, a, b, aa, ab, bb, ba, \ldots\}$. To make these strings end with '**aab**', we need to concatenate the elements of this set with '*aab*'. Hence, the RE is written as

$$(a+b)^*aab.$$

(ii) On the same alphabet, we can write an RE for all the strings that start with **a** and end with **b** as follows:

$$a(a+b)^*b.$$

Example 4.38: Consider the regular expression defined over the alphabet $\Sigma = \{0, 1\}$ as $01(01)^*$. Find the language generated by it.

The language generated by $(01)^*$ is the set of all strings consisting of zero or more occurrences of 01 and is $\{\in, 01, 0101, 010101, \ldots\}$. Since the expression has 01 preceding it, i.e. $01\,(01)^*$, the resulting language set is obtained by concatenating each string of the set corresponding to $(01)^*$. Thus, the resulting language set is $\{01, 0101, 010101, \ldots\}$. We call it a regular set.

Example 4.39: The regular expression $(01 + 10)^*\,11$ defined over the alphabet $\Sigma = \{0,1\}$ comprises the strings that are formed by concatenating a finite number of 01 and 10 substrings followed by 11. The resulting set of sentences is $\{11, 0111, 1011, 011011, \ldots\}$.

Example 4.40: Find the regular expression for the language of all the strings of odd length over the alphabet $\Sigma = \{a, b\}$.

- The regular expression for the strings of length two can be obtained by the following RE:

$$(a+b)(a+b)$$

- All strings of even length can be obtained by applying the * closure to the above expression

$$((a+b)(a+b))^*$$

- Finally, we can find all strings of odd length by concatenating $(a+b)$ with the above expression

$$((a+b)(a+b))^*(a+b)$$

Example 4.41: Find the regular expression for the language defined as $L = \{a^n b^m : n \geq 4, m \leq 3\}$.

In the above language, the minimum number of *a*'s is 4, and the maximum number of *b*'s is 3. The regular expression for the minimum number of *a*'s is *aaaa*; we require strings with the number of *a*'s as 4 and more. Hence, concatenate a^* to the above expression. The regular expression will then be $aaaaa^*$. To this regular expression, we concatenate all strings of *b*'s with maximum length 3 and get the final regular expression as follows:

$$aaaaa^*(\in + b + b\,b + b\,b\,b)$$

Example 4.42: Find the regular expression for the language defined as $L = \{vwv : v, w \in \{a, b\}^* \text{ and } |v| = 2\}$.

Regular expression for the substring *v* with length 2 is $(a+b)(a+b)$.
Regular expression for the substring *w* is $(a+b)^*$
The final regular expression will be $(a+b)\,(a+b)\,(a+b)^*(a+b)(a+b)$.

4.13.1 Algebraic Laws of Regular Expressions

Let P, Q and R be regular expressions. In this section, λ is used to represent the null string as $\in$ and λ are used interchangeably. We can manipulate the regular expressions as follows:

(i) $R + R = R$
(ii) $PQ + PR = P(Q + R)$
(iii) $PQ + RQ = (P + R)Q$
(iv) $R^*R^* = R^*$
(v) $RR^* = R^*R$
(vi) $(R^*)^* = R^*$
(vii) $\lambda + RR^* = R^*$
(viii) $(PQ)^*P = P(QP)^*$

Example 4.43: Prove that the set of strings in which every 0 is immediately followed by at least two 1's can be described by the following regular expressions R_1 and R_2:

$$R_1 = \lambda + 1^* (011)^* (1^*(011)^*)^*$$
$$R_2 = (1 + 011)^*$$

Proof:

$$R_1 = \lambda + 1^* (011)^* (1^*(011)^*)^*$$

Using the manipulating rules, we can write the RE as follows:

$$R_1 = (1^*(011)^*)^*$$
$$R_1 = (1 + 011)^* = R_2$$

Consider the REs $(PQ)^*P$ and $P(QP)^*$. In both the cases we can write the regular set as

$$\{P, PQP, PQPQP, \ldots\}$$

Consider $(P + Q)^*$, which consists of all the strings that are formed by concatenating Ps and Qs, including the null string $\in$. This is also obtained by $(P^* + Q^*)^*$ and $(P^*Q^*)^*$; therefore,

$$(P + Q)^* = (P^* + Q^*)^* = P^*Q^*$$
$$(P + Q)^* = (QP^*)^* = (P^*Q)^*P^*$$
$$\lambda + (P + Q)^*Q = (P^*Q)^*$$
$$(P^*Q)^* = \lambda + (P + Q)^*Q.$$

It is also common to call the set of strings generated by the regular expressions as regular sets, as mentioned earlier. Thus, regular sets, regular expressions, regular grammar and regular language are all analogous to each other.

ADDITIONAL SOLVED EXAMPLES

Example W4.1: For sets A and B, prove that $A \cap B = A$ if and only if $A \subseteq B$.

Solution:

Let us assume that $A \cap B = A$ and prove that $A \subseteq B$. Suppose $x \in A$, which means $x \in A \cap B$. This is because $A \cap B = A$. Therefore x belongs to both A and B. More particularly x is in B. This proves that $A \subseteq B$.

Now we will assume that $A \subseteq B$ and prove that $A \cap B = A$. To prove the equality of $A \cap B$ and A, we shall prove that each set is a subset of the other. By the definition of intersection, $(A \cap B) \subseteq A$. Suppose $x \in A$. Since $A \subseteq B$, x is in B too. Thus x is in both A and B. Therefore, $A \cap B = A$.

Example W4.2: Consider the following statements with respect to sets, and state whether true or false.

(i) Let S denote the set {{a}, b, c}.
  (a) $a \notin S$
  (b) $\{a\} \in S$
(ii) $\{\varnothing\} = \varnothing$
(iii) $\{\varnothing\} \in \{\{\varnothing\}\}$
(iv) $\varnothing \subseteq \{\{\varnothing\}\}$
(v) $\{\varnothing\} \subseteq \{\{\varnothing\}\}$

Solution:

(i) Both the statements are true. The set S contains {a} as an element but not the element a.
(ii) This statement is false because, the left hand side is a set containing an empty set as an element and the right hand side is just an empty set.
(iii) It is true because the left hand side is a set containing an empty set as an element and the right hand side is a set containing a set containing an empty set.
(iv) It is true because an empty set is a subset of every set.
(v) It is false because the set containing an empty set is a member, not the subset.

Example W4.3: If A = {1, 2, 3}, B = {4, 5} and C = {1, 2, 3, 4, 5} find $A \times B$, $C \times B$ and $B \times B$. Hence prove that $(C \times B) - (A \times B) = (B \times B)$.

Solution:

$$A \times B = \{(1, 4), (1, 5), (2, 4), (2, 5), (3, 4), (3, 5)\}$$
$$C \times B = \{(1, 4), (1, 5), (2, 4), (2, 5), (3, 4), (3, 5), (4, 4), (4, 5), (5, 4), (5, 5)\}$$
$$B \times B = \{ (4, 4), (4, 5), (5, 4), (5, 5)\}$$
$$(C \times B) - (A \times B) = \{(4, 4), (4, 5), (5, 4), (5, 5)\} = B \times B$$

Example W4.4: If $A_i = [\,0, I\,]$, where $i \in Z$, the set of integers. Find (i) $A_1 \cup A_2$ (ii) $A_3 \cap A_4$ and (iii) $\bigcup_{i=5}^{10} Ai$

Solution:

(i) $A_1 \cup A_2$ consists of real numbers in the intervals [0, 1] and [0, 2]. The set $A_1 \cup A_2$ consists of real numbers in the interval [0, 2], which is A_2.
(ii) $A_3 \cap A_4$ consists of all real numbers which lie in both the intervals [0, 3] and [0, 4]. The set $A_3 \cap A_4$ consists of real numbers in the range [0, 3], which is A_3.
(iii) $\bigcup_{i=5}^{10} Ai$ denotes the union of sets A_5, A_6, A_7, A_8, A_9, A_{10}, which means the union of [0, 5], [0, 6], [0, 7], [0, 8], [0, 9], [0, 10]. Therefore, $\bigcup_{i=5}^{10} Ai = [0, 10] = A_{10}$

Example W4.5: In a class of 25 students, 12 have taken mathematics, 8 have taken mathematics but not biology, Find the number of students who have taken mathematics and biology and those who have taken biology but not mathematics.

Solution:

Let n(A) denote the cardinality of the set A. Let A and B be sets of students who have taken mathematics and biology respectively. (A – B) is the set of students who have taken mathematics but not biology.

$$n(A) = 12, n(A - B) = 8 \text{ and } n(A \cup B) = 25$$

We know that $n(A - B) + n(A \cap B) = n(A)$. $8 + n(A \cap B) = 12$. That is, $n(A \cap B) = 12 - 8 = 4$

Therefore 4 students have taken both mathematics and biology. Using addition principle we write as under

$$n(A \cup B) = n(A) + n(B) - n(A \cap B)$$

$$25 = 12 + n(B) - 4\text{, Therefore, } n(B) = 17$$

$n(B - A) + n(A \cap B) = n(B)$, $n(B - A) = n(B) - n(A \cap B) = 17 - 4 = 13$, Hence, 13 students have taken biology, but not mathematics.

SUMMARY

Sets are important concepts used in computer science. They are carried forward into other topics of discrete mathematics, such as relations, functions, regular expressions, coding theory etc. Students need to remember set identities and algebraic rules of set theory. The coverage of the topics such as laws of set theory, operations on sets, Venn diagrams and regular expressions is more lucid. A sound knowledge of set theory fundamentals is essential for computer science students. This chapter fairly covers the essentials of set theory for reading the subsequent chapters in this book.

EXERCISES

1. Determine which of the following sets are equal and why?
   (a) $\{a, b, c, d, e\}$ (b) $\{c, b, a, c, d, e\}$
   (c) $\{c, a, b, c, a, d, e\}$ (d) $\{a, b, b, c, d, e\}$
2. Let $S = \{1, \{1\}, \{2\}\}$. With respect to S, which of the following statements are true?
   (a) $1 \in S$ (b) $\{1\} \in S$
   (c) $\{1\} \subseteq S$ (d) $\{\{1\}\} \subseteq S$
   (e) $\{2\} \in S$ (f) $\{2\} \subseteq S$
   (g) $\{\{2\}\} \subseteq S$ (h) $\{\{2\}\} \subset S$
3. Find which of the following statements are true?
   (a) $\Phi \in \Phi$ (b) $\Phi \subset \Phi$
   (c) $\Phi \subseteq \Phi$ (d) $\Phi \in \{\Phi\}$
   (e) $\Phi \subset \{\Phi\}$ (f) $\Phi \subseteq \{\Phi\}$
4. Determine the elements in each of the following sets.
   (a) $\{1 + (-1)^n \mid n \in \mathbb{N}\}$
   (b) $\{n + (1/n) \mid n \in \{1, 2, 3, 5, 7\}\}$
   (c) $\{n^3 + n^2 \mid n \in \{0, 1, 2, 3, 4\}\}$
5. Let $\mathbb{Z}$ be the universal set, and the following are six subsets defined on $\mathbb{Z}$
   $A = \{2m + 1 \mid m \in \mathbb{Z}\}$ $B = \{2n + 3 \mid n \in \mathbb{Z}\}$
   $C = \{2p - 3 \mid p \in \mathbb{Z}\}$ $D = \{3r + 1 \mid r \in \mathbb{Z}\}$
   $E = \{3s + 2 \mid s \in \mathbb{Z}\}$ $F = \{3t - 2 \mid t \in \mathbb{Z}\}$

Which of the following statements are true, and which are false?

(a) A = B
(b) A = C
(c) B = C
(d) D = E
(e) D = F
(f) E = F

6. Which of the following sets are empty?
   (a) $A = \{x \mid x \in N, 2x + t = 3\}$
   (b) $B = \{x \in Z \mid 3x + 5 = 9\}$
   (c) $C = \{x \mid x \in Q, x^2 + 4 = 6\}$
   (d) $D = \{x \in R \mid x^2 + 4 = 6\}$
   (e) $E = \{x \in R \mid x^2 + 3x + 3 = 0\}$
   (f) $F = \{x \mid x \in C, x^2 + 3x + 3 = 0\}$
7. Determine which row of Pascal's triangle contains consecutive entries that are in the ratio 1: 2: 3.
8. Write a computer program (or develop an algorithm) that lists all the subsets of {1, 2, 3,..., *n*}, where $1 \leq n \leq 15$. (The value of *n* should be supplied during the execution of the program.)
9. Let **U** = {1, 2, 3,..., 9, 10}, let A = {1, 2, 3, 4, 5}, B = {1, 2, 4, 8}, C = {1, 2, 3, 5, 7} and D = {2, 4, 6, 8}. Determine each of the following:
   (a) $(A \cup B) \cap C$
   (b) $A \cup (B \cap C)$
   (c) $\overline{C} \cup \overline{D}$
   (d) $\overline{C \cap D}$
   (e) $(A \cup B) - C$
   (f) $A \cup (B - C)$
   (g) $(B - C) - D$
   (h) $B - (C - D)$
   (i) $(A \cup B) - (C \cap D)$

10. If A = [0, 3], B = [2, 7), with **U** = $\mathbb{R}$, determine each of the following:
    (a) $A \cap B$
    (b) $A \cup B$
    (c) $\overline{A}$
    (d) $A \Delta B$
    (e) $A - B$
    (f) $B - A$
11. Let A, B, C, D, E $\subseteq \mathbb{Z}$ be defined as follows:
    $A = \{2n \mid n \in \mathbb{Z}\}$ – that is, A is the set of all (integer) multiples of 2;
    $B = \{3n \mid n \in \mathbb{Z}\}$; $C = \{4n \mid n \in \mathbb{Z}\}$; $D = \{6n \mid n \in \mathbb{Z}\}$; $E = \{8n \mid n \in \mathbb{Z}\}$;
    (a) Which of the following statements are true, and which are false?
       (i) $E \subseteq C \subseteq A$
       (ii) $A \subseteq C \subseteq E$
       (iii) $B \subseteq D$
       (iv) $D \subseteq B$
       (v) $D \subseteq A$
       (vi) $\overline{D} \subseteq \overline{A}$
    (b) Determine each of the following sets.
       (i) $C \cap E$
       (ii) $B \cup D$
       (iii) $A \cap B$
       (iv) $B \cap D$
       (v) $\overline{A}$
       (vi) $A \cap E$
12. Determine which of the following statements are true, and which are false.
    (a) $\mathbb{Z}^+ \subseteq \mathbb{Q}^+$
    (b) $\mathbb{Z}^+ \subseteq \mathbb{Q}$
    (c) $\mathbb{Q}^+ \subseteq \mathbb{R}$
    (d) $\mathbb{R}^+ \subseteq \mathbb{Q}$
    (e) $\mathbb{Q}^+ \cap \mathbb{R}^+ = \mathbb{Q}^+$
    (f) $\mathbb{Z}^+ \cap \mathbb{R}^+ = \mathbb{R}^+$
    (g) $\mathbb{R}^+ \cap \mathbb{C} = \mathbb{R}^+$
    (h) $\mathbb{C} \cup \mathbb{R} = \mathbb{R}$
    (i) $\mathbb{Q}^+ \cap \mathbb{Z} = \mathbb{Z}$

13. Prove or disprove each of the following.
    (a) For sets A, B, C $\subseteq$ **U**, $A \cap C = B \cap C \Rightarrow A = B$.
    (b) For sets A, B, C $\subseteq$ **U**, $A \cup C = B \cup C \Rightarrow A = B$.
    (c) For sets A, B, C $\subseteq$ **U**, $[(A \cap C = B \cap C) \wedge (A \cup C = B \cup C)] \Rightarrow A = B$.
    (d) For sets A, B, C $\subseteq$ **U**, $A \Delta C = B \Delta C \Rightarrow A = B$.
14. If $A = \{a, b, d\}$, $B = \{d, x, y\}$ and $C = \{x, z\}$, how many proper subsets are there for the set $(A \cap B) \cup C$? How many for the set $A \cap (B \cap C)$?
15. Let A, B $\subseteq$ **U**. Use the equivalence $A \subseteq B \Leftrightarrow A \cap B - A$ to show that the dual statement of $A \subseteq B$ is the statement $B \subseteq A$.
16. Provide the justification (selected from the laws of set theory) for the steps that are needed to simplify the set

$$(A \cap B) \cup [B \cap ((C \cap D) \cup (C \cap \overline{D}))],$$

    where A, B, C, D $\subseteq$ **U**.
17. A binary string of length 12 is made up of 12 bits (that is, 12 symbols, each of which is a 0 or a 1). How many such strings either start with three 1's or end in four 0's?
18. A computer company wishes to hire 25 programmers for handling systems programming tasks, and 40 programmers for application programming tasks. 10 of the hired are expected to perform both kinds of tasks. How many programmers need to be hired?
19. A computer science professor has two dozen introductory textbooks on computer science and is concerned about their coverage of the topics (A) compilers, (B) data structures and (C) operating systems. The following data shows the numbers of books that contain material on these topics: $|A| = 8$ $|B| = 13$ $|C| = 13$ $|A \cap B| = 5$ $|A \cap C| = 3$ $|B \cap C| = 6$ $|A \cap B \cap C| = 2$.
    (a) How many of the textbooks include material on exactly one of these topics?
    (b) How many do not deal with any of the topics?
    (c) How many have no material on compilers?

5 Relations and Operations

In Chapter 4, we have dealt with sets and set operations. One of the operations is the Cartesian product of two or more sets. This operation gives a new set wherein the members are not the same as the participating sets but different, and contain tuples of different sizes based on the number of participating sets in the operation. This new set defines the universal relation. This chapter deals with relations and operations on relations.

5.1 CONCEPT OF RELATIONS

This is an extension of set theory that is most widely used in computer science. The concept and theory of relations was introduced by Augustus De Morgan. The digraphs were introduced by Bertrand Russell. In programming languages we deal with relational operators such as ≤, <, >, ≥, ≠, and =, which are commonly used for defining the conditions. In the design of database management systems, a popular data model is the relational model and the software packages built around are called relational database management systems (RDBMS). Relational methods can be found in various areas of computer science, namely, relational semantics of concurrency, analysis of rewriting systems, and modern programming language design, algorithms analysis and the discrete mathematics taught to students of computer science. In addition to these, relationships exist between people, namely, uncle, aunt, brother, sister, father, mother, friend, etc. In set theory, the concept of subset represents the relationship between the two sets. Consider two parallel lines or two perpendicular lines or isomorphic graphs, all representing relations. The said relations are examples of binary relations. Relations may be ternary, quaternary and so on. We shall revisit the concept of the Cartesian product of two sets from the point of view of relations. The Cartesian product is also called the cross product of sets.

5.1.1 Cartesian Product

Let A and B be two sets. The ordered pairs of the type (a, b) are obtained in the new set by taking the Cartesian product of the given two sets. The Cartesian product is denoted by $A \times B$ and is defined as follows:

$$A \times B = \{(a, b) \mid a \in A \text{ and } b \in B\}$$

Example 5.1: Consider A = {1, 2, 3} and B = {x, y}. Obtain the Cartesian product set A × B.

$$A \times B = \{(1, x), (1, y),\\ (2, x), (2, y),\\ (3, x), (3, y)\}$$

Totally there are six pairs, and in each pair the first element is drawn from set A and the second element is drawn from set B. It is seen that the number of elements in A × B is equal to the product of the cardinalities of A and B. That equals 3 × 2, which is equal to 6.

Example 5.2: Prove that for two given non empty finite sets A and B, the cardinality of A × B is equal to the product of the cardinalities of the individual sets. That is, $|A \times B| = |A|\,|B|$.

Let the cardinality of A be *m*, i.e. $|A| = m$, and the cardinality of B be *n*. i.e. $|B| = n$. In order to get the pairs of the type (*a*, *b*), we need to complete two tasks. One task is to choose *a* from A followed by another, to choose *b* from B. There are *m* ways to perform the first task and *n* ways to perform the second task. Therefore, by multiplication principle, these two successive tasks are performed in $(m \times n)$ ways. Thus, $|A \times B| = m \times n = |A|\,|B|$.

Example 5.3: Consider the sets of Example 5.1 and obtain A × B, B × A, B × B, A × A and B × B × B.

$$A \times B = \{(1, x), (1, y), (2, x), (2, y), (3, x), (3, y)\}$$
$$B \times A = \{(x, 1), (x, 2), (x, 3), (y, 1), (y, 2), (y, 3)\}$$
$$B \times B = \{(x, x), (x, y), (y, x), (y, y)\}.\ xy\ xy\ xy$$
$$B \times B \times B = \{(x, x, x), (x, x, y), (x, y, x), (x, y, y),\\ (y, y, x), (y, x, x), (y, y, y), (y, x, y)\}.$$

Example 5.4: Consider the student information database, wherein the students' roll number, name, marks, result and other address details are recorded. A typical entry would look like as follows:

1 Ashok 250 P #20, Brindavan Colony, Delhi, 110001

The roll number is drawn from an integer set, say I_1, the name is drawn from a set of strings representing names, say S_1, marks are drawn from an integer set, say I_2, the result is pass or fail and is drawn from a set {P, F} and the address is drawn from a set, say S_2, that consists of addresses. Each record in the database is considered a 5-tuple and belongs to the set $I_1 \times S_1 \times I_2 \times \{P, F\} \times S_2$.

In certain relations, only one set is involved and such relations are called unary relations. For example, in the following expression the – sign is used in the context of *b* alone. The sign is called unary minus. The minus is used invariably as binary minus sign. Therefore, in case –*b*, the usage is unary and related to only *b*.

$$\frac{-b \pm \sqrt{b^2 - 4ac}}{2a}$$

Even in the real world, we can talk about unary relations. For example, the relations *love* and *friend* are binary in nature. But, to love someone, one should love himself. Similarly, to befriend others, one should be a friend to oneself. Every line is parallel to itself. Every set is a subset of itself.

The Cartesian product of sets defines a new set that contains tuples and this set is called the universal relation R. Other relations are defined over this universal set.

5.2 RELATIONS AS SUBSETS OF CARTESIAN PRODUCTS

In Example 5.3, the set B × B × B has eight elements and each element in the set is called a 3-tuple. The relation is called a ternary relation, as three sets are involved. If four sets are involved in a Cartesian product it is called a quaternary relation, and so on. In general, consider the sets $A_1, A_2, A_3 \ldots A_n$ as the subsets of a universal set U. The Cartesian product of these sets, denoted by $A_1 \times A_2 \times A_3 \times \ldots \times A_n$ is defined as under.

$$A_1 \times A_2 \times A_3 \times \ldots \times A_n = \{(a_1, a_2, a_3, \ldots a_n) \mid a_1 \in A_1, a_2 \in A_2, a_3 \in A_3, \ldots a_n \in A_n\}$$

The element $(a_1, a_2, a_3, \ldots a_n)$ is called an *n*-tuple and the relation is called an *n*-ary relation. Most of the relations are binary in nature. The binary relations on the set A = {1, 2, 3} are defined as under.

$$\begin{aligned} R = A \times A = \{&(1, 1), (1, 2), (1, 3),\\ &(2, 1), (2, 2), (2, 3),\\ &(3, 1), (3, 2), (3, 3)\} \end{aligned}$$

Let us consider the subsets of R as under:

$$R_1 = \{(1, 1), (2, 2), (3, 3)\}$$
$$R_2 = \{(1, 2), (1, 3), (2, 3)\}$$
$$R_3 = \{(1,2), (1,3), (2,1), (2, 3), (3,1), (3,2)\}$$

The relation R_1 represents an *Equality* relation, wherein the first element is equal to the second element. The relation R_2 represents a *Less than* relation, wherein the first element is less than the second element. The relation R_3 represents a *Not equal to* relation, wherein the first element is not equal to the second element. Thus, the relations R_1, R_2, R_3 are subsets of the universal relation R.

$$R_1, R_2, R_3 \subseteq R = A \times A.$$

Example 5.5: Consider two finite sets A and B with $|A| = m$ and $|B| = n$. The Cartesian product A × B and $|A \times B|$ is *mn*. Since every relation is a subset of the Cartesian product, there are 2^{mn} relations possible from A to B. This includes the empty relation and the product A × B itself. The relation AB is called the universal relation. Similarly, there are 2^{mn} relations possible from B to A, one of which is the empty relation and the other is $|B \times A|$.

Example 5.6: Consider a software development company wherein programmers work with Java, C++ and Object Pascal. The operating systems used are Unix and Windows 7. The memory capacities available on the machines are 2MB, 4MB and 8MB. The number of ways in which the programmers have options to work is obtained by the Cartesian product of the three sets.

L = {Java, C++, Object-Pascal}

OS = {Unix, Windows 7}

M = {2, 4, 8}

The Cartesian product L × OS × M gives 18 options for the programmers because the cardinalities of the sets L, OS and M are 3, 2 and 3 respectively. The 3-tuples are listed as follows:

L × OS × M = {(Java, Unix,2), (Java, Unix,4), (Java, Unix,8), (Java, Windows 7, 2),
(Java, Windows 7, 4), (Java, Windows 7, 8), (C++, Unix, 2),
(C++, Unix, 4), (C++, Unix, 8), (C++,Windows 7, 2), (C++,Windows 7, 4),
(C++, Windows 7, 8), (Object-Pascal, Unix, 2), (Object-Pascal, Unix, 4),
(Object-Pascal, Unix, 8), (Object-Pascal, Windows 7, 2),
(Object-Pascal, Windows 7, 4), (Object-Pascal, Windows 7, 8)}

5.2.1 Identity Relation

It is also called an Equality relation, wherein every element of the set is related to itself. If A is the set, then the identity relation I_A is given as follows:

$$I_A = \{(a, a) \mid \text{for all } a \in A\}$$

Given a set A = {1, 2, 3} the relation I_A is {(1,1), (2, 2), (3, 3)}.

5.2.2 Void Relation

Given a set A and its cross product A × A, we know that both the empty relation (Ø) and the universal relation A × A are subsets of A × A. The empty relation is also called a void relation in some books.

Consider a computer program having input set A and output set B. Let A and B be related by the relation R if and only if $b \in B$ is the output for $a \in A$ as the input. This relation gives all possible outputs for all possible inputs to the program, which depicts the universal relation. Also, when there is no input, there is no output, which depicts the void relation.

5.2.3 Inverse Relation

Let R be a relation from set A to set B, R: A → B. The inverse relation of R is denoted by R^{-1} and represents the relation B to A. It is defined as follows:

$$R^{-1} = \{(b, a) \mid (a, b) \in R\}$$

This is equivalent to saying that (a R b) if and only if (b R^{-1} a). It is also equivalent to saying that $(a, b) \in R \rightarrow (b, a) \rightarrow R^{-1}$. The domain of R is identical to the range of R^{-1} and vice versa. Consider a set A = {1, 2, 3} and the relation R = {(1, 2), (2, 3), (3, 3)}. The relation R^{-1} = {(2, 1), (3, 2), (3, 3)}.

5.3 PROPERTIES OF BINARY RELATIONS

If R is a binary relation from A to B then the tuple (x, y) has x belonging to A and y belonging to B. It is concisely represented as follows:

$$R = \{(x, y) \mid x \in A \text{ and } y \in B\}$$

The set A is called the domain of R and denoted as Domain (R). The set B is called the range of R and is denoted as Range (R). Consider the set A = {1, 2, 3, 4, 5} and the relation *less than*. The relation R is given as:

$$R = \{(1, 2), (1, 3), (1, 4), (1, 5), (2, 3), (2, 4), (2, 5), (3, 4), (3, 5), (4, 5)\}$$

The Domain (R) is {1, 2, 3, 4, 5} and Range (R) is {2, 3, 4, 5}. We define the term 'Relative set' for the given relation R. If R is a relation from A to B and $x \in A$, we define the relative set of x, denoted by $R(x)$, as the set of all $y \in B$ such that $x\,R\,y$. It is given as:

$$R(x) = \{y \in B \mid x\,R\,y\}$$

If A_1 is the subset of A, the relative set of A_1, denoted by $R(A_1)$, is the set of all $y \in B$ that are related to $x \in A_1$. This is written as follows:

$$R(A_1) = \{y \in B \mid x\,R\,y \text{ for some } x \in A_1\}$$

Example 5.7: Consider a set $A = \{w, x, y, z\}$ and $R = \{(w, w), (w, x), (x, y), (y, w), (z, y), (y, x)\}$. Now, $R(w) = \{w, x\}$, $R(x) = \{y\}$, $R(y) = \{w, x\}$ and $R(z) = \{y\}$. Consider the subset $A_1 = \{y, z\}$. The relative set $R(A_1) = \{w, x, y\}$.

Example 5.8: Show that given a relation R from A to B, the following facts hold good for A_1 and A_2, the subsets of A.

(i) If $A_1 \subseteq A_2$ then $R(A_1) \subseteq R(A_2)$
(ii) $R(A_1 \cup A_2) = R(A_1) \cup R(A_2)$
(iii) $R(A_1 \cap A_2) = R(A_1) \cap R(A_2)$

Proof:

(i) If $A_1 \subseteq A_2$ then $R(A_1) \subseteq R(A_2)$. If $y \in R(A_1)$ then $x\,R\,y$ for some $x \in A_1$. This x also belongs to A_2 because A_1 is the subset of A_2. Therefore, $y \in R(A_2)$ is true. This proves that $R(A_1) \subseteq R(A_2)$.
(ii) Consider $y \in R(A_1 \cup A_2)$. This means that $x\,R\,y$ for some $x \in (A_1 \cup A_2)$. Therefore, $x \in A_1$, or $x \in A_2$. If $x \in A_1$, then $y \in R(A_1)$. Similarly, If $x \in A_2$ then $y \in R(A_2)$. In either case $y \in R(A_1) \cup R(A_2)$. Therefore, $R(A_1 \cup A_2) = R(A_1) \cup R(A_2)$.
(iii) If $y \in R(A_1 \cap A_2)$ then for some x in $(A_1 \cap A_2)$ there exists y that belongs to both $R(A_1)$ and $R(A_2)$. This means that $y \in R(A_1 \cap A_2)$, which proves that $R(A_1 \cap A_2) = R(A_1) \cap R(A_2)$.

5.3.1 Properties of Binary Relations

In computer science, we mostly deal with relations on a given set rather than the relations across two sets. These relations are discussed in this section and satisfy certain properties including reflexivity, irreflexivity, symmetry, antisymmetry and transitivity.

5.3.1.1 *Reflexivity and irreflexivity*

The relation R on a set A is said to be reflexive if a is related to a ($a\,R\,a$) for all $a \in A$. Similarly, the relation R on a set A is said to be irreflexive if a is not related to a ($a\,R\,a$) for all $a \in A$. Thus, the given relation R is reflexive if every element of A is related to itself and on the other hand, it is irreflexive if no element of A is related to itself.

Example 5.9: Let A = {1, 2, 3} and consider the following relations.

$$R_1 = \{(1, 1), (2, 2), (3, 3)\}$$

$$R_2 = \{(1, 2), (2, 1), (3, 2)\}$$

$$R_3 = \{(1, 1), (2, 3), (3, 1)\}$$

The relation R_1 is reflexive, R_2 is irreflexive and R_3 is not reflexive. R_3 is not reflexive because the element 1 ε A is related to itself, whereas the other elements are not.

Example 5.10: Consider the set of natural numbers ℕ and the relations >, ≥ and divisor. The relation > is irreflexive because every element in ℕ is not greater than itself. However the relation ≥ is reflexive because of equality relation. The equality relation is reflexive. The relation divisor is reflexive because every element of ℕ divides itself.

Example 5.11: Let A be a nonempty set and R be a relation on A × A. The relation R = ∅ is an empty relation and is not reflexive because the tuple $(a, a) \notin R$ for all $a \in A$. But the relation R is irreflexive.

Example 5.12: Consider the set A having n elements, i.e., | A | = n. The universal relation A × A has n^2 elements. The number of relations are 2^{n^2}. The number of relations amongst these reflexives are $2^{(n^2-n)}$. Let $A = \{a_1, a_2, \ldots, a_n\}$ be the set. A relation R on A is reflexive if and only if $R = \{(a_i, a_i) \mid 1 \leq i \leq n\}$. The other $(n^2 - n)$ ordered pairs in A × A, wherein the pairs of the form (a_i, a_j) and $i \neq j$ are included or excluded.

Example 5.13: Consider a set of lines. The relation parallel is reflexive because every line is parallel to itself. Similarly, the relations friend and love defined on a set of persons in the real world are reflexive, as every person loves himself/herself and is a friend of himself/herself.

5.3.1.2 *Symmetric, not symmetric and asymmetric relations*

A relation R on a set A is said to be symmetric, whenever a R b then b R a. The relation R is said to be not symmetric whenever we have some a and b belonging to A and (a R b) but b is not related to a, i.e., (b R a) is not true. The relation R is said to be asymmetric, whenever a R b for all $a, b \in A$, b is not related to a. These are defined as follows:

(i) $(a, b) \in R \rightarrow (b, a) \in R$ for $a, b \in A$, the relation is symmetric
(ii) $(a, b) \in R$ and $(b, a) \notin R$ for $a, b \in A$, the relation is not symmetric
(iii) $(a, b) \in R \rightarrow (b, a) \notin R$ for $a, b \in A$, the relation is asymmetric

Example 5.14: Consider the set A = {a, b, c} and the following relations defined over the universal relation A × A.

(i) $R_1 = \{(a, b), (c, a), (b, c), (b, a), (a, c), (c, b)\}$
(ii) $R_1 = \{(a, b), (c, a), (b, c), (a, c), (c, b)\}$
(iii) $R_1 = \{(a, b), (c, a), (b, c)\}$

The relation R_1 is symmetric. The relation R_2 is not symmetric. The relation R_3 is asymmetric.

Example 5.15: Let $A = \mathbb{Z}$ be the set of integers and $R = \{(a, b) \in A \times A \mid (a < b)\}$. The relation R is asymmetric and not symmetric. Consider the relation divides (|). Let $a, b \in \mathbb{Z}$ and the relation $a \mid b$ is not symmetric.

Example 5.16: Consider the real world examples. The relations cousin, friend, couple and the like are symmetric relations. If *a* is a cousin of *b*, that means *b* is a cousin of *b*. The relations like father, brother, mother are asymmetric. We can also say that the relations which are asymmetric are also not symmetric.

5.3.1.3 *Antisymmetry and transitive*

The relation R on a set A is said to be antisymmetric whenever a R b and b R a implies that $a = b$. It follows that R is not antisymmetric, if we have both a R b and b R a but $a \neq b$.

Example 5.17: Consider the relation less than or equal to (≤) or greater than or equal to (≥) on a set of integers $\mathbb{Z}$. Let $a, b \in \mathbb{Z}$. Then by the property of antisymmetry, we can write as follows:

(i) $(a \geq b)$ and $(b \geq a) \rightarrow (a = b)$
(ii) $(a \leq b)$ and $(b \leq a) \rightarrow (a = b)$

Example 5.18: Consider the relation divides (|) on a set of integers $\mathbb{Z}$. Let $a, b \in \mathbb{Z}$. Then by the property antisymmetry, we can write that:

i) $(a \mid b)$ and $(b \mid a) \rightarrow (a = b)$

A relation R is said to be transitive if (a R b) and (b R c) implies that (a R c). In other words, the relation is said to be not transitive if (a R b) and (b R c) but (a R c).

Example 5.19: Consider the sets A, B and C and the relation subset defined over a universal set. If $A \subseteq B$ and $B \subseteq C$ then $A \subseteq C$. Therefore, the relation $\subseteq$ is transitive.

Example 5.20: Consider the relation *parallel* defined over a set of lines. If line A is parallel to line B and line B is parallel to C then line A is parallel to line C. Therefore, the relation *parallel* is transitive. The same is not true for the relation *perpendicular*.

Example 5.21: Consider the relation ≤ defined over $\mathbb{Z}$. If $a \leq b$ and $b \leq c$ then $a \leq c$ for a, b and $c \in \mathbb{Z}$. Therefore, the relation ≤ is transitive.

If we consider real world relations, like brother, sister, elder, they are examples of transitive relations. The relations such as friend and love need not be transitive. The relations mother and father are not transitive.

5.3.2 Equivalence Relation and Partitions

A given relation is said to be an Equivalence relation when it is reflexive, symmetric and transitive. The relation given in Example 5.19 is an equivalence relation.

Example 5.22: Let A = {*a*, *b*, *c*} and consider that the following relations defined over A × A are equivalence relations.

$$R_1 = \{(a, a), (b, b), (c, c)\}$$

$$R_2 = \{(a, a), (b, b), (c, c), (b, c), (c, b)\}$$

$$R_3 = \{(a, a), (b, b), (c, c), (a, b), (a, c), (b, c), (b, a), (c, b), (c, a)\}$$

Consider the relation R_1, which is reflexive because every element of the set is related to itself. The symmetry property is also satisfied because $a\,R_1 b$ is not true. Since the antecedent is not true, irrespective of the consequence truth value, the implication is true. The same argument holds good in the case of the transitive property too. From the relations R_1, R_2 and R_3, the relation R_1 is called the smallest equivalence relation and the relation R_3 is called the largest equivalence relation, mainly by the number of elements in the relation.

Example 5.23: Consider the set A = ℤ and the following relation R, which is defined as under
R = {(*a*, *b*) ∈ A × A | *a* and *b* yield the same remainder when divided by 2}

In this case, 2 is called modulus and it is written as a ≡ (*b* mod 2). This is read as a is congruent to *b* mod 2 and is an equivalence relation. The relation is reflexive as *a* ≡ *a* (mod 2). The relation is symmetric because *a* ≡ (*b* mod 2) is same as *b* ≡ (*a* mod 2). Similarly, the transitive property also holds good. Hence, it is an equivalence relation.

Example 5.24: Consider the set of triangles in a plane. The relation *similarity of triangles* is an equivalence relation. Let A, B and C represent the triangles in a plane. From the property of similarity of triangles, we can ascertain the following:

(i) A, B and C are similar to themselves.
(ii) If triangle A is similar to triangle B then triangle B is similar to triangle A.
(iii) If triangle A is similar to triangle B and triangle B is similar to triangle C then triangle A is similar to triangle C.

Example 5.25: Consider the arithmetic operation = on Zℤ is an equivalence relation. Let *a*, *b*, *c* ∈ ℤ and then the following properties hold good.

(i) $a = a$
(ii) If $(a = b)$ then $(b = a)$
(iii) If $(a = b)$ and $(b = c)$ then $(a = c)$

5.3.2.1 *Partitions*

Given a set (A) and an equivalence relation (R), we define a partition (P) on A induced by the equivalence relation R and vice versa. Let $P_1 = \{\{a\}, \{b\}, \{c\}\}$ and the set A = {*a*, *b*, *c*}. The equivalence relation induced by P_1 is {(*a*, *a*), (*b*, *b*), (*c*, *c*)}. If $P_2 = \{\{a, b, c\}\}$, then the equivalence relation induced by P_2 is {(*a*, *a*), (*b*, *b*), (*c*, *c*), (*a*, *b*), (*a*, *c*), (*b*, *c*), (*b*, *a*), (*c*, *b*), (*c*, *a*)}. The partition P_1 is called the smallest partition and P_2 is called the largest partition. This convention is basically dependent on the number of elements in the relation. The sets in the partition are also called blocks. The blocks in the partition are also called equivalence classes.

Alternatively, we define the partition of a set A as a set of all subsets of A denoted by $\{A_1, A_2, A_3 \ldots A_n\}$ such that their union is A, and the intersection of A_i and Aj is empty for any distinct A_i and Aj. In other words, the partition of a set A is the division of A into disjoint subsets, as depicted in Fig. 5.1.

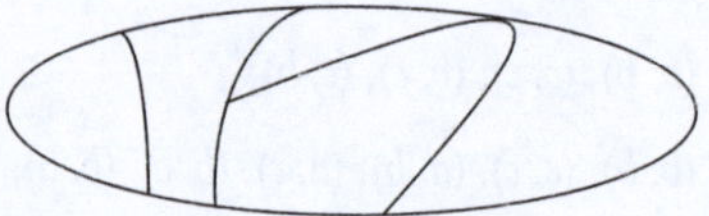

Figure 5.1 Partition of a set into subsets

Let $A = \{a, b, c, d, e, f, g\}$ then $\{a\}, \{b, c, d\}, \{e, f\}, \{g\}$ are partitions of A. Consider the playing cards, wherein the cards are partitioned into four suits, namely, spades, hearts, clubs and diamonds. Another way of partitioning is by dividing into 13 ranks, namely, aces, twos, threes, fours, fives, sixes, sevens, eights, nines, tens, jacks, queens and kings. In the real world, we talk about the partitioning of property amongst siblings. A country is partitioned into states, states into districts and districts into sub-districts or *taluks*.

Example 5.26: Consider a set $A = \{a, b, c\}$ and partition $P = \{\{a, b\}, \{c\}\}$. The equivalence relation induced by P is $\{(a, a), (a, b), (b, a), (b, b), (c, c)\}$. From the relation, it is clear that *a* and *b* are members of the same block if and only if *a* R *b*. The partitions are shown using the set symbols and also using a bar over the partition elements or members. In this example, P is alternatively denoted as $P = \{\overline{a, b}\ ; \overline{c}\}$. The partitions are separated by either a comma or a semicolon.

Theorem 5.1: Let P be a partition on a set A. Then *a* R *b* if and only if *a* and *b* are members of the same block. Then R is an equivalence relation on A.

Proof:

(i) If $a \in A$, then clearly *a* is in the same block. So *a* is related to *a*.
(ii) If *a* R *b* then *a* and *b* are in the same block. So *b* is related to *a*.
(iii) If *a* R *b* and *b* R *c* then *a*, *b* and *c* must belong to the same block of P and *a* R *c*.

Since R is reflexive, symmetric and transitive, it is an equivalence relation. R is induced or determined by P.

Example 5.27: Consider a relation $R = \{(a, a), (a, b), (b, a), (b, b), (c, c), (c, d), (d, c), (d, d)\}$ defined on $A = \{a, b, c, d\}$. The partition P induced by R is $P = \{\{a, b\}, \{c, d\}\}$. We can define R(*a*) as the set of all elements related to *a*. $R(a) = \{a, b\}$. $R(b) = \{a, b\}$. Therefore, the elements *a* and *b* belong to the same block.

Consider two partitions of the set $A = \{a, b, c, d, e, f, g, h, i, j, k\}$ as $P_1 = \{\{a, b, c, d\}, \{e, f, g\}, \{h, i\}, \{j, k\}\}$ and $P_2 = \{\{a, b, c, h\}, \{d, i\}, \{e, f, j, k\}, \{g\}\}$. The corresponding relations are R_1 and R_2. The relations are given as follows:

$$R_1 = \{(a, a), (a, b), (a, c), (a, d), (b, a), (b, b), (b, c), (b, d), (c, a), (c, b), (c, c), (c, d), (d, a), (d, b), (d, c), (d, d), (e, e), (e, f), (e, g), (f, e), (f, f), (f, g), (g, e), (g, f), (g, g), (h, h), (h, i), (i, h), (i, i)\}$$

$$R_2 = \{(a, a), (a, b), (a, c), (a, h), (b, a), (b, b), (b, c), (b, h), (c, a), (c, b), (c, c), (c, h), (h, a), (h, b), (h, c), (h, h), (e, e), (e, f), (e, j), (e, k), (f, e), (f, f), (f, j), (f, k), (j, e), (j, f), (j, j), (j, k), (k, e), (k, f), (k, j), (k, k), (d, d), (d, i), (i, d), (i, i), (g, g)\}$$

The partition P_1 is said to be a refinement of P_2, denoted by $P_1 \leq P_2$ if $R_1 \subseteq R_2$. This means that any two elements that are in the same block or partition of P_1 must also be in the same block of P_2. We define the product of two partitions as $P_1.P_2$, which corresponds to the equivalence relation $R_1 \cap R_2$. The product of P_1 and P_2 is a partition of the set A such that two elements *a* and *b* are in the same block of $P_1.P_2$ if *a* and *b* are in the same block of P_1 and also in the same block of P_2. Therefore, $P_1.P_2$ is a refinement of P_1 and also a refinement of P_2. Similarly, we define the sum of P_1 and P_2, denoted by $P_1 + P_2$, which represents the equivalence relation $(R_1 \cup R_2)^*$ and is Kleene's closure of $(R_1 \cup R_2)$. The sum $P_1 + P_2$ is a partition of the set A such that two elements a and b are in the same block of $P_1 + P_2$ if there exist elements $c_1, c_2, c_3, \ldots c_K$ such that *a* and c_1 are in the same block of P_1 or P_2, c_2 and c_3 are in the same block of P_1 or P_2 and so on, finally c_k and *b* are in the same block of P_1 or P_2. This is called a chain connection. That is, the two elements are in the same block of $P_1 + P_2$ if they are chain connected. One can give some physical interpretation for these two operations on partitions. Consider students partitioned based on income groups, say P_1 and social status, say P_2. One can find a given student in one of the blocks of P_1, given his or her income group and in one of the blocks of P_2 given the social status, without any ambiguity. It is basically, given an object, classification of the given object into one of the classes.

Example 5.28: Consider the set $A = \{a, b, c, d, e, f, g, h, i, j, k\}$ and $P_1 = \{\{a, b, c, d\}, \{e, f, g\}, \{h, i\}, \{j, k\}\}$ and $P_2 = \{\{a, b, c, h\}, \{d, i\}, \{e, f, j, k\}, \{g\}\}$. The following are $P_1.P_2$ and $P_1 + P_2$:

$$P_1.P_2 = \{\{a, b, c\}, \{e, f\}, \{j, k\}, \{h\}, \{g\}, \{d\}, \{i\}\}$$
$$P_1 + P_2 = \{\{a, b, c, d, h, i\}, \{e, f, g, j, k\}\}$$

5.4 OPERATIONS ON RELATIONS

Operations on relations are similar to those on sets, and the major operations include Union, Intersection, Complement, Closure and Composition of relations. These operations on relations give new relations. Let R_1 and R_2 be two relations defined over $A \times B$. We know that R_1 and R_2 are subsets of $A \times B$. The complement of R_1, denoted by $\overline{R_1}$ is defined as under.

$$\overline{R_1} = \{(a, b) \mid (a, b) \notin R_1\}$$

Consider $A = \{a, b, c\}$ and consider the following relations R_1 and R_2 defined over $A \times A$ are given as follows:

$$R_1 = \{(a, a), (b, b), (c, c)\}$$
$$R_2 = \{(a, a), (b, b), (c, c), (b, c), (c, b)\}$$

The compliment of R_1 denoted by $\overline{R_1} = \{(a, b), (b, a), (a, c), (c, a), (b, c), (c, b)\}$. Similarly, $\overline{R_2} = \{(a, b), (b, a), (a, c), (c, a)\}$.

Example 5.29: Consider $A = \{1, 2, 3, 4\}$ and $B = \{x, y, z\}$. The following are the relations R_1 and R_2.

$$R_1 = \{(1, x), (1, y), (2, y), (2, z), (3, y), (4, x)\}$$
$$R_2 = \{(1, y), (2, z), (3, y), (4, y)\}$$

The complements of R_1 and R_2 are obtained by first determining $R_1 \times R_2$, which is given as follows:

$$R_1 \times R_2 = \{(1, x), (1, y), (1, z), (2, x), (2, y), (2, z), (3, x), (3, y), (3, z), (4, x), (4, y), (4, z)\}$$

$$\overline{R_1} = \{(1, z), (2, x), (3, x), (3, z), (4, y), (4, z)\}$$

$$\overline{R_1} = \{(1, x), (1, z), (2, y), (2, z), (3, x), (3, z), (4, x), (4, z)\}$$

The other operations have same meaning as discussed in topic on set theory. $R_1 \cup R_2$, $R_1 \cap R_2$, and $R_1 - R_2$ are defined as follows:

The union is defined as $R_1 \cup R_2 = \{(a, b) \mid (a, b) \in R_1 \text{ or } (a, b) \in R_2\}$. The $R_1 \cup R_2$ for Example 5.27 is obtained as follows:

$$R_1 \cup R_2 = \{1, x), (1, y), (2, y), (2, z), (3, y), (4, x), (4, y)\}$$

The intersection is defined as $R_1 \cap R_2 = \{(a, b) \mid (a, b) \in R_1 \text{ and } (a, b) \in R_2\}$. The $R_1 \cap R_2$ for Example 5.27 is obtained as follows:

$$R_1 \cap R_2 = \{(1, y), (2, z), (3, y)\}$$

The difference is defined as $R_1 - R_2 = \{(a, b) \mid (a, b) \in R_1 \text{ and } (a, b) \notin R_2\}$. The $R_1 - R_2$ for Example 5.27 is obtained as follows:

$$R_1 - R_2 = \{(1, x), (2, y), (4, x)\}$$

The composition of two relations R_1 and R_2 given by $R_1 = \{(1, x), (1, y), (2, y), (2, z), (3, y), (4, x)\}$, where $R_1 \subseteq A \times B$ and $R_2 = \{(y, 1), (z, 2), (y, 3), (y, 4)\}$, where $R_2 \subseteq B \times A$, is a new relation denoted by $R_1 \bullet R_2$. It is defined as $R_1 \bullet R_2 = \{(a, b) \mid \exists z\ (a, z) \in R_1 \text{ and } (z, b) \in R_2\}$. It is important to note that the composition operation is not commutative. The $R_1 \bullet R_2$ for Example 5.27 is obtained as given below.

$$R_1 \bullet R_2 = \{(1, 1), (2, 2), (1, 3), (3, 1), (1, 4), (2, 1), (2, 3), (2, 4), (3, 1), (3, 3), (3, 4)\}$$

Example 5.30: Consider a set $A = \{1, 2, 3, 4\}$, $R_1 = \{(1, 2), (1, 1), (1, 3), (2, 4), (3, 2)\}$ and $R_2 = \{(1, 4), (1, 3), (2, 3), (3, 1), (4, 1)\}$. $R_1 \bullet R_2 = \{(1, 3), (1, 4), (1, 3), (1, 1), (2, 1), (3, 3)\}$. We can determine $R_2 \bullet R_1 = \{(1, 2), (2, 2), (3, 1), (3, 2), (3, 3), (4, 1), (4, 2), (4, 3)\}$. It is evident that $(R_1 \bullet R_2) \neq (R_2 \bullet R_1)$.

Consider the sets A, B and C. Let $R_1 \subseteq A \times B$ and $R_2 \subseteq B \times C$. The composition relation $(R_2 \bullet R_1)$ is a relation from A to C, defined as follows. If $a \in A$ and $c \in C$ then $a\ (R_2 \bullet R_1)\ c$ if and only if, for some $b \in B$, there exist relations $a\ R_1\ b$ and $b\ R_2\ c$. Even though R_1 follows R_2 in $(R_2 \bullet R_1)$, it is to be read as R_2 follows R_1. This interpretation is based on the fact that function is a special kind of relation.

Closure is another operation that can be performed on relations. There are three kinds of closure operations, namely, reflexive closure, symmetric closure and transitive closure. Let R be a relation from A to A. The reflexive closure of R is denoted as r(R) and is defined as follows. Given a relation R, which is not reflexive, we add only those tuples from $A \times A$ that make the relation R just reflexive. The reader should remember that not a single extra tuple is added beyond those required.

Example 5.31: Consider the set $A = \{a, b, c, d\}$ and $R = \{(a, a), (a, b), (b, d), (c, c), (c, d)\}$. The relation R is not reflexive because the tuples (b, b) and (d, d) are missing in this relation. By adding these tuples to R, the relation R becomes reflexive. It is expressed as a union of R and the equality relation E on A, as given below.

$$\text{Reflexive closure of R, } r(R) = (R \cup E)$$
$$= \{(a, a), (a, b), (b, d), (c, c), (c, d), (b, b), (d, d)\}.$$

Another operation is symmetric closure, which is denoted by s(R). If the given relation R is not symmetric then we add only those tuples from A × A such that the relation becomes just symmetric. It is expressed as a union of the given relation R and its inverse (R^{-1}), given as follows:

$$\text{Symmetric closure of R,} \quad s(R) = (R \cup R^{-1})$$

Example 5.32: Consider the set A = {*a, b, c, d*} and R = {(*a, a*), (*a, b*), (*b, d*), (*c, c*), (*c, d*)}. The relation R is not symmetric because the tuples (*b, a*), (*d, c*) and (*d, b*) are missing in this relation. By adding these tuples to R, the relation R becomes symmetric.

$$R = \{(a, a), (a, b), (b, d), (c, c), (c, d)\}.$$

$$R^{-1} = \{(a, a), (b, a), (d, b), (c, c), (d, c)\}.$$

$$s(R) = (R \cup R^{-1})$$

$$= \{(a, a), (a, b), (b, d), (c, c), (c, d), (b, a), (d, b), (d, c)\}.$$

The transitive closure of a given relation is somewhat involved operation. If a given relation is not transitive, we add only those tuples which make the relation just transitive. It is denoted by t(R) and given as follows:

$$t(R) = \bigcup_{i=0}^{\infty} R^i, \text{ where } R^i \text{ represents different powers of R.}$$

R^0 represents an equality relation and R^1 represents the given relation. The other powers of R are obtained by composition operation are:

$$R^2 = R \bullet R$$

$$R^3 = R \bullet R^2$$

$$R^4 = R^2 \bullet R^2$$

$$R^5 = R \bullet R^4$$

and so on. In general, $R^{n+1} = R \bullet R^n$. We stop obtaining the powers of R as soon as the repetition starts. Union of all the powers of R gives the transitive closure of R. This operation is further visited in the discussion on matrix representations of relations (Section 5.5).

Example 5.33: Consider the set A = {*a, b, c, d*} and R = {(*a, a*), (*a, b*), (*b, d*), (*c, c*), (*c, d*)}. This relation is not transitive. The different powers of R are determined as follows:

$$R^0 = \{(a, a), (b, b), (c, c), (d, d)\}$$

$$R^1 = R = \{(a, a), (a, b), (b, d), (c, c), (c, d)\}$$

$$R^2 = R \bullet R = \{(a, a), (a, b), (a, d), (c, c), (c, d)\}$$

$$R^3 = R \bullet R^2 = \{(a, a), (a, b), (a, d), (c, c), (c, d)\}$$

Since R^3 is a repetition of R^2, we stop obtaining further powers of R and take the union of R^0, R^1, and R^2.

$$t(R) = \{(a, a), (b, b), (c, c), (d, d), (a, b), (b, d), (c, d), (a, d)\}$$

5.5 MATRIX REPRESENTATION OF RELATIONS

The relations are also represented alternatively by zero-one matrices. Let R be the given relation on a set A. Let M_R denote its matrix representation. The entry $M_R[i, j]$ is 1 if the tuple (i, j) is in R, otherwise the entry is 0. Thus, M_R is a zero-one matrix. Consider the relation R= {(a, a), (a, b), (b, d), (c, c), (c, d)}, given in Example 5.31. Its matrix representation is as follows:

$$M_R = \begin{bmatrix} 1 & 1 & 0 & 0 \\ 0 & 0 & 0 & 1 \\ 0 & 0 & 1 & 1 \\ 0 & 0 & 0 & 0 \end{bmatrix}$$

Similarly, the matrices for the relations $R_1 = \{(1, 2), (1, 1), (1, 3), (2, 4), (3, 2)\}$ and $R_2 = \{(1, 4), (1, 3), (2, 3), (3, 1), (4, 1)\}$, in Example 5.28 are given below.

$$M_{R1} = \begin{bmatrix} 1 & 1 & 1 & 0 \\ 0 & 0 & 0 & 1 \\ 0 & 1 & 0 & 0 \\ 0 & 0 & 0 & 0 \end{bmatrix} \text{ and } M_{R2} = \begin{bmatrix} 0 & 0 & 1 & 1 \\ 0 & 0 & 1 & 0 \\ 1 & 0 & 0 & 0 \\ 1 & 0 & 0 & 0 \end{bmatrix}$$

It becomes more convenient to work with the matrices instead of directly working with the relations. The operations on relations can be realised as operations on matrices.

Example 5.34: Consider the relations $R_1 = \{(1, 2), (1, 1), (1, 3), (2, 4), (3, 2)\}$ and $R_2 = \{(1, 4), (1, 3), (2, 3), (3, 1), (4, 1)\}$. The composition of these two relations, namely, $R_1 \bullet R_2 = \{(1, 3), (1, 4), (1, 1), (2, 1), (3, 3)\}$. The matrix equivalent of $R_2 \bullet R_1$ is obtained by multiplying the matrices corresponding to R_1 and R_2 in that order. Following are the matrices representing individual matrices R_1 and R_2.

$$M_{R1} = \begin{bmatrix} 1 & 1 & 1 & 0 \\ 0 & 0 & 0 & 1 \\ 0 & 1 & 0 & 0 \\ 0 & 0 & 0 & 0 \end{bmatrix} \text{ and } M_{R2} = \begin{bmatrix} 0 & 0 & 1 & 1 \\ 0 & 0 & 1 & 0 \\ 1 & 0 & 0 & 0 \\ 1 & 0 & 0 & 0 \end{bmatrix}$$

The matrix equivalent to $R_2 \bullet R_1 = \{(1, 3), (1, 4), (1, 3), (1, 1), (2, 1), (3, 3)\}$ is as follows:

$$M_{R1} \times M_{R2} = \begin{bmatrix} 1 & 0 & 1 & 1 \\ 1 & 0 & 0 & 0 \\ 0 & 0 & 1 & 0 \\ 0 & 0 & 0 & 0 \end{bmatrix}$$

The zero-one matrices are multiplied using logical operators namely $\wedge$ and $\vee$ as given below.

$M_{R1} \times M_{R2} = M_{R1} \underset{\wedge}{\vee} M_{R2}$, where we perform a logical AND operation first, followed by a logical OR operation.

Exercise 5.35: Consider the set A = {1, 2, 3} and the relations R and S defined on A. Let R and S be represented by matrices as given below.

$$M_R = \begin{bmatrix} 1 & 0 & 1 \\ 1 & 1 & 1 \\ 0 & 1 & 0 \end{bmatrix} \quad \text{and} \quad M_s = \begin{bmatrix} 1 & 0 & 0 \\ 0 & 1 & 1 \\ 1 & 0 & 1 \end{bmatrix}$$

The composition of S • R is given by the matrix multiplication. $M_R \times M_s = M_R \overset{\vee}{\wedge} M_s$, which is obtained as shown in the following. The transpose of M_s is obtained as shown. Row-wise AND is carried out on the corresponding elements. The OR is performed on the vectors thus obtained. The matrix corresponding to S • R is obtained.

$$M_{S \bullet R} = \begin{bmatrix} 1 & 0 & 1 \\ 1 & 1 & 1 \\ 0 & 1 & 1 \end{bmatrix}$$

The following manipulations are performed. The transpose of M_s is obtained by interchanging the rows and columns.

$$M_R = \begin{bmatrix} 1 & 0 & 1 \\ 1 & 1 & 1 \\ 0 & 1 & 0 \end{bmatrix} \qquad (M_s)^T = \begin{bmatrix} 1 & 0 & 1 \\ 0 & 1 & 0 \\ 0 & 1 & 1 \end{bmatrix}$$

$$(M_R)_1 = [1\ 0\ 1] \qquad (M_s)^{T1} = [1\ 0\ 1]$$

$$(M_R)_2 = [1\ 1\ 1] \qquad (M_s)^{T2} = [0\ 1\ 0]$$

$$(M_R)_3 = [0\ 1\ 0] \qquad (M_s)^{T3} = [0\ 1\ 1]$$

We perform the two binary operations on the vectors of two matrices as given below. The AND operation is first performed, followed by the OR operation, to get the resultant matrix.

$$(M_R)_1 \wedge (M_s)^{T1} = [1\ 0\ 1] \wedge [1\ 0\ 1] = [1\ 0\ 1]$$

$$(M_R)_1 \wedge (M_s)^{T2} = [1\ 0\ 1] \wedge [0\ 1\ 0] = [0\ 0\ 0]$$

$$(M_R)_1 \wedge (Ms)^{T3} = [1\ 0\ 1] \wedge [0\ 1\ 1] = [0\ 0\ 1]$$

The first row of elements of the matrix corresponding to $M_{S \bullet R}$ is obtained by taking the OR of the vectors resulting from the AND operation.

$$(M_{S \bullet R})_{11} = [1 \vee 0 \vee 1] = 1$$

$$(M_{S \bullet R})_{12} = [0 \vee 0 \vee 0] = 0$$

$$(M_{S \bullet R})_{13} = [0 \vee 0 \vee 1] = 1$$

Similarly, other two rows are obtained as under.

$$(M_{S \bullet R})_{21} = [1 \vee 0 \vee 1] = 1 \qquad (M_{S \bullet R})_{31} = [0 \vee 0 \vee 0] = 0$$

$$(M_{S \bullet R})_{22} = [0 \vee 1 \vee 0] = 1 \qquad (M_{S \bullet R})_{32} = [0 \vee 1 \vee 0] = 1$$

$$(M_{S \bullet R})_{23} = [0 \vee 1 \vee 0] = 1 \qquad (M_{S \bullet R})_{33} = [0 \vee 1 \vee 0] = 1$$

The final matrix corresponding to

$$M_{S \bullet R} = \begin{bmatrix} 1 & 0 & 1 \\ 1 & 1 & 1 \\ 0 & 1 & 1 \end{bmatrix}$$

It is important to note that there is a difference between the interpretation of notations used in the composition of two relations and the composition of two functions. In the case of relations, the composition of two relations S and R, denoted by (S • R), is interpreted as tuples (also mean pairs) of R composed with tuples of S. In case of functions, however, the composition of two functions S and R, denoted by (S • R), is interpreted as function R being performed first, followed by S.

5.5.1 Matrices for other Operations

Matrices are also used to represent other operations on relations, namely, union, intersection, complement and inverse. Let A = {1, 2, 3}. Let the relation R be represented by the matrix M_R and the relation S be represented by the matrix M_S as shown below.

$$R = \{(1, 1), (1, 3), (2, 1), (2, 2), (2, 3), (3, 2)\} \text{ and } S = \{(1, 1), (2, 2), (2, 3), (3, 1), (3, 3)\}$$

$$M_R = \begin{bmatrix} 1 & 0 & 1 \\ 1 & 1 & 1 \\ 0 & 1 & 0 \end{bmatrix} \quad \text{and} \quad M_s = \begin{bmatrix} 1 & 0 & 0 \\ 0 & 1 & 1 \\ 1 & 0 & 1 \end{bmatrix}$$

Complement and Inverse are:

$$\overline{R} = \{(1, 2), (3, 1), (3, 3)\} \qquad R^{-1} = \{(1, 1), (2, 2), (2, 3), (3, 1), (3, 2)\}$$

$$M_{\overline{R}} = \begin{bmatrix} 0 & 1 & 0 \\ 0 & 0 & 0 \\ 1 & 0 & 1 \end{bmatrix} \qquad M_{R^{-1}} = \begin{bmatrix} 1 & 1 & 0 \\ 0 & 1 & 1 \\ 1 & 1 & 0 \end{bmatrix}$$

Union and Intersection are:

$$R \cup S = \{(1, 1), (1, 3), (2, 1), (2, 2), (2, 3), (3, 2), (3, 1), (3, 3)\}$$
$$R \cap S = \{(1, 1), (2, 2), (2, 3)\}$$

$$M_{R \cup S} = \begin{bmatrix} 1 & 0 & 1 \\ 1 & 1 & 1 \\ 1 & 1 & 1 \end{bmatrix} \qquad M_{R \cap S} = \begin{bmatrix} 1 & 0 & 0 \\ 0 & 1 & 1 \\ 0 & 0 & 0 \end{bmatrix}$$

Consider R and S as two relations defined on A × B. The following properties hold good:

(i) If $(R \subseteq S)$ then $R^{-1} \subseteq S^{-1}$

(ii) If $(R \subseteq S)$ then $\overline{S} \subseteq \overline{R}$

(iii) $\overline{R \cap S} = \overline{S} \cup \overline{R}$ and $\overline{R \cup S} = \overline{S} \cap \overline{R}$

(iv) $(R \cap S)^{-1} = R^{-1} \cap S^{-1}$ and $(R \cup S)^{-1} = R^{-1} \cup S^{-1}$

Consider an element $(a, b) \in R^{-1}$. It follows from the definition of inverse that the element $(b, a) \in R$. Since R is a subset of S, the element $(b, a) \in S$. Since $(b, a) \in S$, $(a, b) \in S^{-1}$. Therefore, If $(R \subseteq S)$ then

$R^{-1} \subseteq S^{-1}$. Consider $(a, b) \in (R \cap S)^{-1}$. It follows that $(b, a) \in (R \cap S)$, $(b, a) \in R$ and $(b, a) \in S$. This implies that $(a, b) \in R^{-1}$ and $(a, b) \in S^{-1}$. Therefore, $(a, b) \in R^{-1} \cap S^{-1}$. Similarly, Consider $(a, b) \in (R \cup S)^{-1}$. It follows that $(b, a) \in (R \cup S)$, $(b, a) \in R$ or $(b, a) \in S$. This implies that $(a, b) \in R^{-1}$ or $(a, b) \in S^{-1}$. Therefore, $(a, b)\ R^{-1} \cup S^{-1}$.

Consider that $R \subseteq S$. Any element (a, b) that belongs to R also belongs to S, from the definition of proper subset. Since R is proper subset of S, the cardinality of R is less than the cardinality of S. This implies that the cardinality of $\bar{S}$ is less than the cardinality of $\bar{R}$. This is because, if $(a, b) \in$ R then $(a, b) \notin$ R. Therefore $\bar{S} \subseteq \bar{R}$. We leave other proofs on similar lines to the reader.

Exercise 5.36: Consider the set A = {1, 2, 3} and the relations R defined on A. Let R be represented by the matrix as given below. Obtain the transitive closure of R, i.e., t(R).

$$M_R = \begin{bmatrix} 1 & 0 & 1 \\ 1 & 1 & 1 \\ 0 & 1 & 0 \end{bmatrix}$$

The corresponding relation R = {(1, 1), (1, 3), (2, 1), (2, 2), (2, 3), (3, 2)}. The different powers of R are computed using matrices.

R^0 = {(1, 1), (2, 2), (3, 3)} and matrix is $M_{R^0} = \begin{bmatrix} 1 & 0 & 0 \\ 0 & 1 & 0 \\ 0 & 0 & 1 \end{bmatrix}$

R^1 = R = {(1, 1), (1, 3), (2, 1), (2, 2), (2, 3), (3, 2)} and $M_{R^1} = \begin{bmatrix} 1 & 0 & 1 \\ 1 & 1 & 1 \\ 0 & 1 & 0 \end{bmatrix}$

$R^2 = R \bullet R$ = {(1, 1), (1, 3), (2, 1), (2, 2), (2, 3), (3, 2), (3, 1), (3, 2), (1, 2), (1, 3)}

The corresponding matrix for R^2 is $M_{R^2} = \begin{bmatrix} 1 & 1 & 1 \\ 1 & 1 & 1 \\ 1 & 1 & 1 \end{bmatrix}$

$R^3 = R \bullet R^2$ = {(1, 1), (1, 3), (2, 1), (2, 2), (2, 3), (3, 2), (3, 1), (3, 2), (1, 2), (1, 3)}

The corresponding matrix for R^3 is $M_{R^3} = \begin{bmatrix} 1 & 1 & 1 \\ 1 & 1 & 1 \\ 1 & 1 & 1 \end{bmatrix}$

Since the relation R^3 is a repetition of R^2, we stop computing the different powers of R and take the union of the computed matrices, as follows:

$$t(R) = M_{R^0} \cup M_{R^1} \cup M_{R^2}$$

$$= \begin{bmatrix} 1 & 0 & 0 \\ 0 & 1 & 0 \\ 0 & 0 & 1 \end{bmatrix} \cup \begin{bmatrix} 1 & 0 & 1 \\ 1 & 1 & 1 \\ 0 & 1 & 0 \end{bmatrix} \cup = \begin{bmatrix} 1 & 1 & 1 \\ 1 & 1 & 1 \\ 1 & 1 & 1 \end{bmatrix}$$

$$= \begin{bmatrix} 1 & 1 & 1 \\ 1 & 1 & 1 \\ 1 & 1 & 1 \end{bmatrix}$$

Therefore, using matrices is an alternative representation for relations useful in computing the different powers to obtain transitive closure of the relation.

5.6 GRAPHICAL REPRESENTATION OF RELATIONS

In the graphical representation of relations, the elements of the set are denoted by nodes and the relation between the nodes by the directed edges. Hence, it's a diagraph or directed graph. Consider the relation given in Example 5.34 and its digraph representation given in Fig. 5.2. The relation R= {(1, 1), (1, 3), (2, 1), (2, 2), (2, 3), (3, 2)}. The element 1R1 is shown by self loop at the node 1. Similarly, we have a self loop at node 2 that corresponds to the edge (2, 2). The directed edge from 2 to 1 refers to the edge (2, 1). Similarly, the directed edge from 3 to 2 refers to the edge (3, 2).

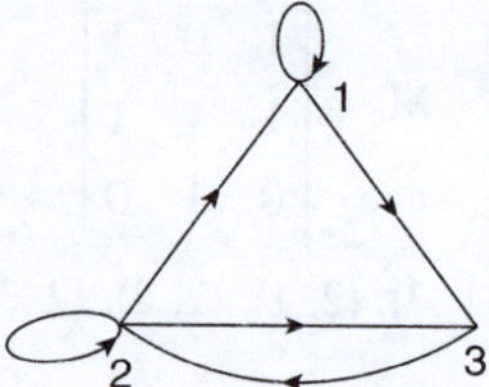

Figure 5.2 Digraph of example 5.34

5.6.1 Representation of Types of Relations

The graphical representation of the reflexive relation R = {(1, 1), (2, 2), (3, 3)} defined on the set A = {1, 2, 3} is given in Fig. 5.3. All the nodes will have self loops.

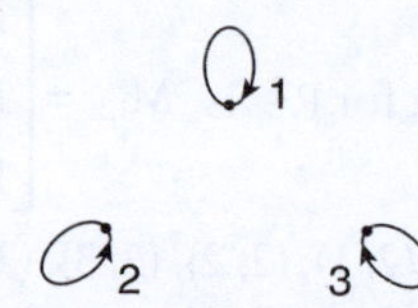

Figure 5.3 Graphical representation of reflexive relations

The graphical representation of the symmetric relation R = {(3, 1), (1, 2), (2, 1), (2, 3), (3, 2), (1, 3)} defined on the set A = {1, 2, 3} is given in Fig. 5.4.

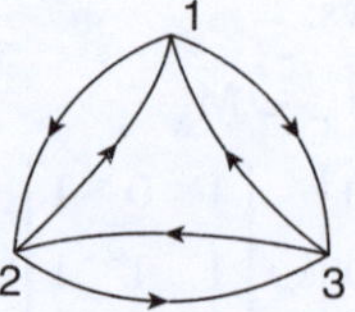

Figure 5.4 Graphical representation of symmetric relations

The graphical representation of an asymmetric relation R = {(3, 1), (2, 1), (2, 3)} defined on the set A = {1, 2, 3} is given in Fig. 5.5.

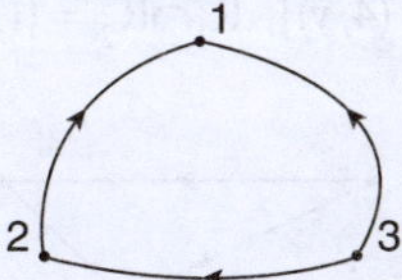

Figure 5.5 Graphical representation of asymmetric relations

The graphical representation of a transitive relation R = {(3, 1), (1, 2), (2, 1), (2, 3), (3, 2), (1, 3), (1, 1), (2, 2), (3, 3)} defined on the set A = {1, 2, 3} is given in Fig. 5.6.

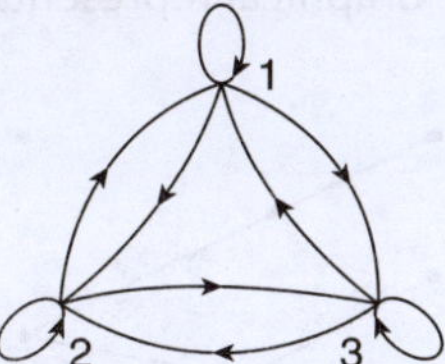

Figure 5.6 Graphical representation of transitive relations

5.6.2 Representations of Operations

The operations on relations are also represented graphically. The union of the relations R_1 and R_2 defined over A × A is given as follows:

$$R_1 = \{(a, a), (b, b), (c, c)\}$$
$$R_2 = \{(a, a), (b, b), (c, c), (b, c), (c, b)\}$$

The complement of R_1 denoted by $\overline{R_1} = \{(a, b), (b, a), (a, c), (c, a), (b, c), (c, b)\}$. Similarly, $\overline{R_2} = \{(a, b), (b, a), (a, c), (c, a)\}$. The graphical representations of $\overline{R_1}$ and $\overline{R_2}$ are given in Figs. 5.7 and 5.8 respectively.

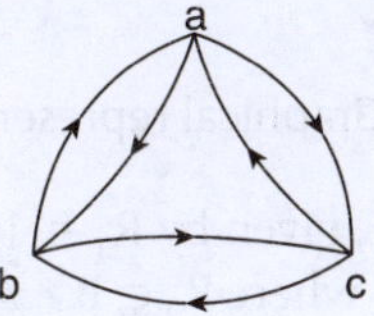

Figure 5.7 Graphical representation of $\overline{R_1}$

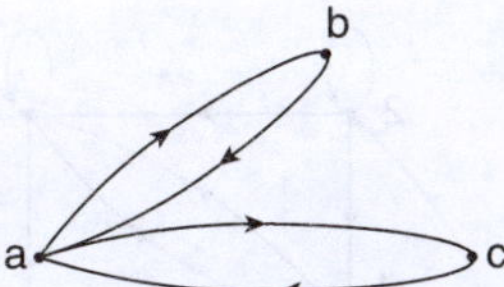

Figure 5.8 Graphical representation of $\overline{R_2}$

The other operations have the same meanings as discussed in the topic on set theory. $R_1 \cup R_2$, $R_1 \cap R_2$, and R_1-R_2 are graphically represented as shown in Figs. 5.9, 5.10 and 5.11 respectively. The relations are

$R_1 \cup R_2 = \{(1, x), (1, y), (2, y), (2, z), (3, y), (4, x), (4, y)\}$, $R_1 \cap R_2 = \{(1, y), (2, z), (3, y)\}$ and $R_1 - R_2 = \{(1, x), (2, y), (4, x)\}$.

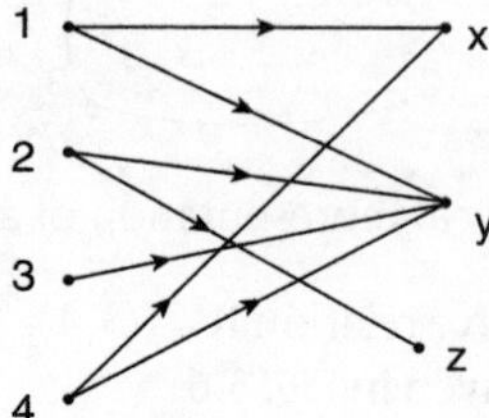

Figure 5.9 Graphical representation of $R_1 \cup R_2$

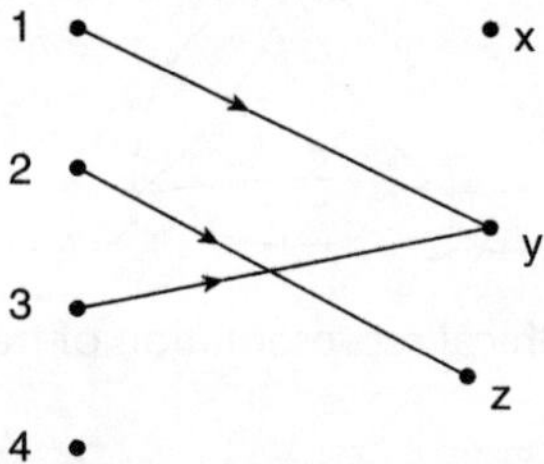

Figure 5.10 Graphical representation of $R_1 \cap R_2$

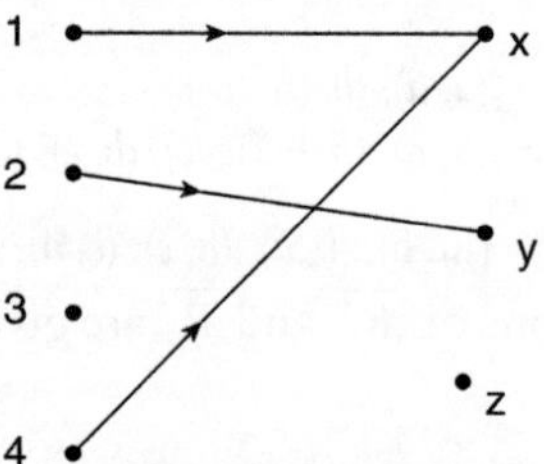

Figure 5.11 Graphical representation of $R_1 - R_2$

The composition of two relations R_1 and R_2 given by $R_1 = \{(1, x), (1, y), (2, y), (2, z), (3, y), (4, x)\}$ where $R_1 \subseteq A \times B$ and $R_2 = \{(y, 1), (z, 2), (y, 3), (y, 4)\}$ where $R_2 \subseteq B \times A$, is a new relation denoted by $R_1 \bullet R_2$. The graphical representation of $R_1 \bullet R_2 = \{(1, 1), (2, 2), (1, 3), (3, 1), (1, 4), (2, 1), (2, 3), (2, 4), (3, 1), (3, 3), (3, 4)\}$ is given in Fig. 5.12.

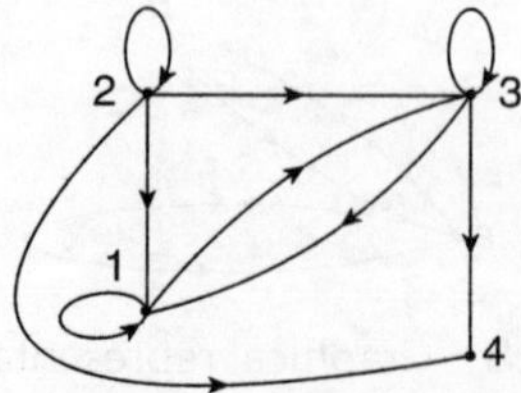

Figure 5.12 Graphical representation of $R_1 \cdot R_2$

5.7 PARTIALLY ORDERED SET (POSET)

A relation R on a given set A is said to be partially ordered, if it is reflexive, antisymmetric and transitive. The set along with the partial order relation R is called a partially ordered set, also abbreviated as POSET. It is denoted by (A, R). Consider the following examples of POSETs.

Example 5.37: The relation subset and the set of all subsets of a given set A, called the power set, is a POSET. It is denoted by (P(A), $\subseteq$).

Example 5.38: Consider the set of positive integers $\mathbb{Z}^+$. The relation $\leq$ together with $\mathbb{Z}^+$.is a POSET.

Example 5.39: Consider the set of positive integers $\mathbb{Z}^+$. The relation divides (|), together with $\mathbb{Z}^+$.is a POSET.

Example 5.40: Consider the set of positive integers $\mathbb{Z}^+$. The relation > together with $\mathbb{Z}^+$. is a not a POSET because the relation > is not reflexive. Therefore the relation is not partially ordered.

Consider the POSETs (A, $\leq$) and (A, $\geq$). The relations $\geq$ and $\leq$ are duals because if the relation R is $\leq$ then the inverse of R denoted as R^{-1} represents the relation $\geq$. The two POSETs are called duals. Given a POSET (A, $\leq$), consider the two elements a and b belonging to A and said to be comparable if $a \leq b$ or $b \leq a$. In a POSET, every pair of elements need not be comparable. This is obvious from Example 5.37 as 2 does not divide 5 and 5 does not divide 2. Thus, the word 'partial' means that some elements need not be comparable. If every pair of elements in a POSET is comparable then the set is said to be linearly ordered and the set A forms a chain. The data ordered in ascending order or in descending order are examples of linear ordering. We also call such an ordering total ordering.

Theorem 5.2: Show that if (A, $\leq$) and (B, $\leq$) are POSETs then (A $\times$ B, $\leq$) is a POSET with partial order $\leq$ defined by

$$(a, b) \leq (a', b') \text{ if } (a \leq a') \text{ is in A and } (b, b') \text{ is in B.}$$

Proof: If $(a, b) \in$ (A $\times$ B) then $(a, b) \leq (a, b)$ since $a \leq a$ in A and $b \leq b$ in B. This means that the relation $\leq$ fulfils the reflexivity property in (A $\times$ B). Consider that $(a, b) \leq (a', b')$ and $(a', b') \leq (a, b)$. Here, a' and a belong to A, and b and b' belong to B. It follows that

$$a' \leq a \text{ and } a \leq a' \text{ in A and } b' \leq b \text{ and } b \leq b' \text{ in B}$$

Since A and B are POSETS, the antisymmetry of the partial orders on A and B implies that

$$a = a' \text{ and } b = b'$$

Therefore the relation $\leq$ satisfies the antisymmetry property in (A $\times$ B). Now consider that

$$(a, b) \leq (a', b') \text{ and } (a', b') \leq (a'', b'')$$

Where a, a' and a'' belong to A, and b, b' and b'' belong to B. Then we can say that

$$a \leq a' \text{ and } a' \leq a''$$

By the transitivity property of partial order on A, we can say that $a \leq a''$. Similarly, we can say that $b \leq b''$, using partial order on B. Hence $(a, b) \leq (a'', b'')$

Consequently, the transitive property holds for the partial order on (A $\times$ B) and we conclude that (A $\times$ B) is a POSET.

Example 5.41: Let S = $\{a, b, c, \ldots, z\}$ be an English alphabet in the usual order. The Kleene's closure of the set S, denoted by S^* consists of all possible words of any length, whether the words are meaningful or not. The word help < helpful and is listed on an earlier page in the dictionary. This type of ordering is called lexicographic ordering.

5.7.1 Graphical Representation of POSET

The graphical representation of POSET is called a Hasse diagram. In Hasse diagrams, we do not show reflexivity and transitivity explicitly. Consider the diagrapm shown in Fig. 5.13.The transitivity is eliminated by showing the three related elements *a, b* and *c*, such that $a \leq b \leq c$ is represented showing *a* at a level lower than *b* and *b* lower than *c*, as shown in Fig. 5.13(d). The elements that are not related are not connected.

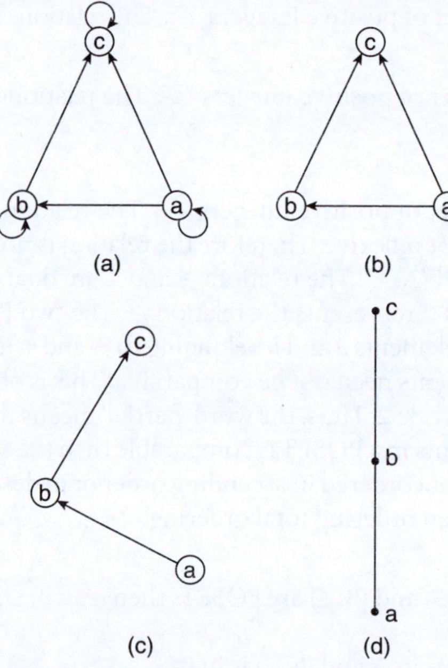

Figure 5.13 (a) Partial order relation (b) Implicit reflexivity (c) Implicit transitivity (d) Levels

Example 5.42: Consider the set A = {1, 2, 3, 4, 12} and the relation as divisibility on A. This means that $a \leq b$ if and only if $a \mid b$, for $a, b \in$ A. The POSET is (A, ≤) and the corresponding Hasse diagram is shown in Fig. 5.14.

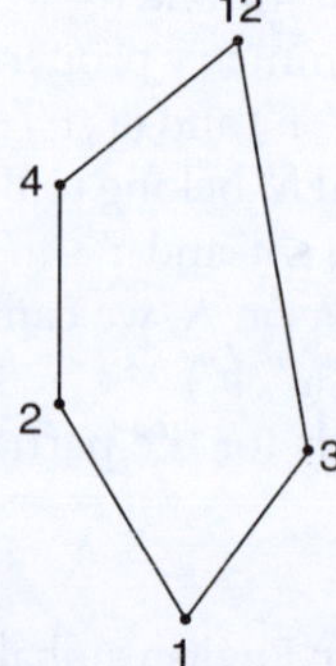

Figure 5.14 Hasse diagram of Example 5.40

Example 5.43: Consider the set A = {0, 1} and its power set P(A)= {Ø, {0}, {1}, {0, 1}} along with the partial order subset (⊆). The POSET is (P(A), ⊆). The corresponding Hasse diagram is given in Fig. 5.15. If the set B = {*a, b, c*}, then the power set of B is denoted as P (B) = {Ø, {*a*}, {*b*}, {*c*}, {*a, b*}, {*a, c*}, {*b, c*}, {*a, b, c*}}. The POSET is (P (B), ⊆) and is shown in Fig. 5.16.

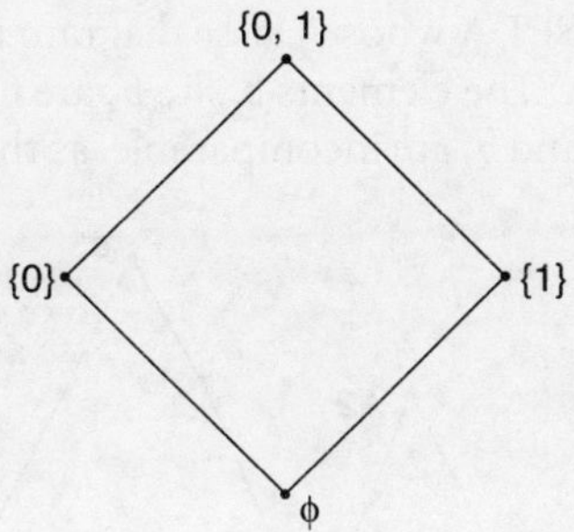

Figure 5.15 Hasse diagram for (P(A), ⊆)

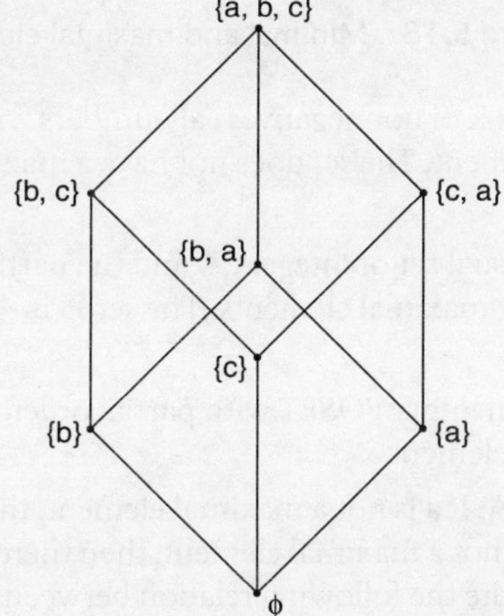

Figure 5.16 Hasse diagram for (P(B), ⊆)

The Hasse diagram for a linearly ordered data set {*a, b, c, d, e*} is shown in Fig. 5.17.

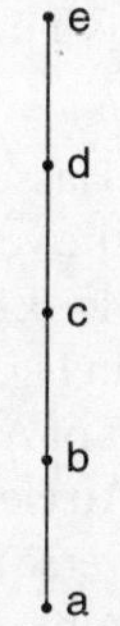

Figure 5.17 Hasse diagram for linear ordering

5.7.2 Maximal and Minimal Elements

An element $a \in$ A is called the maximal element of A if there is no element c in A such that $(a < c)$. Similarly, an element $b \in$ A is called the minimal element of A if there is no element c in A such that $(c < b)$.

Example 5.44: Consider the POSET A whose Hasse diagram is shown in Fig. 5.18. The elements a_1, a_2, a_3 are the maximal elements of A. The elements b_1, b_2, b_3 are called the minimal elements of A. The elements b_1 and b_2, b_1 and b_3, and b_2 and b_3 are incomparable, as there is no line between these elements.

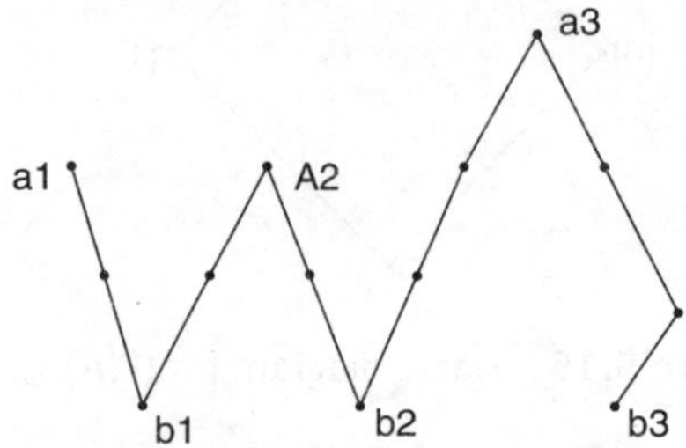

Figure 5.18 Minimal and maximal elements

Example 5.45: Consider the set R of nonnegative real numbers with the partial order ≤. It is a POSET. The set R has 0 as the minimal element. The set does not have a maximal element, as the set is infinite.

Example 5.46: Consider the usual set of integers $\mathbb{Z}$ and the partial order ≤. It is a POSET. The set $\mathbb{Z}$ has neither a minimal element nor maximal elements. The set $\mathbb{Z}$ is $-\infty$ to $+\infty$.

Theorem 3: Let A be a finite nonempty POSET with partial order ≤. Then A has at least one maximal element and at least one minimal element.

Proof: Consider an element $a \in$ A. If a is not a maximal element, then there exists an element $a_i \in$ A such that $(a < a_1)$. Similarly, if a_1 is not a maximal element, then there exists an element $a_2 \in$ A such that $(a_1 < a_2)$ and so on. We can write the following relation between the elements.

$$a < a_1 < a_2 < a_3 < a_4 < a_5 \ldots < a_{k-3} < a_{k-2} < a_{k-1} < a_k$$

This cannot be further extended because A is finite. Hence, we cannot have $a_k < b$ for $b \in$ A. Therefore, a_k is a maximal element of the POSET (A, ≤). One can extend the same argument and say that the dual POSET (A, ≤) has a maximal element, so (A, ≥) has a minimal element.

Example 5.47: Let A = {1, 2, 3, 4, 5} and the Hasse diagram of partial order ≤ on A be as shown in Fig. 5.19 (a). A minimal element of this POSET is the vertex 4 or 5. We shall put 4 in the sorted-list[1], the remaining Hasse diagram of A – {4} is shown in Fig. 5.19(b). The new minimal element in the new A is 5. Let this be placed in the sorted-list[2], as shown in Fig. 5.19(c). The final chain is shown in Fig. 5.19(f).

An element $a \in$ A is called the greatest element of A if $x \le a$ for all $x \in$ A. Similarly, an element $a \in$ A is called the least element of A if $a \le x$ for all $x \in$ A. An element $a \in$ A is called the greatest element of (A, ≤) if and only if a is the least element of (A, ≥). The reverse is also true. Consider Example 5.43, which has 0 as the least element and no greatest element. In Example 5.41, the empty set is the least element and the set itself is the greatest element. In Example 5.44, there is no least element and no greatest element.

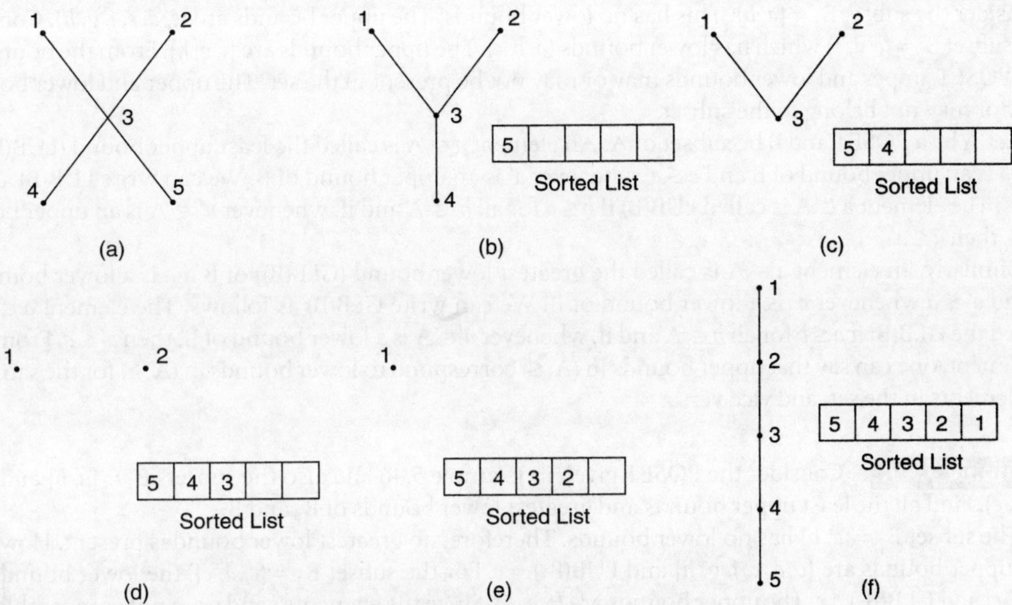

Figure 5.19 Hasse diagram for Example 5.45

Theorem 5.3: A POSET has at the most, one greatest element and one least element.

Proof: Consider that a and b are the two greatest elements of a POSET A. Suppose b is the greatest element. It follows from the definition that $a \leq b$. Similarly, suppose a is the greatest element, we can say that $b \leq a$. By the antisymmetry property, we can write $a = b$. Therefore, the greatest element is at the most one for the given POSET. Since this fact is true for all POSETs, the dual POSET (A, ≥) has at the most one least element. The greatest element, if it exists, is denoted by I and is often called the unit element. Similarly, the least element of the POSET, if it exists, is denoted by 0 and is often called the zero element.

Consider a POSET A and its subset, say, B. An element $a \in$ A is called the upper bound of B if $b \leq a$, for all $b \in$ B. An element $a \in$ A is called the lower bound of B if $a \leq b$, for all $b \in$ B.

Example 5.48: Consider the POSET (A ≤) and A = {a, b, c, d, e, f, g, h}. The Hasse diagram is shown in Fig. 5.20. Determine the upper and lower bounds of the subsets $B_1 = \{a, b\}$ and $B_2 = \{c, d, e\}$.

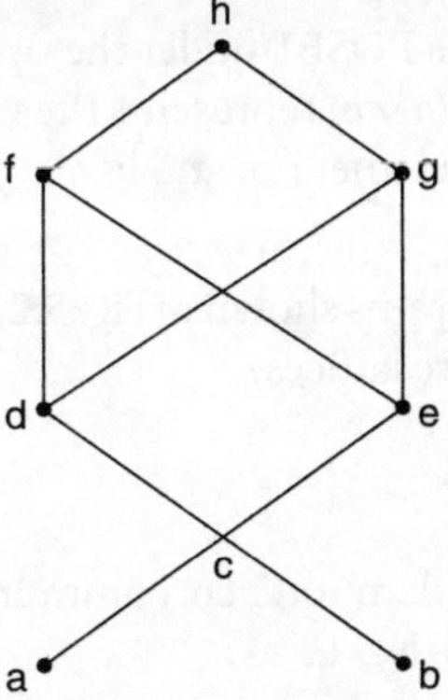

Figure 5.20 Hasse diagram for Example 5.46

Consider the subset $B_1 = \{a, b\}$. This has no lower bounds. The upper bounds are $\{c, d, e, f, g, h\}$. Consider the subset $B_2 = \{c, d, e\}$ which has lower bounds $\{a, b, c\}$. The upper bounds are $\{f, g, h\}$. From the bounds of the POSET, upper and lower bounds may or may not be present in the set. The upper and lower bounds may or may not belong to the subset.

Let A be a POSET and B be subset of A. An element $a \in$ A is called the least upper bound (LUB(B)) of B if a is an upper bound of B and $a \leq a'$ whenever a' is an upper bound of B. We can write LUB(B) as follows. The element a $\in$ A is called LUB(B) if $b \leq a$ for all $b \in$ A and if, whenever $a' \in$ A is an upper bound of B, then $a \leq a'$.

Similarly, an element $a \in$ A is called the greatest lower bound (GLB(B)) of B if a is a lower bound of B and $a' \leq a$ whenever a' is a lower bound of B. We can write GLB(B) as follows. The element $a \in$ A is called the GLB(B) if $a \leq b$ for all $b \in$ A and if, whenever $a' \in$ A is a lower bound of B, then $a' \leq a$. From this argument, one can say that upper bounds in (A, $\leq$) correspond to lower bounds in (A, $\geq$) for the same set of elements in the set, and vice versa.

Example 5.49: Consider the POSET given in Example 5.46 and also the subsets $B_1 = \{a, b\}$ and $B_2 = \{c, d, e\}$. Find all the least upper bounds and greatest lower bounds of B_1 and B_2.

The subset $B_1 = \{a, b\}$ has no lower bounds. Therefore, no greatest lower bound is present. However, the upper bounds are $\{c, d, e, f, g, h\}$ and LUB(B_1) = c. For the subset $B_2 = \{c, d, e\}$, the lower bounds are $\{a, b, c\}$ and GLB(B_2) = c. The upper bounds are $\{f, g, h\}$. Since the elements f and g are not comparable, the subset B_2 has no least upper bound.

5.8 LATTICES

A lattice is a partial order set (L, $\leq$) wherein every pair of elements in the set has a unique least upper bound and greatest lower bound. Consider a pair $\{a, b\}$ of the POSET. We shall denote LUB ($\{a, b\}$) by $a \vee b$ and is also called the join of a and b. Similarly, the GLB($\{a, b\}$) is denoted by $a \wedge b$ and is also called the meet of a and b. A lattice is a mathematical structure and often appears in computing. Any chain is a lattice.

Example 5.50: Consider the set S and its power set P(S). The POSET under the operation $\subseteq$ is (P(S), $\subseteq$), which is a lattice. Let the sets A and B belong to P(S). $A \vee B$ represents the union of the two sets. Similarly, $A \wedge B$ represents the intersection of the two sets.

Example 5.51: Consider the set Z^+. The POSET under the operation $\leq$ is (Z^+, $\leq$), which is a lattice. Let the elements a and b belong to Z^+. Then $(a \vee b)$ represents the LCM of the two elements a and b. Similarly, $(a \wedge b)$ represents the GCD of the two elements a and b.

Example 5.52: The following Hasse diagrams shown in Fig. 5.21 (a) (P(A), $\subseteq$) and (b) A = {1, 2, 3, 4, 12} and the relation as divisibility on A (A, $\leq$), are lattices.

5.8.1 Properties of Lattices

Lattices satisfy algebraic properties such as idempotence, commutativity, associativity, and absorption. Consider the lattice (L, $\leq$) and the elements $a, b \in$ L.

(i) Idempotent property: $(a \vee a) = a$ and $(a \wedge a) = a$

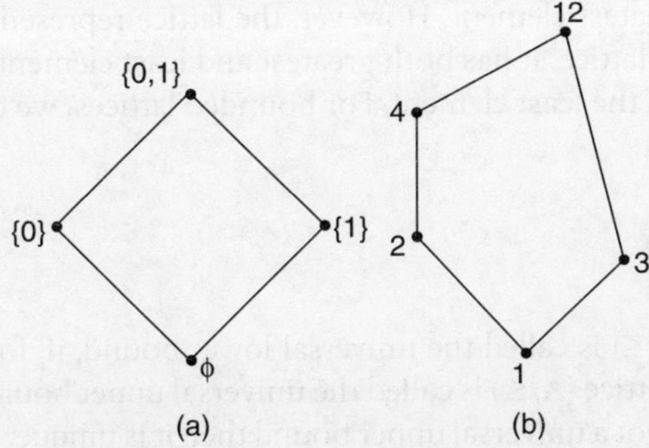

Figure 5.21 Examples of Lattices

(ii) Commutativity property : $(a \vee b) = (b \vee a)$ and $(a \wedge b) = (b \wedge a)$
(iii) Associative property : $(a \vee b) \vee c = (a \vee (b \vee c))$ and $(a \wedge (b \wedge c)) = ((a \wedge b) \wedge c \wedge a))$
(iv) Absorption property : $(a \vee (a \wedge b)) = a$ and $(a \wedge (a \wedge b)) = a$

Some of the observations about the least upper bound $(a \vee b)$ and the greatest lower bound $(a \wedge b)$ are as given as follows:

(a) If $a \leq (a \vee b)$ and $b \leq (a \vee b)$ then $(a \vee b)$ is an upper bound of a and b.
(b) If $(a \leq c)$ and $(b \leq c)$ then $(a \vee b) \leq c$. It follows that $(a \vee b)$ is the least upper bound of a and b.
(c) If $a \leq (a \wedge b)$ and $b \leq (a \wedge b)$ then $(a \wedge b)$ is an lower bound of a and b.
(d) If $(c \leq a)$ and $(c \leq b)$ then $c \leq (a \wedge b)$. It follows that $(a \wedge b)$ is the greatest lower bound of a and b.

Theorem 5.4: If L is a lattice, then for every a and b ∈ L the following relations hold good.

(i) $(a \vee b) = b$ if and only if $(a \leq b)$
(ii) $(a \wedge b) = a$ if and only if $(a \leq b)$
(iii) $(a \wedge b) = a$ if and only if $(a \vee b) = b$

Proof:

(i) Suppose that $(a \vee b) = \text{b}$. Since $a \leq (a \vee b)$, we can write $(a \leq b)$. If $(a \leq b)$ then since $(b \leq b)$, b is an upper bound of a and b. From the definition of LUB, we can write $(a \vee b) \leq b$. Since $(a \vee b)$ is an upper bound, $b \leq (a \vee b)$ and hence $(a \vee b) = b$.
(ii) Suppose that $(a \wedge b) = a$. Since $a \leq (a \vee b)$, we can write $(a \leq b)$. If $(a \leq b)$ then since $(b \leq b)$, b is an upper bound of a and b. From the definition of LUB, we can write $(a \vee b) \leq b$. Since $(a \vee b)$ is an upper bound, $b \leq (a \vee b)$ and hence $(a \vee b) = b$.
(iii) Combining (i) and (ii) we get the proof for (iii)

Example 5.53: Consider L as a linearly ordered set. If a and $b \in$ L then either $a \leq b$ or $b \leq a$. It follows that L is a lattice as every pair of elements has an LUB and a GLB.

5.8.2 Special Types of Lattices

Under this section, we discuss two concepts, namely, complemented lattice and distributive lattice.

5.8.2.1 *Complemented lattice*

A lattice is said to be bounded if it has the greatest element 1 and the least element 0. Consider the lattice represented by the divisibility relation and the set Z^+. We know that the lattice is not bounded as it has the least element but no greatest element. However, the lattice represented by the power set and the subset relation is a bounded lattice. It has both greatest and least elements. The set itself is the greatest element and the empty set is the least element. For bounded lattices, we can write the following for all elements belonging to the set.

(i) $0 \leq a \leq 1$
(ii) $a \vee 0 = a$ and $(a \wedge 0) = 0$
(iii) $a \vee 1 = 1$ and $(a \wedge 1) = a$

An element a in a lattice (A, ≤) is called the universal lower bound, if, for every element $b \in A$, $a \leq b$. Similarly, an element a in a lattice (A, ≤) is called the universal upper bound, if, for every element $b \in A$, $b \leq a$. Further, if a lattice has got a universal upper bound then it is unique. The lattice represented by the power set and the subset relation has both universal lower bound and universal upper bound defined.

Let (A, ≤) be a lattice with universal lower and upper bounds as 0 and 1 respectively. Consider the elements a and b in A. The element b is said to be a complement of a if $(a \vee b) = 1$ and $(a \wedge b) = 0$. Using the commutative property, one can say that when b is a complement of a, then a is also a complement of a. An element may have more than one complement or no complement at all.

Example 5.54: Consider the lattice shown in Fig. 5.22. The element d is a complement of e and e is a complement of d. The elements b and e are complements of d. The element c has no complement. However, 0 is the unique complement of 1 and vice-versa.

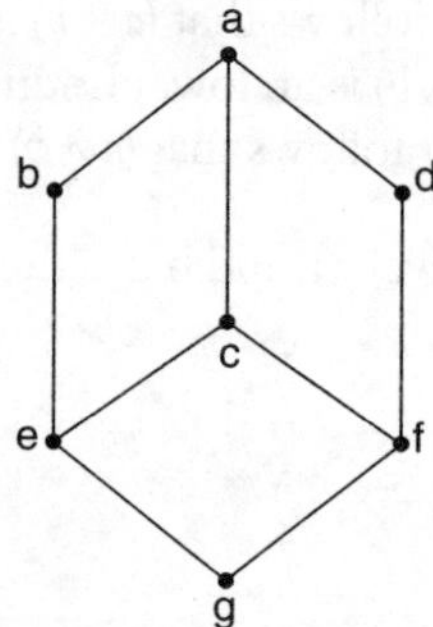

Figure 5.22 Example of a lattice

A lattice is said to be a complemented lattice if every element in the lattice has a complement. The complemented lattice must have universal lower and upper bounds.

Example 5.55: Consider the complemented latticeshown in Fig. 5.23. The complement of b is c and the complement of c is b. The complements of the element c are the elements a and b.

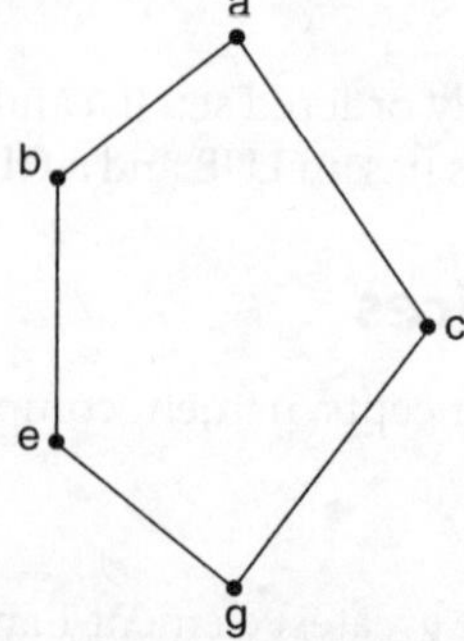

Figure 5.23 Example of a lattice

5.8.2.2 *Distributive lattice*

A lattice is said to be a distributive lattice if the meet operation distributes over the join operation and vice-versa. This holds good because of the commutativity property satisfied by the meet and join operations.

(i) $(a \vee (b \wedge c)) = ((a \vee b) \wedge (b \vee c))$
(ii) $(a \wedge (b \vee c)) = ((a \wedge b) \vee (b \wedge c))$

The lattice shown in Fig. 5.24 is a distributive lattice. However, the lattice given in Fig. 5.25 is not a distributive lattice. The reasons are as follows:

(i) $(b \wedge (c \vee d)) = (b \wedge a) = b$
(ii) $(b \wedge c) \vee (b \wedge d) = (e \vee e) = e$

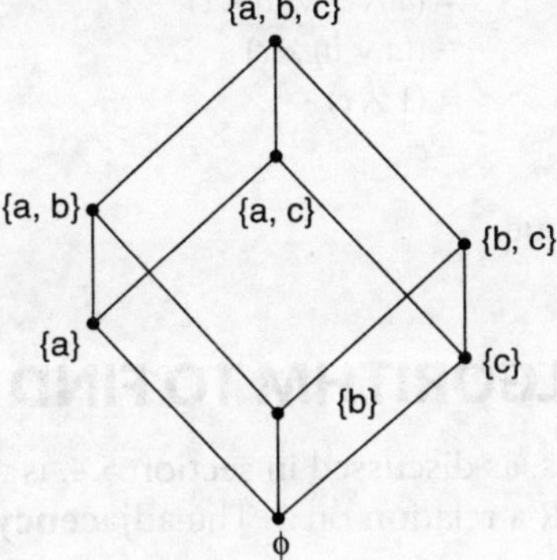

Figure 5.24 Example of a distributive lattice

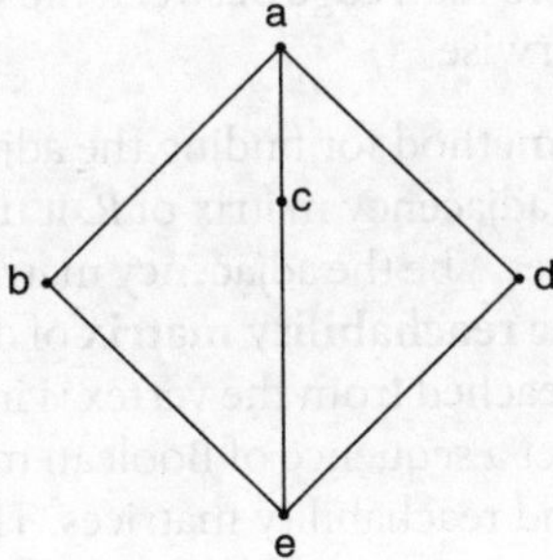

Figure 5.25 Example of a non-distributive lattice

Theorem 5.5: If the meet operation is distributive over join operation in a lattice then the join operation also distributes over the meet operation. The vice-versa holds good.

Proof: Consider the expression $(a \vee (b \wedge c)) = (a \vee b) \wedge (a \vee c)$. Let us consider right-hand-side.

$$\begin{aligned}(a \vee b) \wedge (a \vee c) &= ((a \vee b) \wedge a) \vee ((a \vee b) \wedge c)\\ &= a \vee ((a \vee b) \wedge c)\\ &= (a \vee (a \wedge c)) \vee (b \wedge c)\\ &= a \vee (b \wedge c)\end{aligned}$$

Using the principle of duality, one can prove that $(a \wedge (b \vee c)) = (a \wedge b) \vee (a \wedge c)$. Thus the theorem.

Theorem 5.6: If an element has a complement in a distributive lattice, then the complement is unique.

Proof: Let us assume that an element has two complements b and c. We can write the following expressions.

$$(a \vee b) = 1 \qquad (a \vee c) = 1$$
$$(a \wedge b) = 0 \qquad (a \wedge c) = 0$$

Consider the complement b and the expression $b = (b \wedge\wedge 1)$

$$\begin{aligned} &= (b \wedge (a \vee c)) \\ &= (b \wedge a) \vee (b \wedge c) \\ &= 0 \vee (b \wedge c) \\ &= (a \wedge c) \vee (b \wedge c) \\ &= ((a \vee b) \wedge c) \\ &= (1 \wedge c) \\ &= c \end{aligned}$$

Therefore, the complement is unique.

5.9 WARSHALL'S ALGORITHM TO FIND TRANSITIVE CLOSURE

One of the operations on relations, as discussed in section 5.4, is transitive closure of a given relation. Let S be the finite set $\{v1, ..., vn\}$, R a relation on S. The **adjacency matrix** A of R is an $n \times n$ Boolean (zero-one) matrix defined by

$$\begin{aligned} A\,[i, j] &= 1 \text{ if there is an edge between the vertices vi and vj} \\ &= 0 \text{ otherwise} \end{aligned}$$

Warshall's algorithm is an efficient method for finding the adjacency matrix of the transitive closure of relation R on a finite set S from the adjacency matrix of R. It uses properties of the digraph D, in particular, walks of various lengths in D. Let A be the adjacency matrix of R and T be the adjacency matrix of the transitive closure of R. T is called the **reachability matrix** of digraph D due to the property that $T[i, j] = 1$ if and only if the vertex vj can be reached from the vertex vi in D, by a sequence of traversal of edges. In **Warshall's algorithm,** we construct a sequence of Boolean matrices $A = W^{[0]}, W^{[1]}, W^{[2]}, ..., W^{[n]} = T$, where A and T represent adjacency and reachability matrices. This can be done from the digraph D as follows.

$[W^{[1]}]\ i, j = 1$ if and only if there is a walk from vi to vj, with elements of a subset of $\{v1\}$ as interior vertices.

$[W^{[2]}]\ i\,j = 1$ if and only if there is a walk from vi to vj, with elements of a subset of $\{v1, v2\}$ as interior vertices. Continuing this process, we generalise to the following.

$[W^{[k]}]\ i, j = 1$ if and only if there is a walk from vi to vj, with elements of a subset of $\{v1, v2, \ldots, vk\}$ as interior vertices. It is to be noted that in constructing $W^{[k]}$ from $W^{[k-1]}$ we shall either keep 0's or change some 0's to 1's. No 1's ever get changed to 0's.

Example 5.56: Consider the transitive closure of the relation represented by the digraph shown in Fig. 5.26. Use the aforesaid method. Indicate what edges must be added to this digraph to get the digraph of the transitive closure, and draw the digraph of the transitive closure.

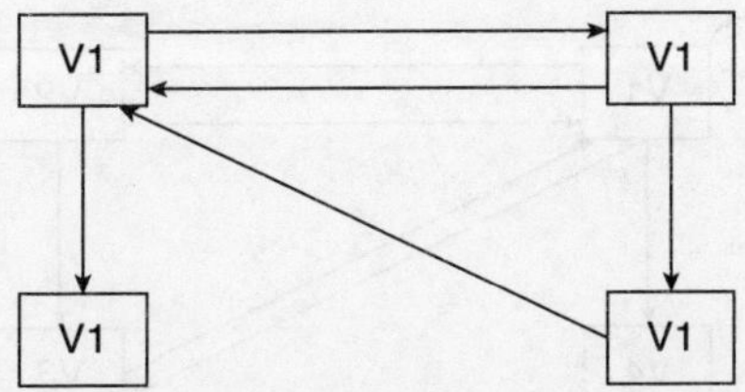

Figure 5.26 Digraph showing the relation

$$A=\begin{bmatrix}0&0&0&0\\1&0&0&0\\1&0&1&0\\0&1&0&1\end{bmatrix}$$

Walks {*v*2, *v*1, *v*2}, {*v*2, *v*1, 4*v*} and {*v*3, *v*1, *v*2} have elements of {*v*1} as interior vertices. Therefore, $[W^{[1]}]_{2,2}=1$, $[W^{[1]}]_{2,4}=1$, and $[W^{[1]}]_{3,2}=1$

$$W^{[1]}=\begin{bmatrix}0&0&0&0\\1&1&0&0\\1&1&1&1\\0&1&0&1\end{bmatrix}$$

We need to consider only walks with *v*2, or *v*1 and *v*2 as interior vertices, since walks with *v*1 as an interior vertex have already been considered. Walks {*v*1, *v*2, 3*v*}, {*v*1, *v*2, *v*1}, and {3*v*, *v*1, *v*2, 3*v*} have elements of {*v*1, *v*2} as interior vertices. Therefore, $[W^{[2]}]_{1,1}=1$, $[W^{[2]}]_{1,3}=1$, and [*W*[2]] 3, 3 = 1. There are other walks with elements of subsets of {*v*1, *v*2} as interior vertices, but they do not contribute any new 1's to $W^{[2]}$.

$$W^{[2]}=\begin{bmatrix}0&0&0&0\\1&1&1&0\\1&1&1&1\\0&1&1&1\end{bmatrix}$$

There are walks with elements of subsets of {*v*1, *v*2, *v*3} as interior vertices. We need to consider only walks with *v*3 (possibly along with *v*1 or *v*2 or both) as interior vertices since walks with *v*1, *v*2 (but not *v*3) as interior vertices have already been considered. However, none of these walks create any new 1's in $W^{[3]}$. We continue this process to obtain $W^{[4]}$. However, any walks we construct with *v*4 as an interior vertex contributes no new 1's. Therefore, $T = W^{[4]} = W^{[3]} = W^{[2]}$. Therefore,

$$T=\begin{bmatrix}1&1&1&1\\1&1&1&1\\1&1&1&0\\0&0&0&0\end{bmatrix}$$

One must add edges from *v*1 to *v*1, *v*1 to *v*3, *v*2 to *v*2, *v*3 to *v*2, and *v*3 to *v*3. The graph of the transitive closure is given in Fig. 5. 27.

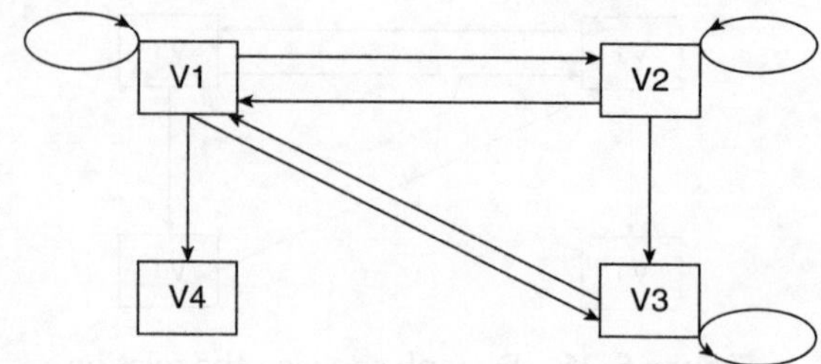

Figure 5.27 Transitive closure of the diagraph in Fig. 5.26

Algorithm: Warshall's algorithm

Input: Adjacency matrix A of relation R on a set of *n* elements

Output: Adjacency matrix T of the transitive closure of R.

```
Start
Step1: T:=A          /* initialize T to A */
Step2: for j := 1 to n
     Step 2.1 for i := 1 to n
        Step 2.2.1 if i j T , = 1 then
           Step 2.2.2 ai :=  ai v aj /* Boolean OR of row i and row j, store it in ai */
        next i
           next j
Stop
```

Here, the matrix *T* at the end of each iteration of *j* is the same as $W^{[j]}$ in the digraph implementation of Warshall's algorithm.

Example 5.57: Consider the adjacency matrix A as given below. We shall trace the pseudocode implementation of Warshall's algorithm on A, giving the details of each Boolean OR between rows

$$A = \begin{bmatrix} 0 & 0 & 0 \\ 0 & 1 & 1 \\ 0 & 1 & 0 \end{bmatrix}$$

When $j=1, i=1,\ T_{ij'}=0$ — No action takes place

$i=2,\ T_{ij'}=0$ — No action takes place

$i=3\ T_{ij'}=0$ — No action takes place, Therefore, $W^{[1]}=T=A$

$j=2, i=1\ T_{ij}=1$ — $(0\,1\,0)$ OR $(0\,1\,1) = (0 \vee 0, 1 \vee 1, 0 \vee 1) = (0\,1\,1)$
Therefore, row 1 of T becomes $(0\,1\,1)$

$i=2\ T_{ij}=1,$ — row 2 is ORed with row 2 and put into row 2.
However, row 2 OR row 2 is equal to row 2.

$i=3\ T_{ij'}=0$ — No action takes place. At this stage we have

$$T = W^{[2]} = A = \begin{bmatrix} 0 & 0 & 0 \\ 0 & 1 & 1 \\ 0 & 1 & 1 \end{bmatrix}$$

$j=3\ i=1,\ T_{ij}=0$ — No action takes place

$i=2\ T_{ij'}=1$ — $(0\,1\,1)$ OR $(0\,0\,0) = (0 \vee 0, 1 \vee 0, 1 \vee 0) = (0\,1\,1)$

The result is put into row 2, which is unchanged

$i=3$ $T_{ij'}=0$ No action takes place.

At this stage $T = [3]\ W = [2]\ W$ above. We now have the transitive closure.

ADDITIONAL SOLVED EXAMPLES

Example W5.1:

(i) If A is the set of students who were registered at the University of Michigan for the fall 2001 semester and B is the set of subjects registered {Mathematics, Biology, History, English} then R is the binary relation = $\{(a, b) \mid a \in A$ is registered for the subject $b \in B\}$.

(ii) Let A be the set of surnames listed in the telephone directory. Then R = $\{(a, n) \mid a$ appears on page $n\}$. Then R is a binary relation from set A to the set of natural numbers.

(iii) $\{(a, b) \mid a, b \in \mathbb{N}$, and a/b is an integer$\}$ and $\{(a, b) \mid a, b \in \mathbb{N}, a - b = 2\}$ are binary relations on $\mathbb{N}$.

(iv) $\{(a, b) \mid a, b \in \mathbb{N}$, and $b = a^2\}$ is a binary relation on R and the corresponding graph is called a parabola.

Example W5.2: Prove or disprove that every asymmetric relation is also an antisymmetric relation.

Solution:

By the definition of asymmetric relation one can say:
if there is no element where, if (x, y), then (y, x);
this "sounds" like :

$$\neg\exists x, y(x \,R\, y \rightarrow y \,R\, x).$$

Moving "inward" $\neg$, it is equivalent to

$$\forall\exists x, y(\neg(x \,R\, y \rightarrow y \,R\, x)).$$

Now, we need the equivalence between $p \rightarrow q$ and $(p \wedge \neg q)$ to get

$$\forall\exists x, y(x \,R\, y \wedge \neg\ (y \,R\, x)).$$

This is not the same as the formula in the definition. According to the definition of *asymmetric*, **if** *a* is the *father of* b, then b is **not** the *father of a*, which is reasonable. Therefore, an asymmetric relation is not antisymmetric.

Example W5.3: Let R and S be two binary relations on a set A. Prove or disprove the following.

(i) If R and S are asymmetric then $R \cup S$ and $R \cap S$ are asymmetric.

(ii) If R and S are antisymmetric then $R \cup S$ and $R \cap S$ are antisymmetric.

Solution:

(i) By the definition of asymmetry, we can write $\forall x, y((x \,R\, y \rightarrow \neg\ (y \,R\, x)))$.
When R and S are asymmetric, the tuples (x, y) are present but not (y, x) in both R and S. Their union will contain tuples of the type (x, y) present in R or S. Therefore their union is also asymmetric.

(ii) On similar lines, $\forall x, y((x \,R\, y \wedge (y \,R\, x)) \rightarrow (x = y))$ can be solved. If R is antisymmetric, then it contains the tuples of the type (x, y) and (y, x). If S is antisymmetric it also contains the tuples (x, y) and (y, x) along with the tuples (x, x) and (y, y). Therefore $R \cup S$ is antisymmtric and $R \cap S$ is not.

Example W5.4: In ANSI FORTRAN, a nonexecutable specification statement called EQUIVALENCE statement allows two or more variables in a given program to refer to the same memory location. For example:

EQUIVALENCE (X, Y, Z)

Conveys information to the compiler that the variables X, Y, Z will share a common memory location. Here, the set of all program variables is partitioned by the equivalence relation.

Example W5.5: Let R be an equivalence relation defined on the set A = {1, 2, 3, 4, 5, 6, 7} which induces a partition A = {1, 2} ∪ {3} ∪ {4, 5, 7} ∪ {6}. Find R.

Solution:

Consider the subset {1, 2}, which represents the sub relation {(1, 1), (1, 2), (2, 1), (2, 2)}. In similar manner, the other subsets {3}, {4, 5, 7} and {6} represent the sub relations {(3, 3)}, {(4, 4), (5, 5) (7, 7), (4, 5) (4, 7), (5, 4), (5, 7), (7, 4), (7, 5)}, and {(6, 6}. The reflexive relations, corresponding to the partition is {(1,1), (1, 2), (2, 1), (2, 2), (3, 3), (4, 4), (5, 5), (7, 7), (4, 5) (4, 7), (5, 4), (5, 7), (7, 4), (7, 5), (6, 6)}.

SUMMARY

We have dealt with the properties and types of relations, in reasonable detail. Relations are important mathematical concepts and find numerous applications in computer science. Partitions of a given set and the equivalence relation defined over the set are alternate representations. Digraphs and matrices are alternate representations for relations. Partial ordered relations are different from totally ordered relations. Hasse diagrams are pictorial representations for POSETS. Lattices and different types of lattices are useful concepts. Boolean algebra is a lattice. Certain theorems are also dealt with in this chapter.

EXERCISES

1. Find the domain and range of the following relations
   (a) {(Alabama, 7), (California, 53), (Colorado, 7), (Florida, 25), (Kansas, 4)}.
   (b)

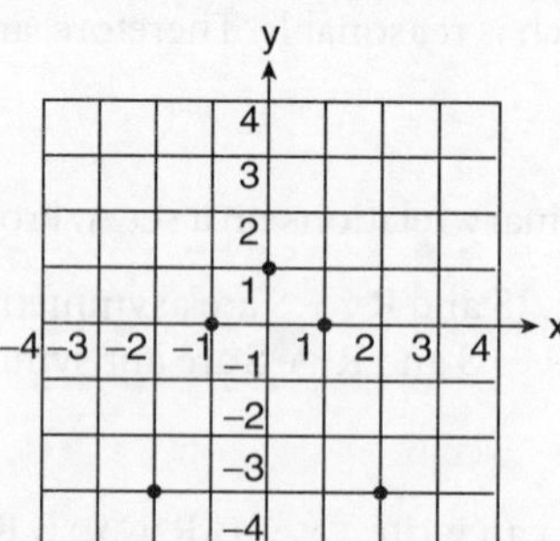

(i) Define the key terms.
   (*a*) Relation in *x* and *y* (*b*) Domain of a relation (*c*) Range of a relation

(ii) Use a mathematical equation to define a relation whose second component *y* is 1 less than 2 times the first component *x*.

3. Given the function defined by $g(x) = \frac{1}{2}X - 1$ find the function values.
   (a) g(0) (b) g(2) (c) g(4) (d) g(−2)
4. Given the functions defined by $f(x) = x^2 - 2x$ and $g(x) = 3x + 5$, find the function values.
   (a) F(t) (b) g(w + 4) (c) f(−t).
5. Let A = {1, 2, 3, 4} and R = {(1, 1), (1, 2) (2, 3), (3, 3), (3, 4)} be a relation on A. Give graph representations of R, R^2, R^3 and R^4.
6. A ball is dropped from a 50m building. The height (in meters) after *t* sec is given by $-4.9t^2 + 50$
   (a) Find *h*(1) and *h*(1.5).
   (b) Interpret the meaning of the function values found in part (a).
7. Which of the following relations on the set S = {1, 2, 3, 4} are reflexive, irreflexive, symmetric, antisymmetric or transitive?
   (a) {(1, 3), (2, 4), (3,1), (4, 2)}
   (b) {(1, 3), (2, 4)}
   (c) {(1,1), (2, 2), (3,3), (4, 4), (1, 3), (2, 4), (3, 1), (4, 2)}
   (d) {(1, 1), (2, 2), (3, 3), (4, 4)}
   (e) {(1,1), (2, 2), (3, 3), (4, 4), (1, 2), (2, 3), (3, 4), (4, 3), (3, 2), (2,1)}
8. Suppose f and g are maps from R to R defined by $f(x) = x^2$ and $g(x) = x + 1$ for all $x \in$ R. What is $g \circ f$ and what is $f \circ g$?
9. If $(P, \subseteq)$ is a POSET and $A \subseteq P$, then an upperbound for A is an element u with $a \subseteq u$ for all $a \in A$.
10. In the figure below, you see three diagrams. Which of these diagrams are Hasse diagrams?

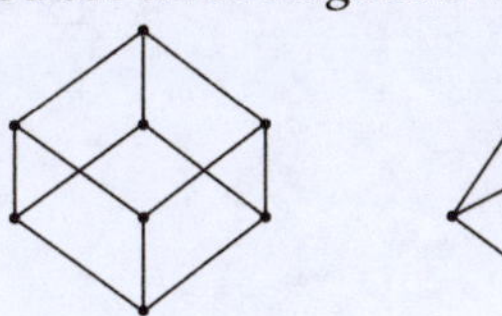
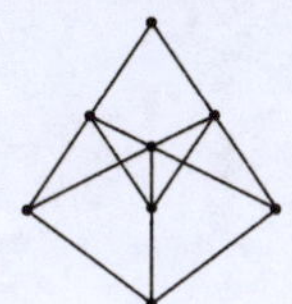
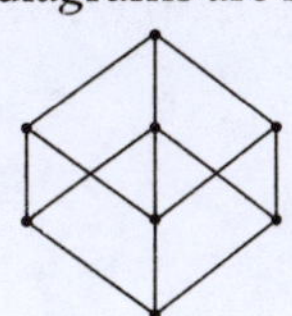

11. Let a = {1, 2, 3, 6, 9,18} and let us define R on A by *x* R *y* if *x*/*y*. Draw a Hasse diagram for the POSET (A, R) and topologically sort (A, R).
12. Let A = {*a*, *b*, *c*, *d*, *e*, *f*, *g*, *h*} and let A1 = {*a*, *b*, *c*, *d*}, A2 = {*a*, *c*, *e*, *f*, *g*, *h*}, A3 = {*a*, *c*, *e*, *g*}, A4 = {*b*, *d*}, A5 = {*f*, *h*}. Determine which of the following are partitions of A.
    (a) {A1, A2} (b) {A1, A5} (c) {A3, A4, A5} (d) {A1, A3, A4}
13. Let A = {1, 2, 3, 4} and R = {(1,1)(1, 2)(2,1)(2, 2)(2, 3)(2, 4)(3, 4)(4,1)}. Draw the digraph and the find the relation matrix.
14. Define (a) Path (b) Cycle (c) R^n (d) R^∞.
15. Find R, R^2, R^3 and R^∞ for the following digraph.

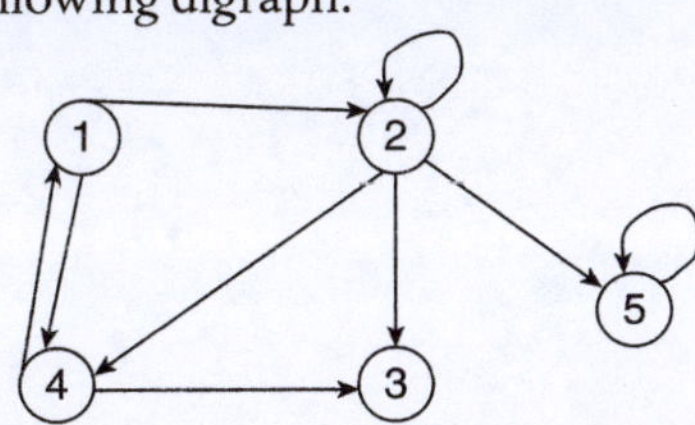

16. Let A = $\mathbb{Z}$, the set of integers, let R be the relation given by R = {(*a*, *b*) ε A × A| *a* < *b*}. Verify the relation of R.
17. Let A = {1, 2, 3, 4}, B = {*a*, *b*, *c*},
    R = {(1, *a*), (1, *b*), (2, *b*), (2, *c*), (3, *b*), (4, *a*)}
    S = {(1, *b*), (2, *c*), (3, *b*), (4, *b*)}
    Compute (a) $\overline{R}$ (b) $R \cap S$ (c) R^{-1} (d) $R \cup S$.

18. Let A = {2, 3, 4, 6, 8, 12, 24, 48} and let R be the relation 'divides'. Let B = {4, 6, 12}. Determine
    (a) All upper bounds of B
    (b) All lower bounds of B
    (c) The LUB of B.
    (d) The GLB of B.
    (e) The maximal and minimal elements.
19. Let A = {*a*, *b*, *c*, *d*, *e*}. Consider the partition P = {{*a*, *b*}, {*c*, *d*}, {*e*}} of A. Find the equivalence relation inducing this partition.
20. Show that a subset of a linearly ordered POSET is also a sub lattice.
21. Show that a sub lattice of a distributive lattice is also a distributive lattice.
22. For each of the adjacency matrices A given below,
    (a) Draw the corresponding digraph.
    (b) Find the matrix T of the transitive closure using the digraph implementation of Warshall's algorithm. Show all working (see Example 5.56).
    (c) Indicate what arcs must be added to the digraph for A to get the digraph of the transitive closure, and draw the digraph of the transitive closure.

(i) $A = \begin{bmatrix} 0 & 0 & 1 \\ 1 & 1 & 0 \\ 0 & 1 & 0 \end{bmatrix}$ (ii) $A = \begin{bmatrix} 0 & 0 & 1 \\ 0 & 1 & 1 \\ 1 & 1 & 1 \end{bmatrix}$ (iii) $A = \begin{bmatrix} 1 & 0 & 0 & 0 \\ 0 & 0 & 0 & 1 \\ 0 & 0 & 1 & 0 \\ 0 & 1 & 0 & 0 \end{bmatrix}$

6 General Functions and Growth Functions

In Chapter 5, we dealt with relations. Every relation involves sets and combinations of the elements of the sets. One can map the elements of one set to another. This mapping is also called a function. All functions are relations, but every relation is not a function. In this chapter, we shall discuss functions and their applications in computer science. In a sense, every program is viewed as a function. The input to a program is a set and the output of the program is another set.

6.1 OVERVIEW OF FUNCTIONS

In mathematics, a **function** is considered as a relation between a set of inputs and a set of permissible outputs with the property that each input is related to exactly one output. Consider as an example, the function that relates each real number (x) to its square (x^2). The output of a function (f) corresponding to an input x is denoted by $f(x)$, which is read as 'f of x'. In this example, if the input is (−3), then the output is 9. We write this as $f(-3) = 9$. The input variable(s) are sometimes referred to as the argument(s) of the function. The functions are central objects of investigation in most of the fields of modern mathematics. There are many ways to describe or represent functions. Some functions are described by formulae or algorithms. Another representation could be a pictorial representation, also called a graph of the function. In science, functions are sometimes represented by tables that give the corresponding outputs for the inputs. A function is described implicitly, for example, the inverse of a function or a solution of a differential equation. The input and output of a function is expressed as an ordered pair, or a tuple. In the example $f(x) = x^2$, we have the ordered pair (−3, 9). If both input and output are real numbers, this ordered pair is viewed as the Cartesian coordinates of a point in the graph of the function. But no graph is able to define exactly, every point in an infinite set.

In modern mathematics, a function is defined by its set of inputs, called the domain of the function and a set of the outputs, called the co-domain or range of the function. The pictorial representation of a set of all pairs of inputs and outputs is called its graph. In case of the function $f(x) = x^2$, the domain and co-domain are real numbers, and the ordered pairs are of the form (x, x^2). The collection of functions with the same domain and the co-domain are called function spaces. The study of properties within the function spaces are called real analysis or complex analysis. In arithmetic, one visualises operations such as addition, subtraction, multiplication and division, as functions. In evaluation of arithmetic functions, the output is a number. Another important operation defined on functions is function composition, where the output of one function becomes the input to another function.

Consider two sets A and B and let *f* denote a rule that associates each member of A with a unique member of B. Then, we say that *f* is a function from A into B. It is denoted as $f:A \rightarrow B$. A function may also be defined as, given two non-empty sets A and B, a function $f:A \rightarrow B$. is subset of A X B with the following properties.

(i) No – two different ordered pairs in *f* have the same first element.
(ii) All the elements of A occur a first elements in *f*, i.e, $f:A \rightarrow B$. is a relation from A to B satisfying the properties (i) and (ii).

It is important to note that every function is a relation but the converse is not true. Let *x* denotes a member of the set A, then the member of set B, which the function (*f*) associates to $x \in A$, is denoted by *f*(*x*) called the value of the function (*f*) for *x*. The function may be described as $y = f(x)$, where $x \in A$ and $y \in B$. Here *x* and *y* are called respectively independent and dependent variables.

6.2 DOMAIN AND RANGE OF A FUNCTION

For the function $f: A \rightarrow B$. A– is called the domain of f and B– is called the co-domain of *f*. The set of all the function values is called the range of *f*.

Thus, range of $(f) = \{f(x): x \in A\}$ and is depicted as shown in Fig. 6.1.

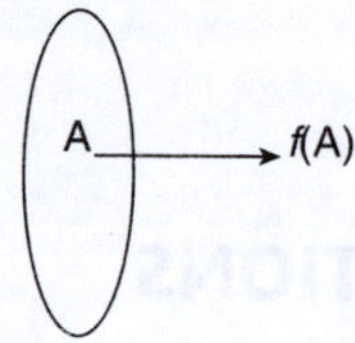

Figure 6.1 Domain and co-domain of a function

We are concerned exclusively with functions, for which the sets A and B are some subsets of the set R, which is a set of real numbers. The function notation is also used in non-numeral situations, as given in the following examples.

(i) Consider a set of employees in a firm. Now, we shall associate each employee of a firm with the salary obtained. This association is denoted by the function *f*. If *x*- denotes an employee, then *f*(*x*) denotes the salary obtained by the employee.
(ii) Consider an electronics experiment to study the effect of temperature on a semiconductor whose resistance decreases as temperature increases. Let *x* denote the temperature T and the resistance $R = a\,e^{\frac{b}{T}}$ denotes the value of the function *f*(*x*). The resistance depends on temperature.
(iii) Consider a mechanical engineering experiment of load and bending moment of a metal rod. Let *x* denote load applied (*w*) and *f*(*x*) denotes the bending moment $B = \frac{wL}{12}$, (L/12 is constant for given set), where bending depends on the load applied.
(iv) Consider the following Java statement:

```
int max ( float real)
{------}
```

and the following Pascal statement,
Function max (*x*: real) : integer

Both statements indicate that the domain of the max function is the set of real numbers and its co-domain is the set of integers. It is interesting to note that it is possible to add and multiply the two real valued functions with the same domain. Consider two functions as given below:

$f_1 : A \rightarrow R$; $f_2 : A \rightarrow R.$, where R is a set of real numbers. Then, $f_1 + f_2$ and $f_1 f_2$ are also functions from A to R, defined by

$$(f_1 + f_2)(x) = f_1(x) + f_2(x)$$

$$(f_1 f_2)(x) = f_1(x) f_2(x).$$

Example 6.1: Consider the functions $f_1 : A \rightarrow R$ and $f_2 : A \rightarrow R.$ such that $f_1(x) = x^3, f_2(x) = x^3 - x^4$. Find the functions $(f_1 + f_2)$ and $(f_1 f_2)$.

Solution:

$$(f_1 + f_2)(x) = f_1(x) + f_2(x)$$
$$= x^3 + x^3 - x^4 = 2x^3 - x^4.$$

and

$$(f_1 f_2)(x) = f_1(x) f_2(x).$$
$$= x^3 (x^3 - x^4) = x^6 - x^7$$

Example 6.2: Consider the sets A = {1, 2, 3, 4, 5} , B = { *a, b, c, d, e* }. Determine whether *f* is a function.

$$f = \{(1, a), (2, b), (3, b), (4, c), (5, c)\}$$

Solution:

1 is associated with *a*, 2 is associated with *b*, 3 is associated with *b*, 4 is associated with *c*, 5 is associated with *d*. It is also denoted as $f(1) = a, f(2) = b, f(3) = b, f(4) = c, f(5) = d$. In this process, it should be noted that every element of A is associated with a unique element of B and all elements of A are used. Hence, it is a function.

Example 6.3: Consider the sets A = {*p, q, r*}, and B = {*a, b*}. Let R be the relation from A to B. The function *f* is given as in Fig. 6.2.

$f = \{(p, a)\ (q, a)\ (q, b)\ (r, b)\}$. Determine whether R – is a valid function from A to B.

Solution:

The function $f = \{(p, a), (q, a), (q, b), (r, b)\}$. The element *q* of A is related to two different elements *a* and *b*, of B, Hence, *f* is not a function.

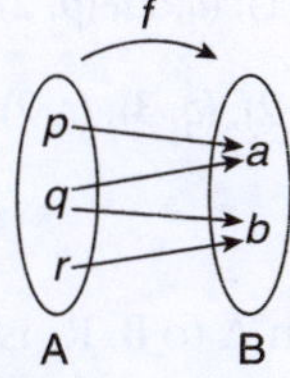

Figure 6.2 A typical function

Example 6.4: Consider the sets A = {1, 2, 3, 4, 5}, B= {*p*, *q*, *r*, *s*} and the function *f* given in Fig. 6.3. Find out whether *f* is a function.

Solution:
Here, $f(1) = p$, $f(2) = p$, $f(3) = q$, and $f(4) = r$. The element 5 of A is not associated with any element of B. Therefore, function *f* is not a function from A to B.

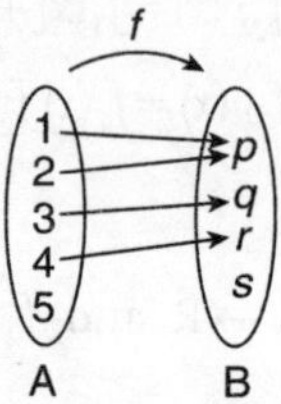

Figure 6.3 Typical function of Example 6.4

Example 6.5: A = {1, 2, 3, 4 }, B = { *x*, *y*, *z* } and the function *f* is as given in Fig. 6.4. Find out whether *f* is a function.

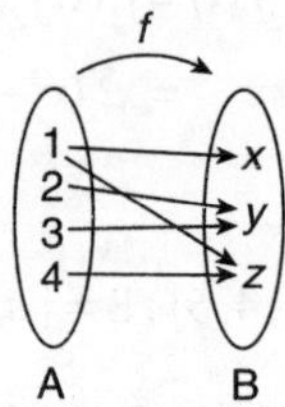

Figure 6.4 Typical function of Example 6.5

Solution:
The relation corresponding to $f = \{(1, x), (2, y), (3, y), (1, z), (4, z)\}$. The element 1 is not uniquely associated. There are two different pairs (1, *x*), (1, *z*) in *f*, having the same first element and different values. Therefore, *f* is not a function from A to B.

Example 6.6: If A = {*p*, *q*, *r*, *s*} and B = {1, 2, 3} are the sets, which of the following relations from A to B are not functions?

$$R_1 = \{(p, 1), (q, 2), (r, 1), (s, 2)\}$$

$$R_2 = \{(p, 1), (q, 1), (r, 1), (s, 1)\}$$

$$R_3 = \{(p, 1), (q, 2), (p, 2), (s, 3)\}$$

$$R_4 = \{(p, 2), (q, 3), (r, 2), (s, 2)\}$$

Solution:
The relations R_1, R_2, and R_4 are functions from A to B. R_3 is not a function from A to B. There is a repetition of (p, 1) and (p, 2).

Example 6.7: If A = {−1, 1, 0, 2, 5} and f: A→Z, where Z is the set of integers, is defined by $f(x) = x^3 - x^2 - x + 1$ then find the range of f.

Solution:

The range of $f = f(\text{A}) = \{f(-1), f(1), f(0), f(2), f(5)\}$

$$f(x) = x^3 - x^2 - x + 1, f(-1) = 0, f(1) = -2, f(0) = 1, f(2) = 3, f(5) = 96.$$

Hence the range of $f = f(\text{A}) = \{0, -2, 3, 1, 96\}$.

6.3 DIFFERENT TYPE OF FUNCTIONS

There are different types of functions, namely, constant functions, into functions, onto functions, one-to-one functions, one-to-one and onto functions, inverse functions, and hashing functions.

6.3.1 Constant Function

A function f: A →B is such that $f(a) = c$ for every $a \in$ A where c is a fixed element of B, is called a constant function. The elements of A have the same image in B and $f(\text{A}) = \{c\}$ as shown in Fig. 6.5.

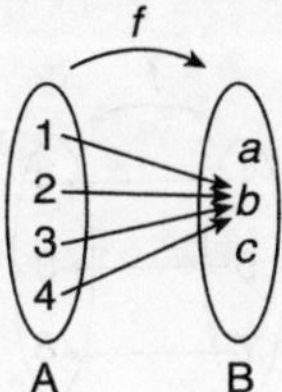

Figure 6.5 Constant function

6.3.2 Into Function

A function f: A →B is called an into function, if some elements of the co-domain are not images under the function f, as shown in Fig. 6.6. Let the sets be A = {1, 2, 3, 4, 5} and: B = {a, b, c, d, e}. The function is defined as follows:

$$f(1) = a, f(2) = b, f(3) = b, f(4) = c, \text{ and } f(5) = d$$

Figure 6.6 Into function

6.3.3 Onto Function

A function f: A →B is called an onto function, if all the elements of the co-domain are images under the function. Consider the relation on the two sets A = {1, 2, 3, 4, 5} and, B = {a, b, c, d}, is as shown in Fig. 6.7.

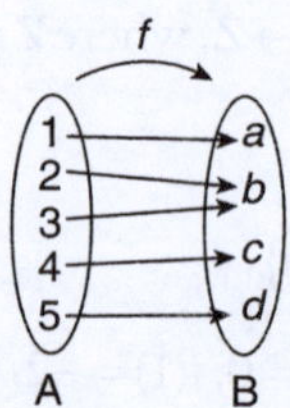

Figure 6.7 Onto function

The same is written as $f(1) = a, f(2) = b, f(3) = b, f(4) = c$, and $f(5) = d$. The **onto** function is also called Surjection.

6.3.4 One-One Function

A function f: A →B is called a one-to-one function, if different elements of the domain have different images. That is $x_1 \neq x_2, \Rightarrow f(x_1) \neq f(x_2)$. The one-to-one function is also called an Injective function.

Example 6.8: Consider two sets A = {1, 5, 7}, and B = {2, 3, 8}. The relation defined as R = {(1,2), (5,3), (7,8)} is clearly a one-to-one function and is shown in Fig. 6.8.

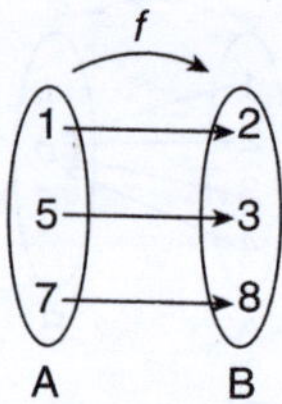

Figure 6.8 One-to-one function

It is to be noted that if there exists more than one member of A, say x_1, x_2 such that $x_1 = x_2$, then we have $f(x_1) = f(x_2)$.

Example 6.9: Consider the function $y = x^2$ with the domain as the set of real numbers. Show that the function is not one-one.

Solution:
Consider $x_1 = 1$ and $x_2 = -1$. It is clear that $x_1 \neq x_2$, but $f(x_1) = f(x_2) = 1$. Therefore, the function is not one-to-one.

6.3.5 One-to-One and Onto

A function f: A →B is said to be one-to-one and onto, if it is both a one-to-one and an onto function. It is also called a Bijective function.

Example 6.10: Consider the two sets A = {1, 2, 3, 4} and B = {p, q, r, s}. The mappings are as given in Fig. 6.9. $f(1) = p, f(2) = q, f(3) = r, f(4) = s$.

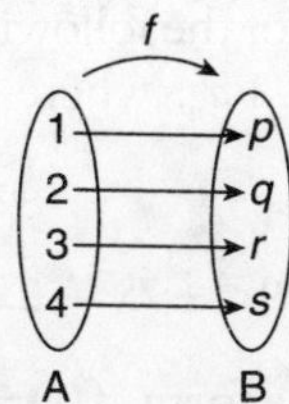

Figure 6.9 Bijective function

Whenever a function is to be bijective across finite sets A and B, the domain and co-domain must have the same number of elements. The sets A and B are said to have a one-to-one correspondence with each other.

Example 6.11: Consider the function f: R → R and $f(x) = 2x + 11$ for all x in R. For $x_1, x_2 \in$ R, show that $f(x_1) = f(x_2)$.

Solution:

The function $f(x_1) = 2x_1 + 11$ and $f(x_2) = 2x_2 + 11$. Therefore,

$2x_1 = 2x_2$.

$x_1 = x_2$. Thus the function is one-one.

On the other hand, if h: R →R, and $h = x^3 - x^2$, then $h(0) = 0$, $h(1) = 0$. Therefore, h is not one-to-one. Since $h(0) = h(1)$, but $0 \neq 1$. The function h is not one-one because there exist real numbers x_1 and x_2, $h(x_1) = h(x_2)$ and $x_1 \neq x_2$.

Example 6.12: Consider the two sets A = {a, b, c}, and B = {a, b, c, d, e}. The function is defined as follows:

f = {(a, a), (b, c), (c, d)} is a one-to-one function from A to B.

and g = {(a, a), (b, c), (c, c)} is a function from A to B, but it fails to be a one-one function because $g(b) = c$, $g(c) = c$. and $g(b) = g(c)$ but $b \neq c$.

Example 6.13: Consider the two sets A = {a_1, a_2,----a_n} and B = { b_1, b_2,-------,b_n} with $m \leq n$. The function f: A →B is a one-one function and has the form {(a_1, x_1), (a_2, x_2),------(a_m, x_m)},where there are n– choices for x_1 (any element of B), $n - 1$ choices for x_2 (any element except that chosen for x_1), $n - 2$ choices for x_3, and so on. So we have $n - (m - 1) = (n - m + 1)$ choices for x_m. Hence, the number of one-one functions from A to B is given by

N$(n - 1)(n - 2)$-------$(n - m + 1) = {}^{n}p_m$. Consequently in the earlier example there are ${}^{5}p_3 = 5.4.3 = 60$ one- one functions f: A →B.

6.4 ALTERNATIVE TYPES OF FUNCTIONS

Functions can also be classified into two groups, namely, Algebraic functions and Transcendental functions.

6.4.1 Algebraic Functions

These consist of a finite number of terms, which contain integral or fractional powers of the independent variable (argument) x. The function is operated on by the four basic arithmetic operators +, –, ×, and ÷. A few sub-classes of algebraic functions are as follows.

(a) Polynomial function: A function is of the following form $f(x)= a_0x^n + a_1x^{n-1} + a_2x^{n-1} + ---- + a_n$, where a_0, a_1, a_2,---a_n are constants and n is non-negative integer and $a_0 \neq 0$.

Example 6.14: Consider the polynomial $9x^4 + 7x^3 - 2x^2 + x + 18 = 0$ of degree 4.

(b) Rational function: A function of the form $f(x) = \frac{p(x)}{q(x)}$, $q(x) \neq 0$ $p(x)$ and $q(x)$ are polynomials, known as rational functions.

Example 6.15: Consider the rational function $f(x) = \frac{2x^3 - 5x + 6}{x^2 + 5}$, where $p(x) = 2x^3 - 5x + 6$ and $q(x) = x^2 + 5$.

(c) Absolute value function: For any real number x, the absolute function f defined as under.

$$f(x) = |x| = \begin{matrix} x & \text{if } x \geq 0 \\ -x & \text{if } x < 0 \end{matrix}$$

(d) Exponential Function: An exponential function with base a is defined by $f(x) = a^x$ where $a > 0$, $a \neq 1$, and x is any real number.

6.4.2 Transcendental Function

A function which is not algebraic is called a transcendental function. Trigonometric functions, inverse trigonometric functions, logarithmic functions, hyperbolic functions and inverse hyperbolic functions come under this category.

Example 6.16: The following are a few examples of transcendental functions:

$$\sin x + \cos x = 1,\ f(x) = \frac{\cosh ax}{\cos^3 x},\ \sin^{-1}x + \cos^{-1}x = \frac{\pi}{2},\ x\log_e x - 1.2 = 0.$$

6.4.3 Hashing Functions

Hash function is any function that is used to map data of arbitrary size to data of fixed size. The values returned by the hash function are called hash values, hash codes etc. One usage of hash function is in computer science where it is called hash table and used for rapid data lookup.

In computer science, there is a common requirement to store large volumes of data and yet provide random access to it. The same exercise is more challenging if the expected volume of data is completely unknown. One of the possible solutions is to go for linked list representations. While searching for the given data, in the worst case, the entire list needs to be searched. Instead, one can think of a collection of linked list representations of data so as to reduce the search time. One can store the data items by converting into a three digit integer, in the range 000 to 999. Then choose a prime number, say 7919, nearer to 1000. Assume that one can define an array of size 7919 of pointers to the list of data. More than one data item can be reduced to the same integer in the range 0–7919. We say that there were resultant collisions. These collisions are resolved by chaining the colliding data.

In the best case, each data item gets hashed to different locations, when the data size is less than 7179. The retrieval becomes easy, $\Omega(1)$. Otherwise, there is a need to chain the data items. The worst case is when all the data items collide and form a chain. In such a case, the retrieval requires a linear search, O(n). In compiler implementation, we represent the symbol table by a hash function. In Cobol, we represent indexed sequential file implementation using hashing function. Typical hashing is depicted in Fig. 6.10 and illustrated through Example 6.16.

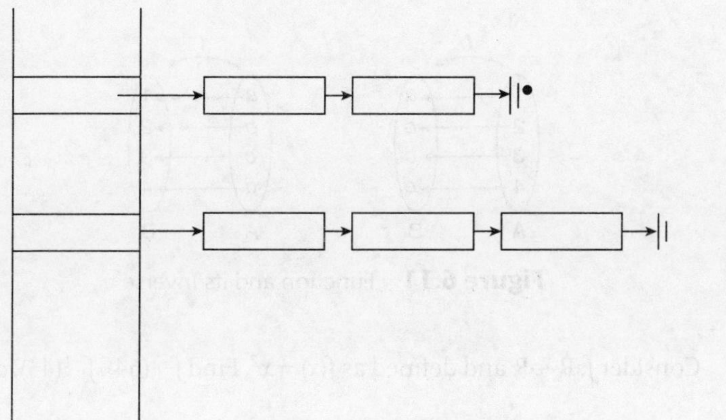

Figure 6.10 A typical hashing table with chained nodes

Example 6.17: Consider the names of 1000 students in a class to be stored in a hash table. Let the ASCII values (asv) of the characters of the name be added and divided by the prime number, say 7179. The remainder is used to store the name in the table. Thus the hashing function for the name 'Rama' is obtained as under.

$$\begin{aligned} H(\text{Rama}) &= \text{asv}(R) + \text{asv}(a) + \text{asv}(m) + \text{asv}(a)/7179 \\ &= (82 + 97 + 109 + 97)/7179 \\ &= 385 \end{aligned}$$

The array location 385 is made to point to the node containing the name 'Rama'. Similarly, the name 'Ashok' is stored in the location 502. The name 'Babz' is also 385 and is located next to the node containing 'Rama' as there is collision with the name 'Rama'. The collision is resolved by chaining.

There are many other functions in computer science, namely, characteristics function, mod, floor and ceiling functions, factorial function, logarithmic function, exponential functions, Boolean functions, string functions, matrix functions, trigonometric functions and the like. The permutation functions are dealt with in chapter 7 under Section 7.7. Readers will usually come across these functions in some context or the other, in their curriculum.

6.5 INVERSE OF AN ELEMENT

Consider the two sets A = {1, 2, 3, 4, 5}, and B = {*a*, *b*, *c*, *d*}. A function *f*: A →B is defined by *f*(1) = *a*, *f*(2) = *a*, *f*(3) = *b*, *f*(4) = *b*, and *f*(5) = *c*. The element *a* is the image of the elements 1 and 2. Now the elements 1, 2 are called inverse images (or pre images) of '*a*'. Symbolically it is written as $f^{-1}(a) = \{1, 2\}$. Similarly $f^{-1}(b) = \{3, 4\}$, $f^{-1}(d) = \Phi$ (null set). No element has *d* as its image. In general, it can be stated that for *f*: A →B, if b ∈ B, and f^{-1}(b) is defined as, $f^{-1}(b) = \{x \in A / f(x) = b\}$. Here, $f^{-1}(b)$ is called the pre image set of B under *f*.

6.5.1 Inverse Function

The inverse of a function exists only if the function is bijective, also called one-one and onto. If *f*: A →B is a bijection, then the inverse function exists, f^{-1}: A →B. It is also true that the inverse of every element of B is a single element of A. f^{-1}: A →B. is also a bijection. The function *f*: A →B is defined by $f(1) = a$, $f(2) = b, f(3) = c, f(4) = d$, and A = {1,2,3,4}, B = {*a, b, c, d*}. The inverse function f^{-1}: A →B is defined by $f^{-1}(a) = 1$, $f^{-1}(b) = 2, f^{-1}(c) = 3$, and $f^{-1}(d) = 4$. The inverse of the function exists, which is shown in Fig. 6.11.

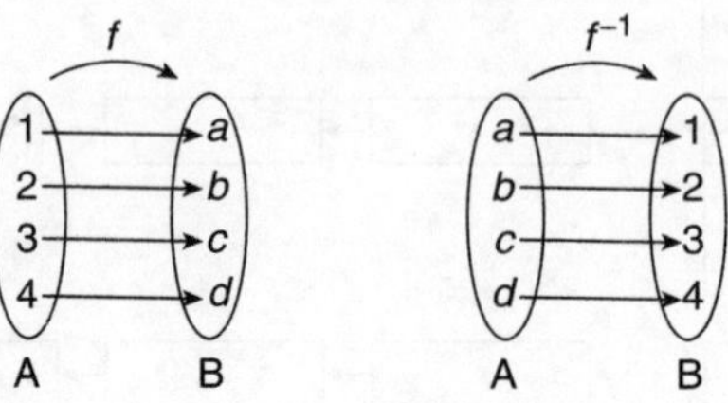

Figure 6.11 Function and its Inverse

Example 6.18: Consider *f*: R→R and defined as $f(x) = x^2$. Find $f^{-1}(64), f^{-1}(45)$, and $f^{-1}(-16)$.

Solution:

Let $f^{-1}(64) = x$, then $f(x) = 64$, given that $f(x) = x^2$.

$$\therefore \quad x^2 = 64,\ x = \pm 8.$$

$$\therefore \quad f^{-1}(64) = \{8, -8\} \in R.$$

Again, let $f^{-1}(45) = x$, then $x^2 = 45$: then $x = \pm 3\sqrt{5} \in R$.

$$f^{-1}(45) = f^{-1}(45) = \{3\sqrt{5}, -3\sqrt{5}\}$$

And Let $f^{-1}(-16) = x, f(x) = -16$. and $x^2 = -16, \Rightarrow x = \pm i4 \notin R$.

$\therefore f^{-1}(-16) = \Phi$. The complex numbers does not belong to set of reals.

Example 6.19: Consider *f*: R→R be defined by $f(x) = 7x + 8$. Find $f^{-1}(x), f^{-1}(1), f^{-1}(-3)$.

Solution:

$$f(x) = 7x + 8 \tag{1}$$

Let $f^{-1}(x) = y$, then $x = f(y)$. But $f(y) = 7y + 8$.

$$\Rightarrow \quad x = 7y + 8,\ 7y = x - 8,\ y = \frac{x-8}{7}$$

i.e.

$$f^{-1}(x) = \frac{x-8}{7} \tag{2}$$

Now using equation (2) put $x = 1,\ f^{-1}(1) = \frac{1-8}{7} = -1$.

$$\text{Put } x = -3,\ f^{-1}(-3) = \frac{-3-8}{7} = \frac{-11}{7}$$

Example 6.20: Consider the two sets A = {1, 2, 3, 4, 5}, and B = {5, 6, 7, 8}. A function *f*: A → B is defined as *f* = {(1, 6), (2, 8), (3, 5), (4, 5, (5, 7)}. Determine $f^{-1}(8),\ f^{-1}(7),\ f^{-1}(5)$. If $B_1 = \{6, 7\}, B_2 = \{5, 6, 7\}$. Find $f^{-1}(B_1)$ and $f^{-1}(B_2)$.

Solution:

$$f^{-1}(8) = \{x \in A \mid f(x) = 8\} = \{2\}.$$

$$f^{-1}(7) = \{x \in A \mid f(x) = 7\} = \{5\}.$$

$$f^{-1}(5) = \{x \in A \mid f(x) = 5\} = \{3,4\}.$$

Now $f^{-1}(B_1) = \{x \in A \mid f(x) \in B_1\}$ Now $B_1 = \{6,7\}$

$f(x) \in B_1$ when $f(x) = 6, f(x) = 7$. Now $f(x) = 7$, when $x = 5$,

and $f(x) = 6$ when $x = 1$. $\therefore\ f^{-1}(B_1) = \{1, 5\}$

similarly for $B_2 = \{5,6,7\}$

$$f^{-1}(B_2) = \{x \in A \mid f(x) \in B_2\} = \{1,3,5\}.$$

Example 6.21: As a simple example, if f converts a temperature in degrees Celsius C to degrees Fahrenheit F, the function converting degrees Fahrenheit to degrees Celsius would be a suitable f^{-1}.

$$F = f(C) = \frac{9}{5}C + 32.$$

Then its inverse function converts degrees Fahrenheit to degree Celsius.

$$f^{-1}(F) = \frac{5}{9}(F - 32) = C$$

Since

$$f^{-1}(f(c)) = f^{-1}\left(\frac{9}{5}C + 32\right) = \frac{5}{9}\left[\left(\frac{9}{5}C + 32\right) - 32\right] = C$$

For every C. Such an inverse function exists if and only if f is bijective

6.6 COMPOSITE FUNCTION

This is also called a function of a function. Consider three non-empty sets A, B, C and the functions $f: A \to B$ and $g: B \to C$ The composition or product of these functions is defined as the function $(g \circ f)$. The same is shown in Fig. 6.12.

$$g \circ f: A \to C \text{ with } (g \circ f)(x) = g[f(x)], \text{ for all } x \in A.$$

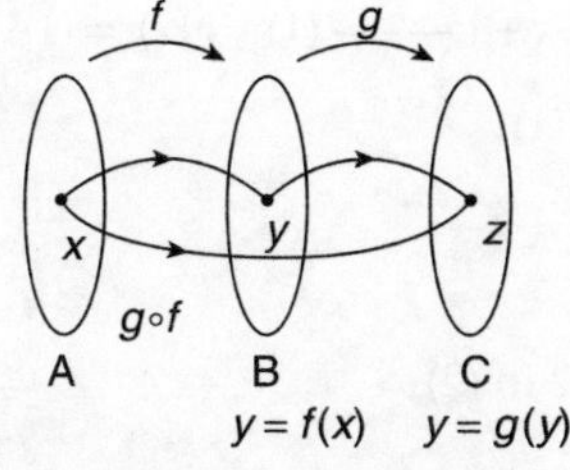

Figure 6.12 Composite function

A composite function $g(f(x))$ can be visualized as the combination of two 'machines'. The first takes input x and gives output $f(x)$. The second takes $f(x)$ and gives output $g(f(x))$. The composition of two functions takes the output of one function as the input to the second function. More specifically, the composition of f with a function g: B→ C is the function.

$$(g \circ f)(x): A \to C \text{ defined by } (g \circ f)(x) = g(f(x))$$

The value of x is obtained by first applying f to x to obtain $y = f(x)$ and then applying g to y to obtain $z = g(y)$. In the notation $g \circ f$, the function on the right, f, acts first and the function on the left, g acts next. The notation can be memorised by reading the notation as "g of f" or "g after f".

It is to be noted that the composition of functions $(g \circ f)$ is only defined when the co domain of f is the domain of g, and $f \circ g$ need not be defined. The required condition for $f \circ g$ is that the co domain of g should be equal to the domain of f. It is not generally true that $g \circ f = f \circ g$. That is, the order of the composition is important.

If f and g are treated as relations, then the composition of f and g denoted by $f \circ g$, whereas the composition of the functions f and g is denoted by $g \circ f$. Suppose that an airplane's elevation at time t is given by the function $h(t)$, and that the oxygen concentration at elevation x is given by the function c(x). Then $(c \circ h)(t)$ describes the oxygen concentration around the plane at time t.

Example 6.22: Consider the functions f: R→R and g: R→R defined by $f(x) = x^2 + 1$ and $g(x) = x + 1$ find $f \circ g$ and $g \circ f$.

Solution:

Let $f(x) = x^2 + 1$ ------- (1) : $g(x) = x + 1$ --------- (2)

$$(f \circ g)(x) = f[g(x)]$$
$$= f[x+1] = (x+1)^2 + 1$$
$$= x^2 + 2x + 2.$$

And consider

$$(g \circ f)(x) = g[f(x)]$$
$$= g[x^2 + 1] = (x^2 + 1) + 1$$
$$= x^2 + 2$$

Example 6.23: Let f and g be real valued functions defined by (1) $f(x) = x^2 + x + 1$ and (2) $g(x) = 1 - x^2$. Find $(g \circ f)(2)$ and $(f \circ g)(-2)$.

Solution:

$f(x) = x^2 + x + 1$ ------- (1) $g(x) = 1 - x^2$ ------- (2)

(i) $(g \circ f)(2) = g[f(2)]$ put $x = 2$ in (1).

$= g[7]$ $\because f(2) = 7$

$= -48.$

(ii) $(f \circ g)(-2) = f[g(-2)]$ put $x = -2$ in (2)

$= f(-3)$ $\because g(-2) = 1 - 4 = -3$

$= 7.$

Example 6.24: Let f: Z →Q and g: Q →Q given by

$$f(x)=\frac{1}{6}x, \quad \forall x \in Z$$

$g(x) = x^2 + x + 1 \quad \forall x \in Q$. Then find $(g \circ f)$: Z→Q

Solution:

$$(g \circ f)(x) = g[f(x)] = g\left(\frac{x}{6}\right)$$

$$= \left(\frac{x}{6}\right)^2 + \frac{x}{6} + 1$$

$$= \frac{x^2 + 6x}{36} + 1 \quad \forall x \in Z$$

Example 6.25: If A = {1, 2, 3, 4, 5}, B = {1, 2, 3, 8, 9} and the functions f: A →B and g: A →A are defined by f = {(1, 8), (3, 9), (4, 3), (2,1), (5, 2)} and g = {(1, 2), (3, 1), (2, 2), (4, 3), (5, 2)}, find $f \circ g, g \circ f, f \circ f$ and $g \circ g$ if they exist.

Solution:

$$(f \circ g)(1) = f\{g(1)\} = f(2) = 1$$

$$(f \circ g)(2) = f\{g(2)\} = f(2) = 1$$

$$(f \circ g)(3) = f\{g(3)\} = f(1) = 8$$

$$(f \circ g)(4) = f\{g(4)\} = f(3) = 9$$

$$(f \circ g)(5) = f\{g(5)\} = f(2) = 1$$

$$(f \circ g) = \{(1, 1), (2, 1), (3, 8), (4, 9), (5, 1)\}$$

Now the range $(f) = \{1, 2, 3, 8, 9\}$

domain $(g) = \{1, 2, 3, 4, 5\}$

∴ range $(f) \not\subset$ domain(g). Hence $(g \circ f)(a) = g\{f(a)\}$ is not defined.

i.e., $g \circ f$ is not defined.

Again, the range $(f) = \{1, 2, 3, 8, 9\} \not\subset$ domain$(f) = \{1, 2, 3, 4, 5\}$

Hence, $f \circ f$ is not defined.

range $(g) = \{1, 2, 3\} \subseteq$ domain$(g) = \{1, 2, 3, 4, 5\}$

Hence, $g \circ g$ is defined.

Now $(g \circ g)(1) = g\{g(1)\} = g(2) = 2$ and so on

$$(g \circ g) = \{(1, 2), (2, 2), (3, 2), (4, 1), (5, 2)\}$$

Example 6.26: If S = {1, 2, 3, 4, 5} and if the functions f, g, h: S →S are given by

$$f = \{(1, 2), (2, 1), (3, 4), (4, 5), (5, 3)\}$$

$$g=\{(1, 3), (2, 5), (3,1), (4, 2), (5, 4)\}$$
$$h=\{(1, 2), (2, 2), (3, 4), (4, 3), (5, 1)\}$$

(i) verify $g \circ f = f \circ g$
(ii) Explain why *f* and *g* have inverses but *h* does not.
(iii) Find f^{-1} and g^{-1}
(iv) Show that $(f \circ g)^{-1} = g^{-1} \circ f^{-1} \neq f^{-1} \circ g^{-1}$.

Solution:

(i) $(f \circ g)(1) = f\{g(1)\} = f(3) = 4$
$(f \circ g)(2) = f\{g(2)\} = f(5) = 3$ and so on
$(f \circ g) = \{(1,4),(2,3),(3,2),(4,1),(5,5)\}$ (1)
$(g \circ f)(1) = g\{f(1)\} = g(2) = 5$
$(g \circ f)(2) = g\{f(2)\} = g(1) = 3$ and so on
$(g \circ f) = \{(1,5),(2,3),(3,2),(4,4),(5,1)\}$ (2)
Hence, from (1) and (2) $g \circ f \neq f \circ g$.

(ii) Both *f* and *g* are one-to-one and onto, ∵ they are invertible.
$h(1) = h(2) = 2, 1 \neq 2$
Also, range$(h) = \{1, 2, 3, 4\} \neq S$ ∵ *h* is not one-to-one.
So inverse of *h* does not exist.

(iii) f^{-1} is obtained by reversing the elements in all the ordered pairs of *f*
i.e., $f^{-1} = \{(2, 1),(1, 2),(4, 3),(5, 4),(3, 5)\}$ (3)
It is easily verified that
$f \circ f^{-1} = f^{-1} \circ f = \{(1, 1),(2, 2),(3, 3),(4, 4),(5, 5)\} = I$
Similarly, $g^{-1} = \{(3, 1),(5, 2),(1, 3),(2, 4),(4, 5)\}$ (4)

(iv) From (1) $(f \circ g)^{-1} = \{(4, 1),(3, 2),(2, 3),(1, 4),(5, 5)\}$ (5)
From (3) and (4)
$g^{-1} \circ f^{-1} = \{(2, 3),(1, 4),(4, 1),(3, 2),(5, 5)\}$ (6)
$f^{-1} \circ g^{-1} = \{(3, 2),(5, 1),(1, 5),(2, 3),(4, 4)\}$ (7)
From (5), (6) and (7) it is obvious that
$(f \circ g)^{-1} = g^{-1} \circ f^{-1} \neq f^{-1} \circ g^{-1}$.

Example 6.27: Consider the functions *f* and *g* to be functions from R to R defined by $f(x) = ax + b$ and $g(x) = x^2 - 3x + 5$ If $(g \circ f)(x) = 100x^2 + 30x + 5$, determine *a* and *b*.

Solution:

$$(g \circ f)(x) = g(f(x)) = g(ax + b)$$
$$= (ax+b)^2 - 3(ax+b) + 5$$
$$= a^2x^2 + b^2 + 2abx - 3ax - 3b + 5$$
$$= x^2a^2 + x(2ab - 3a) + b^2 - 3b + 5 \quad (1)$$

Given that

$$(g \circ f)(x) = 100x^2 + 30x + 5$$

$$= x^2a^2 + x(2ab - 3a) + b^2 - 3b + 5 \quad (2)$$

From equations (1) and (2),

$$a^2 = 100,\ a = \pm 10, \quad 2ab - 3a = 30, \quad b^2 - 3b + 5 = 5$$

For $a = 10, b = 3$, and for $a = -10, b = 0$. The value of b also can be obtained by solving

$$b^2 - 3b + 5 = 5.$$

Property 1: Let f: A→B, g: B→C and h: C→D, then $h \circ (g \circ f) = (h \circ g) \circ f$. The composition of functions is associative.

Proof:

Observe that $h \circ (g \circ f)$: A → D and $(h \circ g) \circ f$: A → D.

$$[h \circ (g \circ f)](x) = h[(g \circ f)(x)] \quad \forall x \in A$$

$$= h[g(f(x))] \quad \forall x \in A$$

and

$$[(h \circ g) \circ f](x) = (h \circ g) f(x)$$

$$= h\,[g(f(x))]$$

Hence, $[h \circ (g \circ f)](x) = [(h \circ g) \circ f](x)$. Where x is an arbitrary element of A.

Property 2: Let f: A →B and g: B →C. Then the following conclusions holds true.

(i) If both f and g are one-one, then $(g \circ f)$ is also one-one.

(ii) If f is onto B and g is onto C, then $(g \circ f)$ is also onto C.

(iii) If both f and g are one-to-one correspondences, then $(g \circ f)$ is also a one-to-one correspondence.

Proof:

First we note that $(g \circ f)$: A → C

Let $x_1 x_2 \in A$, suppose that

$$(g \circ f)(x_1) = (g \circ f)(x_2)$$

$$g(f(x_1)) = g(f(x_2))$$

⇒ $f(x_1) = f(x_2)$ because g is one-one

⇒ $x_1 = x_2$ because f is one-one

Hence, $(g \circ f)$ is one-one.

(ii) Let $x_3 \in C$, and g is onto C, there exists $x_2 \in B$, such that $g(x_2) = x_3$. Now $x_2 \in B$, and f is onto B. So there exist $x_1 \in A$ such that $f(x_1) = x_2$. Then
$(g \circ f)(x_1) = g(f(x_1)) = g(x_2) = x_3$ Hence $(g \circ f)$ is onto C.

(iii) Both the functions f and g are one-one and onto and it is also true that $(g \circ f)$ is one-one and onto, Hence $(g \circ f)$ is a one-to-one correspondence, which follows from (i) and (ii).

Property 3: Using the concept of composition of functions we can define the power of a Function. Let f: A →A be a function, f – is a relation on A, so we can define

$$f^2, f^3, f^4, ---f^n.$$

Now $\forall\, a \in A,$

$$f^1(a) = f(a)$$

$$f^2(a) = (f \circ f)(a) = f(f(a))$$

$$f^3(a) = (f \circ f^2)(a) = f(f^2(a))$$

$$f^n(a) = (f \circ f^{n-1})(a) = f(f^{n-1}(a)).$$

If A– is a finite set and f: A →A is a one-one function on A, then f is automatically onto A. Now f: A →A is a one-one and by induction it can be shown that $f^n: A \to A$ is a one-one function for all integers $(n \geq 1)$.

6.7 IDENTITY FUNCTION

A function f: A →A such that $f(a) = a$ for every $a \in A$, is called the identity function. The image of every element of A (under f) is itself. In this case $f(A) = A$. The identity function is usually denoted by I_A. The unique function over a set A that maps each element to itself, is called the identity function for A. Each set has its own identity function, so the subscript cannot be omitted unless the set can be inferred from context. Under composition, an identity function is 'neutral': if f is any function from X to Y, then

$$f \circ I_A = f.$$

6.8 INVERTIBLE FUNCTION

Consider a one-one function with a domain A and range B. Let $y \in B$, the function being one-one and $x \in A$, such that $f(x) = y$. A new function g is defined such that $g(y) = x \leftrightarrow f(x) = y$. The domain of the function g is the range of the given function f and vice versa. The function g is said to be an inverse of the function. Now the function f is an invertible function. One-One functions are invertible. Clearly, if g is the inverse of f, then f is also the inverse of g. This is shown in Fig. 6.13, and can be stated as follows:

A function f: A →B is said to be invertible if there exists a function g: B →A such that $g \circ f = I_A$ and $f \circ g = I_B$, where I_A is the identity function on A, and I_B is the identity function on B.

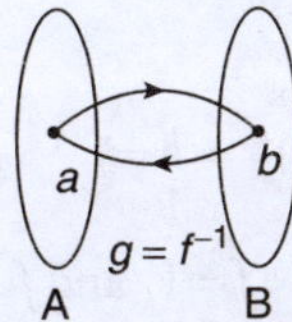

Figure 6.13 Invertible function

In this case, f is called invertible. The notation $(g \circ f)$ and f^{-1} are taken to be multiplication and reciprocal notations. With this analogy, identity functions are like the multiplicative identity, 1, and inverse functions are like reciprocals, hence the notation. That is, the two possible compositions of f and f^{-1} need to be the respective identity maps of A and B. An inverse function for f, denoted by f^{-1}, is a function in the opposite direction, from B to A, satisfying $f \circ f^{-1} = I_B$, $f^{-1} \circ f = I_A$

In the function definition f: A →B and g: B →C, if f is many-to-one, g: B →C will not satisfy the definition of the function. The function must be one-one. If f: A →B is an into function then there must be a point in y for which no x exists. The inverse of a function can be defined within the same sets for x and y only when it is one-one and onto (bijection). A monotonic function i.e., a bijection function, is always invertible.

Example 6.28: Consider the function f: R →R defined as $y = f(x) = x^2$. The function f is a many-to-one function. For $x = a, y = a^2$ is an into function because y– does not take the negative real values. Hence, it is not invertible.

Example 6.29: Consider f: R → [0, a) defined as $y = f(x) = x^2$. The function is not invertible; it is a many-to-one function.

Example 6.30: Consider f: [0, a) → [0, a) defined as $y = f(x) = x^2$. The function f– is invertible, because the function is bijective.

Property 1: If f: A →B and g: B →C are invertible functions, then $g \circ f$: A →C is an invertible function and $(g \circ f)^{-1} = f^{-1} \circ g^{-1}$. Given that f and g are invertible functions, they are both one-to-one and onto. Hence, $(g \circ f)$ is both one-to-one and onto, and is invertible.

Consider $h = f^{-1} \circ g^{-1}$, which is a function from C to A.

$$(g \circ f) \circ h = (g \circ f) \circ (f^{-1} \circ g^{-1})$$

$$= g \circ (f \circ f^{-1}) \circ g^{-1}$$

$$= g \circ (I_B) \circ g^{-1}$$

$$= g \circ g^{-1} = I_C$$

And

$$h \circ (g \circ f) = (f^{-1} \circ g^{-1}) \circ (g \circ f)$$

$$= f^{-1} \circ (g \circ g^{-1}) \circ f$$

$$= f^{-1} \circ (I_B) \circ f$$

$$= f^{-1} \circ f = I_A$$

Hence h is the inverse of $(g \circ f)$: i.e., $h = (g \circ f)^{-1}$.

$$(g \circ f)^{-1} = h = f^{-1} \circ g^{-1}$$

Property 2: If f: A →B is invertible then $f^{-1} \circ f = I_A$ and $f \circ f^{-1} = I_B$

Let us assume that f is invertible. Hence there exists a function f^{-1}: B →A satisfying $x = f^{-1}(y)$ where $y = f(x)$.

$$f^{-1} \circ f(x) = f^{-1}[f(x)] = f^{-1}[y] = x.$$

By definition of I_A, we can write $f^{-1} \circ f = I_A$

Similarly, one can have

$$f^{-1} \circ f(y) = f^{-1}[f(y)] = f[x] = y.$$

By definition of I_B, $f \circ f^{-1} = I_B$.

From this it can be generalised that f: A →B and g: B →A. The function g– is equal to f^{-1} if and only if $g \circ f = I_A$ and $f \circ g = I_B$

6.9 GROWTH FUNCTIONS

In science, engineering and technology, one frequently comes across the asymptotic behavior of functions. One has to estimate the growth of a function in terms of another, usually simpler, function. It is important to study the limiting behavior of a function when argument tends towards a particular value or infinity.

6.9.1 Big-O Notation

In mathematics, the big-O notation, also called the upper bound, is a member of a larger family of notations called the Landau notation or Bachmann–Landau notation, named after Edmund Landau and Paul Bachmann. This is also called the asymptotic notation. In computer science, the big-O notation is used to classify algorithms based on how they respond to changes in the input size. The algorithms are analysed in terms of processing time or working space requirements. In analytic number theory, it is used to estimate the 'error committed', while replacing the asymptotic size or asymptotic mean size of an arithmetic function by the value or the mean value it takes at a large finite argument. A well known example is the problem of estimating the remainder term in the prime number theorem and polynomial approximations.

Big-O notation characterises functions according to their growth rates. The different functions with the same growth rate are represented using the same O notation. The letter O is used because the growth rate of a function is also referred to as the order of the function. A description of a function in terms of big O notation provides an upper bound on the growth rate of the function. In addition to big-O notation, there are several other related notations, namely, o, Ω, ω, and Θ, which describe the other kinds of bounds on asymptotic growth rates. Big-O notation is also used in many other fields to provide similar estimates. The functions used in the estimation often include, $\log(n)$, n, $n\log(n)$, n^2, 2^n, and $n!$.

In computer science, it is useful in the analysis of the complexity of algorithms, or the number of effective operations an algorithm uses as its input grows. In mathematics, it is used to characterise the residual term of a truncated infinite series, especially an asymptotic series. Formally, the Big-O notation is defined as follows:

- **Definition:** A function $f(x)$ is said to be of the order $O(g(x))$ if and only if there exists constants C and k, such that for every $(x > k)$, $|f(x)| \le C|g(x)|$. In principle, $O(g(x))$ is a set and is given as under.

$$O(g(x)) = \{ f(x) | \exists\, C, k \;\; \forall x \ge k, |f(x)| \le C|g(x)| \}.$$

Therefore, from a set theoretical point of view, we have the following subset relation.

$$O(1) \subset O(\log(n)) \subset O(n) \subset O(n \log(n) \subset O(n^2) \subset O(2^n) \subset (n!).$$

Example 6.31: Use the definition of Big-O to prove that $F(x) = 5x^4 - 37x^3 + 13x - 4 = O(x^4)$

Solution:
We must find integers C and k, such that

$$5x^4 - 37x^3 + 13x - 4 \le C\,|\,x^4| \text{ for all } x \ge k.$$

$$|\,5x^4 - 37x^3 + 13x - 4\,| \le |\,5x^4 + 37x^3 + 13x + 4\,| \le |\,5x^4 + 37x^4 + 13x^4 + 4x^4\,| = 59\,|\,x^4|.$$

Where the first inequality is satisfied if $(x \ge 0)$ and the second inequality is satisfied if $x \ge 1$. Therefore, $|\,5x^4 - 37x^3 + 13x - 4\,| \le 59\,|\,x^4\,|$ if $(x \ge 1)$. So we have $C = 59$ and $k = 1$.

Big-O gives us a formal way of expressing asymptotic upper bounds, a way of bounding from the above growth of a function. Knowing where a function falls within the Big-O hierarchy allows one to compare it quickly with other functions, giving an idea of which algorithm has the best time performance.

Example 6.32: Give a Big-O estimate for these functions.

(a) $3n + n^3 + 4$
(b) $1 + 2 + 3 + \text{----} + n + 3n^2$.
(c) $\text{Log}_{10} 2^n + 10^{10} n^2$.

Solution:

(a) Since $3n + n^3 + 4 \le 3n^3 + n^3 + 4n^3 = 8n^3$ for $n > 1$.
Therefore, $3n + n^3 + 4 = O(n^3)$

(b) We have $1 + 2 + 3 + \text{----} + n \le n + n + \text{---} + n = n.n = n^2$
Hence, $1 + 2 + 3 + \text{----} + n + 3n^2 \le n^2 + 3n^2 = 4n^2$
$\therefore 1 + 2 + 3 + --- + n + 3n^2 = O(n^2)$.

(c) $\log_{10} 2^n + 10^{10} n^2 = n \log_{10} 2 + 10^{10} n^2$

$$\le n^2 \log_{10} 2 + 10^{10} n^2$$

$$\le (\log_{10} 2 + 10^{10}) n^2 \quad \text{if } n \ge 1$$

The term $\log_{10} 2 + 10^{10}$ is constant.

$\therefore \log_{10} 2^n + 10^{10} n^2 = O(n^2)$.

It is interesting to note that any linear function $(an + b)$ is of $O(n^2)$, where $C = a + |b|$, $k = 1$ and $an + b \le (a + |b|) n^2$ for $n \ge 1$.

6.9.2 Other Notations

There are other notations used in the analysis of algorithms, namely, big-omega (Ω) and big-theta (Θ). The Ω indicates lower bound and Θ indicates tight bound.

- **Ω-notation:** As was the case for the O-notation, this provides an asymptotic upper bound on a function. Ω -notation gives an asymptotic lower bound. For a given function g(*n*), the set of functions is denoted by $\Omega(g(n))$.
  $\Omega(g(n)) = \{f(n)$: there exists positive constants C and *k* such that $0 \le Cg(n) \le f(n)$, for all $n > k\}$. For all values *n* to the right of *k*, the value of $f(n)$ is on or above $Cg(n)$. In other words, a function is of order $\Omega(g(n))$ if and only if there exists a constant $(C > 0)$ and a constant *k*, such that for every $(n > k)$, $|f(n)| \ge C|g(n)|$.
- **Big-theta (Θ) notation:** For any two functions $f(n)$ and $g(n)$, we have $f(n) = \Theta(g(n))$ if $f(n) = O(g(n))$ and $f(n) = \Omega(g(n))$.
  Figure 6.14 gives the graphical representations of the different notations discussed. A function $f(n) = \Theta(g(n))$ if and only if there exists constants $(C_1 > 0)$, $(C_2 > 0)$ and a constant *k*, such that for all $(n > k)$, $C_1|g(n)| \le |f(n)| \le C_2|g(n)|$ From the above consideration, the big-theta notation is a stricter version of big-oh and big-omega notations. This is because $f(n)$ is Θ $(g(n))$ if and only if $f(n)$ is both $O(g(n))$ and $\Omega(g(n))$, which is equivalent to saying that $\Theta(g(n)) = O(g(n)) \cap \Omega(g(n))$. The graphical representation reveals how the growth functions differ from each other.

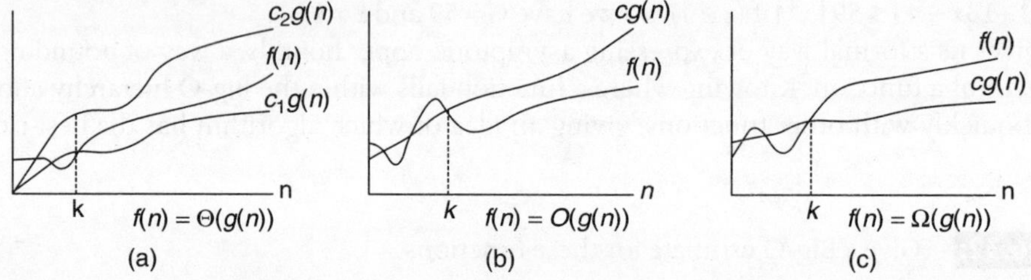

Figure 6.14 Graphical representation of notations

Example 6.33: Consider the expression $an^2 + bn + c = \Theta(n^2)$ for any constants *a* and *c*, such that $(a > 0)$. This implies that $an^2 + bn + c = \Omega(n^2)$ and $an^2 + bn + c = O(n^2)$. To obtain asymptotic upper and lower bounds, it is usual to prove asymptotically tight bounds from asymptotic upper and lower bounds.

Example 6.34: Use the definition of big-theta to prove that $7x^2 + 1 = \Theta(x^2)$.

Solution:

Clearly, we have $7x^2 + 1 = \Omega(x^2)$. We need to show that $7x^2 + 1 = O(x^2)$. We have $7x^2 \le 7x^2 + 1 \le 7x^2 + x^2 \le 8x^2$, We need $(x \ge 1)$ to obtain the second inequality, if $(x \ge 1)$, $7x^2 \le 7x^2 + 1 \le 8x^2$. This means that $7x^2 + 1 = \Theta(x^2)$.

Example 6.35: Show that $\frac{n^2}{2} - 3n = \Theta(n^2)$

Solution:

To solve this, we must determine the positive constants C_1 and C_2 and k such that

$$C_1 n^2 \le \frac{n^2}{2} - 3n \le C_2 n^2 \quad \text{for } n \ge k$$

Dividing by n^2 yields

$$C_1 \le \frac{1}{2} - \frac{3}{n} \le C_2 \quad \text{for } n \ge k$$

$$C_1 \le \frac{1}{2} - \frac{3}{n} \text{ holds for } n \ge 7, \text{by } C_2 \le \frac{1}{14}$$

$$\frac{1}{2} - \frac{3}{n} \le C_2 \text{ holds for } n \ge 1 \text{ by } C_2 \ge \frac{1}{2}$$

Thus, by choosing $C_1 = \frac{1}{14}, C_2 = \frac{1}{2}$ and $k = 7$, we can verify that $\frac{n^2}{2} - 3n = \Theta(n^2)$. Big Ω is very similar to big-O. Big Ω- notation is used to indicate a lower bound on functions, for large values of the independent variables. Note that f is $\Omega(g)$ if and only if g is $O(f)$.

Example 6.36: Show that $2x^3 + x^2 - 3x + 2 = \Omega(x^3)$

Solution:
If $(x \ge 3)$, then $(x^2 - 3x \ge 0)$ and so $x^2 - 3x + 2 \ge 0$. Thus, $|2x^3 + x^2 - 3x + 2| = 2x^3 + x^2 - 3x + 2 \ge 2x^3$. Choosing $C = 2$, and $k = 3$ in the definition of big Ω shows that the above example is $\Omega(x^3)$. The O-notation is an asymptotic upper bound, which may or may not be asymptotically tight. The bound $2n^2 = O(n^2)$ is asymptotically tight, but the bound $2n = O(n^2)$ is not. We use O-notation to denote an upper bound that is not asymptotically tight. The o(g(n)) is formally defined as a set,

$O(g(n)) = \{f(n)$: for any positive constant $c > 0$, there exists a constant $k > 0$ such that

$$0 \le f(n) < c\,g(n) \text{ for all } n \ge k.\}$$

The definitions of O-notation and o-notation are similar. The main difference is $F(n) = O\,g(n)$, the bound $0 \le f(n) \le c\,g(n)$ holds for some constant $(c > 0)$, but in $f(n) = o\,(g(n))$, the bound $0 \le f(n) \le c\,g(n)$ holds for all constants $(c > 0)$. Intuitively, in the o-notation, the function $f(n)$ becomes insignificant relative to $g(n)$ as n approaches infinity: This means that $\lim_{n\to\infty} \frac{f(n)}{g(n)} = 0$.

- ***ω*- notation:** By analogy, ω-notation is to Ω-notation as o-notation is to O-notation. We use ω-notation to denote a lower bound that is not asymptotically tight. One way to define it is as given in the following line,

$$f(n) \in \omega(g(n)) \text{ if and only if } g(n) \in o(f(n))$$

Hence, $\omega\,(g(n))$ (read as 'little omega of g of n') as a set
$\Omega(g(n)) = \{f(n)$: for any positive constant $c > 0$, there exists a constant $k > 0$ such that $0 \le cg(n) < f(n)$ for all $n \ge k\}$.

Example 6.37: $\frac{n^2}{2} = \omega(n)$, but $\frac{n^2}{2} \ne \omega(n^2)$

The relation $f(n) = \omega\,(g(n))$ implies that $\lim_{n\to\infty} \frac{f(n)}{g(n)} = \infty$. if the limit exists. Here $f(n)$ becomes arbitrarily large relative to $g(n)$ as n- approaches infinity.

Example 6.38: Consider the functions $\log_m n = xm$ $\log_k n = xk$ then $n = (m)^{xm}$, $n = (k)^{nk}$ using definition of logarithms

$$(m)^{xm} = (k)^{nk}$$

Taking the logarithm base to m on both sides, we get

$$\log_m m^{xm} = \log_m k^{xk}$$

$$xm \log_m m = xk \log_m k \text{ and } \log_m m = 1$$

$$xm = xk \log_m k$$

$$\log_m n = (\log_m k) * (\log_k n) \quad \because \quad \log_m n = xm \;\&\; \log_k n = xk$$

This is nothing but the identity of logarithmic functions. Hence $\log_m n$ and $\log_k n$ are constant multiples of each other. Thus $\log_m n = O(\log_k n)$ and $\log_k n = O(\log_m n)$ for all m and k. Functions that are $O(n^k)$ for some k are said to be of polynomial order and that of $O(d^m)$ for some $d > 1$ (but not $O(n^k)$ for any k) are said to be of exponential order. Even a small exponential order function, such as 2^n grows more quickly than any polynomial order function such as n^k. Because of the incredible rate of growth of exponential order functions, problems that require exponential time algorithms for solutions are considered to be intractable.

6.9.3 Comparison of Functions

Many of the relational properties of real numbers apply to asymptotic comparisons also. Assuming $f(n)$ and $g(n)$ are asymptotically positive, the following properties hold good.

- **Transitivity:**

  $f(n) = \Theta(g(n))$ and $g(n) = \Theta(h(n)) \Rightarrow f(n) = \Theta(h(n))$

  $f(n) = O(g(n))$ and $g(n) = O(h(n)) \Rightarrow f(n) = O(h(n))$

  $f(n) = \Omega(g(n))$ and $g(n) = \Omega(h(n)) \Rightarrow f(n) = \Omega(h(n))$

  $f(n) = o(g(n))$ and $g(n) = o(h(n)) \Rightarrow f(n) = o(h(n))$

  $f(n) = \omega(g(n))$ and $g(n) = \omega(h(n)) \Rightarrow f(n) = \omega(h(n))$

- **Reflexivity:**

  $f(n) = \Theta(f(n))$

  $f(n) = O(f(n))$

  $f(n) = \Omega(f(n))$

- **Symmetry:**

  $f(n) = \Theta(g(n))$ if and only if $g(n) = \Theta(f(n))$

- **Transpose Symmetry:**

  $f(n) = O(g(n))$ if and only if $g(n) = \Omega(f(n))$

  $f(n) = o(g(n))$ if and only if $g(n) = \omega(f(n))$

Because these properties hold for asymptotic notations, one can draw an analogy between the asymptotic comparison of two functions f and g and the comparison of two real numbers a and b.

$$f(n) = O(g(n)) \approx a \le b$$

$$f(n) = \Omega(g(n)) \approx a \ge b$$

$$f(n) = \Theta(g(n)) \approx a = b$$

$$f(n) = o(g(n)) \approx a < b$$

$$f(n) = \omega(g(n)) \approx a > b$$

We say that $f(n)$ is asymptotically smaller than $g(n)$ if $f(n) = o\, g(n)$, and $f(n)$ is asymptotically larger than $g(n)$ if $f(n) = \omega(g(n))$. One property of real numbers, however, that does not carry over to asymptotic notation is **Trichotomy**, By this property, for any two real numbers a and b, exactly one of these i.e., $(a < b)$, $(a = b)$, or $(a > b)$, must hold. Although any two real numbers can be compared, not all functions are asymptotically comparable. That is for two functions $f(n)$ and $g(n)$, there may be a case where neither $f(n) = O\, g(n))$ nor $f(n) = \Omega(g(n))$ holds good. For example, functions n and $n^{1 + \sin n}$ cannot be compared using asymptotic notation, since the value of the exponent in $n^{1 + \sin n}$ oscillates between 0 and 2, taking on all in-between values. Consider $f(n)$ and $g(n)$ to be asymptotically non-negative functions, then $\max(f(n), g(n)) = \Theta(f(n) + g(n))$. For any real constants a and b, when $(b > 0)$, $(n + a)^b = \Theta(n^b)$.

6.10 APPLICATIONS

The order of an algorithm or growth of its running time characterises the algorithm's efficiency and also allows a comparison of the relative performances of alternative algorithms. While studying the complexity of an algorithm, we are concerned with the increase in the number of operations required by the algorithm as the size of the data increases. We then try to compare the values of this function, for large n, to the values of some known function, such as a power function, exponential function, or logarithm function. Thus, the growth of functions refers to the relative size of the values of two functions for large values of the independent variable.

Although, we can sometimes determine the exact running time of an algorithm, the extra precision is not usually worth the effort of computing it. For large inputs, the multiplicative constants and lower-order terms of exact running time are dominated by the effects of the input size itself. When we look at input sizes large enough to make only the order of growth of the running time relevant, we are studying the asymptotic efficiency of algorithms. That is, we are concerned with how the running time of an algorithm increases with the size of the input *in the limit*, as the size of the input increases without bound. Usually, an algorithm that is asymptotically more efficient will be the best choice for all but very small inputs. This is one area wherein experience with the concept of a limit from calculus is of great help. Most algebra and calculus texts use $\log x$ to denote $\log_{10} x$ (or, perhaps, $\log_e x$), but in computer science, the use of base 2 is more prevalent. So we shall use $\log x$ to denote $\log_2 x$.

6.11 SORTING

In computer based problem solving, searching and sorting are the two activities mainly carried out. Sorting is an important concept in computer science. It provides a systematic arrangement of elements in a given record or a file. A telephone directory is an example of sorted data and it is has lexicographic ordering, which is one way of alphabetical ordering.

6.11.1 Efficiency Considerations

Many well established sorting methods are available in the literature. The important fact is that the programmer must be able to identify the multiple interconnected and sometimes conflicting efficiency considerations, to make the best choice of sorting technique. The most important considerations are listed here.

(i) Time spent by programmer in coding a particular program.
(ii) The amount of machine time required for running the program
(iii) The amount of space required for the program.

However, with reference to the first consideration, a programmer must be able to make the right choice of technique and code it using programming skill. The remaining two considerations need to be optimised, one at the expense of the other. This is often called time-space tradeoff. The techniques are analysed in terms of **best case, worst case and average case.** Once a particular sorting technique has been selected, it is important to make the program as efficient as possible. Often, in the light of skilled programming concepts, efficiency is neglected in consideration of clarity. Not in the case of sorting though, where efficiency is critical. Once the program is debugged, the next step is to improve the performance. Any improvement in sorting time significantly affects overall efficiency, since sorting is used quite frequently. (This kind of tweaking can be achieved for example, by removing function calls from inner loops and using inline functions)

Usually space constraints are less important than time considerations, since most sorting programs require space closer to $O(n)$ than to $O(n^2)$. Programs that require less time usually require more space and vice-versa. Efficient algorithms that need to consider both constraints are of $O(n \log (n))$. Skilled programming is needed to verify the algorithms. In the present scenario, wherein processor speed and memory space are not scarce, an optimal algorithm is still going to take less time and space.

Example 6.39: Bubble sort

In this technique, each data element $x[i]$ is compared with its successor $x[i + 1]$ for a given order. The data elements are interchanged if they are not in the proper order. In general, $(n - 1)$ passes are sufficient to sort a file of size n. A pass is one scan of data elements from beginning to end. If data elements are already partially in order, in such cases, the data is sorted before $(n - 1)$ iterations, making the remaining passes unnecessary. It is important to detect that the data is sorted before every pass. This is achieved by considering the interchanges in a given pass. No interchanges in a pass means that the data is sorted. The analysis of efficiency is done in two ways, a) without consideration of unnecessary passes. b) With consideration of unnecessary passes.

(a) There are $(n - 1)$ passes and $(n - 1)$ comparisons in each pass.
The total no of comparisons $= (n - 1)(n - 1)$
$= n^2 - 2n + 1.$
It is therefore of order $O(n^2)$

(b) The elimination of unnecessary passes improves the speed of the sort. The number of comparisons on iteration i is now $(n - i)$. If there are k-iterations, the total number of comparisons is given by,

$$= (n - 1) + (n - 2) + \text{----}(n - k).$$

$$= nk - (1 + 2 + 3 + \text{------} k)$$

$$= nk - \frac{(k(k+1))}{2}$$

$$= \frac{2nk - k^2 - k}{2}$$

Hence, the average number of iterations is *k* and is of order O(*n*). The entire expression still has O(n^2), but the constant multiplier is smaller than in (a). Therefore, the method requires little additional space and is O(*n*), so that the file is completely sorted and only one pass of (*n* – 1) comparisons (no interchanges) is required to establish that the data is sorted.

Example 6.40: Quick sort

This technique is also known as a partition exchange sort. Consider an array (*x*) containing *n*- data elements. An element *a* is selected from a specific position within the array; it can even be the first element of an array. The data elements of *x*- are partitioned, so that *a* is put in the ith position with the following conditions (i) the data elements from 0 to *i* – 1, which are less than or equal to *a* and (ii) the data elements in the range *i* + 1 to *n* – 1, which are greater than or equal to *a*. Two partitions are created with the ith position as a pivot. When the conditions hold for particular element (*a*) and position (*i*), and if *a* is the smallest data element of *x*, it remains in position *i*, when the array of data is completely sorted. The process repeated with two partitions, namely, sub arrays *x*[0] through *x*[*i* – 1] and *x*[*i* + 1] through *x*[*n* – 1].

Let us consider that the array size is *n*, which is a power of 2, i.e. $n = 2^m$, so that $\log_2 n = m$. The time analysis of the quick sort algorithm is as follows:

$$T(n) = \begin{cases} a & n = 1 \\ cn + T(n/2) + T(n/2) \end{cases}$$

Where *cn*- is the time required to partitioning the array. T(*n*/2) is the time required to sort the left sub array, and T(*n*/2) is the time required to sort the right sub array

Best Case Analysis

The best case analysis is possible when the array is always partitioned in half.

Then $$T(n)) = cn + T(n/2) + T(n/2).$$

$$= cn + 2\,T(n/2).$$

Now *n*- is power 2 i.e. $n = 2^m$.

$$T(2^m) = c(2^m) + 2T(2^{m/2})$$

$$= 2T(2^{m-1}) + c\,2^m$$

$$= 2[2T(2^{m-2}) + c\,2^{m-1}] + c2^m$$

$$= 2^2[T(2^{m-2})] + 2c\,2^m$$

= -----------------------

$$= 2^m[T(2^{m-m})] + mc\,2^m$$

$$= 2^m[T(1)] + mc\,n$$

$$= 2^m\,a + (mc)\log_2 n \quad (n = 2^m,\ m = \log_2 n)$$

$$T(n) = O\,(n\log n).$$

Worst-case Analysis

In this case, upon every function call, the given array is partitioned into two sub arrays. One of them is an empty array. The time analysis takes the form as given below:

$$\begin{aligned}
T(n) &= cn + T(0) + T(n-1).\\
&= cn + T(n-1).\\
&= cn + c(n-1) + T(n-2).\\
&= cn + c(n-1) + c(n-2) + T(n-2).\\
&= \text{-----------------------------------}\\
&= cn + c(n-1) + c(n-2) + ---- + c(1) + T(0).\\
&= cn + c(n-1) + c(n-2) + ---- + c(1) + T(0).\\
&= c[n + (n-1) + (n-2) + ---- + 2 + 1\\
&= c\left(\frac{n(n+1)}{2}\right)\\
&= \frac{cn^2}{2} + \frac{cn}{2}
\end{aligned}$$

Therefore $\quad T(n) = O(n^2).$

Example 6.41: **Selection sort**

The technique consists of selecting a data element and placing it in a proper position and repeating the same till all the data elements get placed in their respective positions, except the last one. The last data element gets automatically placed when the remaining $(n - 1)$ data elements are placed in order. The initial n-element priority queue is reduced by one element after each selection. The entire array is sorted after $n - 1$ iterations. The first pass makes $(n - 1)$ comparisons, the second pass makes $(n - 2)$ comparisons and so on.

$$\text{Total number of comparisons} = (n-1) + (n-2) + (n-3) + \text{---} = \frac{n(n-1)}{2}$$

It is of the order $O(n^2)$. The method is faster than the bubble sort. A little additional storage is required. There is no improvement if the input data is completely sorted or unsorted.

Example 6.42: **Merge sort**

This is a process of combining two or more sorted data sets into a third sorted data set. The technique works as follows. The data is divided into n- sub data sets of size 1, and adjacent (disjoint) pairs of data are merged. Then we have approximately $n/2$ data sets of size 2. The process is repeated until we get only one data set of size n. Let us assume the size of file is an integral power of s i.e. $n = 2^k$ Then

$$T(n) = \begin{cases} 1 \text{ if size} = 1 \\ n + 2^*T(n/2), & \text{otherwise} \end{cases}$$

$$T(n) = 2^*T(n/2) + n \quad \text{where } n = 2^k$$

$$
\begin{aligned}
T(2^k) &= 2*T(2^{k-1})+2^k \\
&= 2*[2*T(2^{k-2})+2^{k-1}]+2^k \\
&= 2^2*T(2^{k-2})+2^k+2^k \\
&= 2^2*[2*T(2^{k-3})+2^{k-2}]+2*2^k \\
&= 2^3*[T(2^{k-3})]+3*2^k \\
&= \text{------------------------} \\
&= 1*2^k+k*2^k
\end{aligned}
$$

Here, $n=2^k$, $k=\log_2 n$. $\therefore$ $T(n)=O(n+n\log_2 n)$.

The amount of time for the execution of merge sort is proportional to $n*\log_2 n$, which is a significant improvement over selection sort.

ADDITIONAL SOLVED EXAMPLES

Example W6.1: Consider the function $f(x)=\dfrac{4x+3}{5x-2}$. Find f^{-1}.

Solution:

Let $f^{-1}(x)=y$ then $x=f(y)$. Given that $f(x)=\dfrac{4x+3}{5x-2}$

Then $$x=\frac{4y+3}{5y-2}$$

Solving for y

$$x(5y-2)=4y+3$$

$$y(5x+4)=2x+3$$

$y=\dfrac{2x+3}{5x+4}$. Hence $f^{-1}(x)=\dfrac{2x+3}{5x+y}$

Example W6.2: If R is the set of real numbers, then show that the function f: R→R defined by $f(x)=5x^3-1$ is one-one and onto.

Solution:

(i) To show that f- is a one-one function.

Let $x_1, x_2 \in$ R, the two different elements

Then $x_1 \neq x_2$

$\Rightarrow 5x_1^3-1 \neq 5x_2^3-1$

$\Rightarrow 5x_1^3 \neq 5x_2^3$

$\Rightarrow f(x_1) \neq f(x_2)$

This shows that different elements in R have different f-images in R.

Hence f- is a one-one function.

(ii) To show that f- is an onto function.
Let $y = 5x^3 - 1$ be any arbitrary element in R.

Then $x = \left[\frac{y+1}{5}\right]^{\frac{1}{3}}$ $x \in R \ \forall \ y \in R$

Now $f(x) = f\left(\frac{y+1}{5}\right)^{\frac{1}{3}} = 5\left(\left(\frac{y+1}{5}\right)^{1/3}\right)^3 - 1$

$= 5\left(\frac{y+1}{5}\right) - 1 = y + 1 - 1 = y$

$f(x) = y.$

This shows that for every element $y \in$ R (range set) there exists a pre-image in the domain set R of f.
Hence f- is an onto function.
Hence f- is a one-one and onto function.

Example W6.3: In the set of natural numbers ℕ, show that the function f: ℕ → ℕ defined by $f(x) = x^2$ is one-one and into.

Solution:

(i) To show f- to be a one-one function.
Let $x_1, x_2 \in$ ℕ (domain set) be two different elements.
Then $x_1 \neq x_2$ i.e. $x^2_1 \neq x^2_2$
$\Rightarrow f(x_1) \neq f(x_2)$. Thus f- is a one-one function.

(ii) To show that f- is an into function.
Given $f(x) = x^3, f(2) = 8, f(3) = 27$,----. However, there are natural numbers in the range set ℕ such as 23, 4, 5, 6, 7, 9, 10,---- etc which are not f-images of any natural number in the domain set ℕ of f. Thus f is an into function. Hence f is a one-one and into function.

Example W6.4: If A, B, C, and D are four sets and f, g, and h are three functions defined as

$$f: A \to B:, \quad g: B \to C, \text{ and } h: C \to A$$

Then prove that $(h \circ g) \circ f = h \circ (g \circ f)$

Solution:

Let x be an arbitrary element of set A, then

$$[(h \circ g) \circ f](x) = (h \circ g) f(x)$$

$$= (h \circ g) y \quad y = f(x), \forall x \in A, y \in B$$

$$= h(g(y)) \text{ since } z = g(y) \ \forall \ y \in B, z \in C$$

$$= h(z) = x \ \in A$$

$$[h \circ (g \circ f)](x) = h[(g \circ f)(x)]$$

$$= h[g(f(x))]$$

$$= h[g(y)]$$
$$= h(z) = x \in A.$$

Hence $[(h \circ g) \circ f](x) = [h \circ (g \circ f)](x)$ and is associative.

Example W6.5: Consider $f(x) = x + 2$, $g(x) = x + 2$ and $h(x) = 3x \ \forall x \in R$. Then show that
(i) $(h \circ g) \circ f = h \circ (g \circ f)$ (ii) $f \circ (f \circ f)$

Solution:

(i) $[f \circ (g \circ h)](x) = f[(g \circ h)(x)]$
$$= f[g(h(x))]$$
$$= f[g(3x)] \quad \because h(x) = 3x.$$
$$= f[3x - 2] \quad \because f(x) = x + 2.$$
$$= 3x - 2 + 2 = 3x \quad \because f(x) = x + 2.$$

Again consider

$$[(f \circ g) \circ h] = (f \circ g) \circ h(x)$$
$$= (f \circ g)(3x)$$
$$= f\ [g(3x)]$$
$$= 3x - 2 + 2 = 3x \quad \because f(x) = x + 2.$$
$$[f \circ (g \circ h)](x) = [(f \circ g) \circ h]$$

Example W6.6: Let the functions f, g and h be defined as follows:

$$f: R \to R \quad f(x) = 2x + 3$$
$$g: R \to R \quad g(x) = 3x + 4$$
$$h: R \to R \quad h(x) = 3x$$

then find (i) $f \circ (g \circ h) = f \circ (g \circ h)(x)$
$$= f(g(h(x)))$$
$$= f(g(3x)) \quad \because h(x) = 3x$$
$$= f(3(3x) + 4)$$
$$= f(9x + 4) = 2(9x + 4) + 3 = 18x + 11$$

(ii) $(f \circ g) \circ h = (f \circ g) \circ h(x)$
$$= (f \circ g)(3x) \quad \because \ h(x) = 3x$$
$$= f(g(3x))$$

$$= f(3(3x)+4) \quad \because \; g(x)=3x+4$$
$$= f(9x+4) \quad \because f(x)=2x+3$$
$$= 2(9x+4)+3$$
$$= 18x+11.$$

This shows that $f\circ(g\circ h)=(f\circ g)\circ h$.

Example W6.7: If f and g are functions defined on the set of real numbers by $f(x)=x+1, g(x)=x^2+2$, then find

(i) $(g\circ f)(-2)$ (ii) $(f\circ g)(-2)$ (iii) $g\circ f(x)$ (iv) $f\circ g(x)$.

Solution:

(i) $(g\circ f)(-2)=g[f(-2)]$

$$= g(-2+1)=g(-1)=(-1)^2+2=3$$

$(\because f(x)=x+1, g(x) = x^2+2)$

(ii) $(f\circ g)(-2)=f(g(-2))=f[(-2)^2+2]=f(6)=6+1=7$

(iii) $(g\circ f)(x)=g(f(x))$

$$= g(x+1)$$
$$= (x+1)^2+2$$
$$= x^2+2x+3$$

(iv) $(g\circ g)(x)=g[g(x)]$

$$= g(x^2+2)$$
$$= (x^2+2)^2+2$$
$$= x^4+4x^2+6$$

Property: Let f: A→B and g: B→C, be two invertible mappings, then prove that if $(g\circ f)$ is invertible then $g\circ f: A\to C$ is also invertible.

Proof:

Since f and g are one-one functions, we have

$$f(x_1)=f(x_2)\Rightarrow x_1=x_2 \quad \forall x_1,x_2\in R$$

$$g(y_1)=g(y_2)\Rightarrow y_1=y_2 \quad \forall y_1,y_2\in R$$

$$(g\circ f)(x_1)=(g\circ f)(x_2)$$

$$\Rightarrow \quad g(f(x_1))=g(f(x_2))$$

$$\Rightarrow \quad (f(x_1))=(f(x_2))$$

$$\Rightarrow \quad x_1=x_2$$

Since f and g are both onto functions, for each $z \in C$, there exists $y \in B$ such that $g(y) = z$ and $f(x) = y$ for each $x \in A$.

Thus $z = g(y) = g[f(x)] = (g \circ f)(x)$

Hence each element $z \in C$ has pre-image under $(g \circ f)$ and therefore $(g \circ f)$ is an onto function. Since $(g \circ f)$ is a one-one and onto function, therefore it is invertible.

SUMMARY

The chapter covers the functions and types of functions. Every function is a relation but vice versa need not be true. The operations on functions are dealt with. Every program is viewed to be a function. Computer science applications in terms of growth functions are given with fair examples. An introduction has been given to the analysis of algorithms with examples on sorting of data. There are many other types of functions, such as hashing functions, characteristic function and the like. The content of this chapter serves as a basis to the course on analysis and design of algorithms.

EXERCISES

1. Determine whether or not each of the following relations is a function with domain {1, 2, 3, 4}. If any function not a function, give reasons.
   (a) $R_1 = \{(1, 1)\ (2, 1)\ (3, 1)\ (4, 1)\ (3, 3)\}$
   (b) $R_2 = \{(1, 2)\ (2, 3)\ (4, 2)\}$
   (c) $R_3 = \{(1, 1)\ (2, 1)\ (3, 1)\ (4, 1)\}$
   (d) $R_4 = \{(1, 4)\ (2, 3)\ (3, 2)\ (4, 1)\}$
2. Let $A = \{a, b, c, d\}$ and $B = \{0, 1, 2, 3\}$. Find whether the following are functions from A to B. If they are functions, find whether they are one-to-one or onto, and indicate their ranges.
   $F = \{(a, 0), (b, 1), (c, 2), (d, 2)\}$ $\quad$ $G = \{(a, 1), (b, 0), (b, 2), (c, 1), (d, 1)\}$
   $H = \{(a, 2), (b, 1), (c, 2), (d, 3)\}$ $\quad$ $I = \{(a, 1), (b, 1), (c, 2), (d, 3)\}$
3. Determine whether or not each of the following relations is a function . If a relation is a function, find its range.
   (e) $R_1 = \{(x, y) / x, y \in z, y = x^2 + 7\}$. Relation from Z to Z.
   (f) $R_2 = \{(x, y) / x, y \in Q, x^2 + y^2 = 1\}$. Relation from Q to Q.
4. Find out whether each of the following functions from Z to Z is one-one
   (a) $f(x) = x + 2$ (b) $f(x) = x^3 + 1$ (c) $f(x) = x^3$
5. Find $(f \circ g)$ and $(g \circ f)$ where $f(x) = x^2 + 1$ and $g(x) = x^2 + 2$ are functions from R to R.
6. Find (a) $f \circ g$ (b) $g \circ f$ (c) $f \circ f$ (d) $g \circ g$, where $f(x) = x^2 + 3x + 2 : g(x) = 2x - 3$ are functions from R to R.
7. If $A = \{1, 2, 3\}$, $B = \{w, x, y, z\}$ and $f: A \to B$
   (a) How many functions f- are there?
   (b) How many of them are one-one?
8. If $A = \{1, 2, 3, 4, 5\}$, $B = \{1, 2, 3, 8, 9\}$ and the functions $f: A \to B$, $g: A \to B$ are defined by $f = \{(1, 8)\ (3, 9)\ (4, 3)\ (2, 1)\ (5, 2)\}$ and $g = \{(1, 2)\ (3, 1)\ (2, 2)\ (4, 3)\ (5, 2)\}$
   find (a) $f \circ g$ (b) $g \circ f$ (c) $f \circ f$ (d) $g \circ g$. if they exist.
9. Consider the function $f: R \to R$, which is defined by $f(x) = 2x - 3$. Find the formula that defines f^{-1}.

10. If $f: Z \rightarrow N$ defined by $f(x)=\begin{cases} 2x-1, & \text{if } x>0 \\ -2x, & \text{if } x\leq 0 \end{cases}$ prove that f –is one-one and onto. Also determine f^{-1}.

11. For the function $f(x) = 6x^4 - 2x^3 + 5$ express it in the form $|f(x)| \leq M|g(x)|$, and show that $|f(x)| \leq 13x^4$.

12. If A set has n elements, how many functions and bijections are there from A to A?

13. If set A has m elements and B has n elements, how many functions are there from A to B?

14. For any given two functions $f(n)$ and $g(n)$, show from the example given below, that

    $f(n) = \Theta(g(n))$ if $f(n) = O(g(n))$ and $f(n) = \Omega(g(n))$, where $f(n) = \frac{n^2}{2} - 3n$.

15. Use the definition of big O-to prove that $f(x) = 2x^2 + 5x + 9 = O(g(x)) = x^2$ and find the value of k such that $|f(x)| \leq 3|g(x)|$ for all $(x > k)$.

16. The two functions $f(n) = 5n^4 - 7n^2$ and $g(n) = n^2$ are defined for positive integer n. Show that f and g have the same order.

17. Consider the function $f(n) = n^5$ and $g(n) = n^7$. Show that if $n \geq 1$, then $n^5 \leq n^7$.

7 Algebraic Structures

In Chapter 6, we have covered functions. The mathematical operations have been visualised as functions. The concept of sets has been discussed in Chapter 4. In this chapter, we present the concept of algebraic structures. A set, together with a number of operations on it, is called an algebraic structure. A group is an algebraic structure and has variants such as permutation group, matrix group, transformation groups etc. In this chapter, the emphasis is on groups.

7.1 HISTORY OF GROUP THEORY

The original motivation for group theory was the quest for solutions to polynomial equations of degree higher than four. The nineteenth century French mathematician, Evariste Galois, extending prior work of Paolo Ruffini and Joseph-Louis Lagrange, gave a criterion for the solvability of a particular polynomial equation in terms of the symmetry groups of its roots (also called solutions). The elements of such a Galois group correspond to certain permutations of the roots. Geometry is a field in which groups are used systematically, especially symmetry groups. After novel geometries such as hyperbolic and projective geometry had emerged, Klein used group theory to organise them in a more coherent way. Another field where group theory finds use is number theory, wherein groups are used to describe factorisation into prime numbers. After contributions from other fields such as number theory and geometry, the notion of group was generalised and firmly established as a theory around 1870. In mathematics, a **group** is viewed as consisting of a set of elements together with an operation to combine any two of its elements to get a new element that also belongs to the same set while satisfying four conditions, namely group axioms, closure, associativity, identity and invertibility. Groups share a fundamental kinship with the notion of symmetry. For example, a symmetry group encodes the symmetry features of a geometrical object. The group consists of the set of transformations that leave the object unchanged in spite of combinations of transformations and being performed in a sequence. There are Lie groups that belong to symmetry groups which find application in the study of standard models of particle physics, and Point groups that are used to understand symmetry phenomena in molecular chemistry.

Modern group theory is considered an active mathematical discipline that studies groups in their own right. To explore groups, mathematicians have devised various notions to break groups into smaller, better-understandable pieces such as subgroups, quotient groups and simple groups. In addition to their abstract properties, group theorists also study the different ways in which a group is expressed concretely, both from hierarchical and computational points of view. The modern concept of an abstract group has developed out of several fields of mathematics.

7.1.1 Main Classes of Groups

Groups are divided into five main classes, namely, permutation groups, matrix groups, transformation groups, abstract groups and topological groups.

7.1.1.1 *Permutation groups*

Given any set A, the collection G of bijections of A are called permutations. The set G is closed under composition and inverse operations. If a set A consists of *n* elements and G consists of all permutations, then G is called a symmetric group S_n. In general, any permutation group G is also a subgroup of the symmetric group of A. In many cases, the structure of a permutation group is studied based on the properties of its action on the corresponding set. For example, one proves that for $n \geq 5$, the alternating A_n is simple. This fact plays a key role in the impossibility of solving a general algebraic equation of degree $n \geq 5$ in radicals.

7.1.1.2 *Matrix groups*

Another class of groups is given by matrix groups or linear groups. Here, the group G consists of a set of invertible matrices of given order *n* over a field *K*, which is closed under the operations product and inverse. Such a group acts on the *n*-dimensional vector space K^n by linear transformations. This property makes matrix groups conceptually similar to permutation groups, and the geometry of the action may be exploited to establish properties of the group *G*.

7.1.1.3 *Transformation groups*

Permutation groups and matrix groups are special cases of transformation groups, the groups that act on a certain space *X* and preserve the inherent structure. In the case of permutation groups, *X* is a set and in case of matrix groups, *X* is a vector space. The concept of a transformation group is closely related to the concept of a symmetry group. The transformation groups frequently consist of all transformations that preserve a certain structure. The theory of transformation groups forms a bridge between group theory and differential geometry. The research originating with Lie and Klein has considered group actions on manifolds by homomorphism or diffeomorphism. The groups themselves may be discrete or continuous.

7.1.1.4 *Abstract groups*

Most of the groups considered in the first stage of the development of group theory are 'concrete', having been realised through numbers, permutations, or matrices. It was not until the late nineteenth century that the idea of an abstract group as a set with operations satisfying a certain system of axioms began to take hold. A typical manner of specifying an abstract group is through a presentation by generators and relations. A significant source of abstract groups is given by the construction of a factor group, or quotient group, *G*/*H*, of a group *G* by a normal subgroup *H*.

The change of perspective from concrete to abstract groups makes it natural to consider the properties of groups that are independent of a particular realisation, or in modern language, are invariant under isomorphism, as well as the classes of groups with such a property: finite groups, periodic groups, simple groups, solvable groups, and so on. Rather than exploring properties of an individual group, one seeks to establish results that apply to a whole class of groups. The new paradigm is of paramount importance for the development of mathematics.

7.1.1.5 *Topological and algebraic groups*

An important elaboration of the concept of a group occurs if G is endowed with additional structure, notably, of a topological space, differentiable manifold, or algebraic variety. If the group operations, m (multiplication) and i (inversion),

$$\text{m: } G \times G \to G, \quad (g,h) \to gh, \quad \text{i:} G \to G, \quad g \mapsto g-1$$

are compatible with this structure, this means that these operations are continuous, smooth or regular (in the sense of algebraic geometry) maps. This follows that G becomes a topological group, a Lie group, or an algebraic group.

Topological groups form a natural domain for abstract harmonic analysis, whereas Lie groups (frequently realised as transformation groups) are the mainstays of differential geometry and unitary representation theory. Certain classification questions that cannot be solved in general can be approached and resolved for special sub-classes of groups. Thus, compact connected Lie groups have been completely classified. There is a fruitful relation between infinite abstract groups and topological groups: whenever a group Γ can be realised as a lattice in a topological group G, the geometry and analysis pertaining to G yield important results about Γ. A comparatively recent trend in the theory of finite groups exploits their connections with compact topological groups.

7.2 BINARY OPERATIONS AND GENERAL PROPERTIES

Let S be a non-empty set and $*$ (read as star) be an operation on S. The operation on the set is a rule, which assigns to each ordered pair of elements of the set, a unique element of S. This means that a binary operation on a non-empty set S is a function * from $S \times S$ to S.

$*: S \times S \to S$. In other words, for all $a, b \in S$, $a * b \in S$. The binary operation defined on set S is usually denoted by a tuple $(S, *)$.

7.2.1 Properties of Binary Operations

Let (S, *) be an algebraic system, where * is a binary operation on S. We shall look at the following binary operations:

7.2.1.1 *Closure property*

Consider a binary operation $*: A \times A \to A$. The operation is said to be closed, if for all $a, b \in S, a * b \in S$. The new element also belongs to S.

Example 7.1: A set of integers Z is closed with respect to the binary operations, namely addition, multiplication and subtraction, but not with respect to division.

$$\forall\, a, b \in Z,\ a+b \in S,\ (a \times b) \in Z, \text{ and } \quad (a/b) \notin Z$$

Example 7.2: The set of odd integers is not closed with respect to addition, since the sum of two odd integers is an even integer, which is not a member of the set.

Example 7.3: The binary operations addition (+) and multiplication (×) are closed on the set of natural numbers $\mathbb{N}$.

$\mathbb{N} = \{1, 2, 3, ---\}$. $\forall\, a, b \in N,\ a+b \in N,\ (a \times b) \in N$. However, the subtraction operation (–) is not a closed binary operation on $\mathbb{N}$. $\forall$ a, b $\in$ N, $(a-b) \notin N$

It is necessary to note that a set S is said to be closed with respect to a given operation, if the given operation on members of S always produces (again) a member of S. Thus, by definition, such a binary operation is said to have the closure property. The binary operations denoted by symbols with standard meaning such as. ,*, +, –, •, ∗, Δ, ⋃, ⋂, ∧, ∨, etc, are used in mathematics.

7.2.1.2 *Associative property*

Consider a binary operation *. For any $\forall\, a, b, c \in S$: $(a*b)*c = a*(b*c)$

Example 7.4: The addition (+) and multiplication (•) are associative in the following sets.

$\mathbb{N}$ = the set of natural numbers.
I or $\mathbb{Z}$ = the set of integers.
Q = the set of rational numbers.
R = the set of real numbers.
C = the set of complex numbers.
$(a+b)+c = a+(b+c)$
$(a \bullet b) \bullet c = a \bullet (b \bullet c)$

7.2.1.3 *Commutative property: (Abelian)*

For $\forall$ a, b $\in$ S, $a*b = b*a$

Example 7.5: Consider the elements *a* and *b*. The operations $a+b = b+a$ and $a \bullet b = b \bullet a$, are commutative.

7.2.2 Existence of an Identity Element

Consider an element $e \in S$, such that $\forall a \in S,\ a*e = e*a = a$. Then the element $e \in S$ is called the identity element of S with respect to the binary operation $*$. For example, 0 and 1 are the identity elements of $\mathbb{Z}$ with respect to the addition and multiplication operations respectively, for any $a \in Z$, $a+0 = 0+a = a$ and $a \times 1 = 1 \times a = a$.

7.2.3 Existence of an Inverse

Consider an element $a \in S$, there exists an element $a^{-1} \in S$ such that $a*a^{-1} = a^{-1}*a = e$. The element, $a^{-1} \in S$ is called the inverse of *a* under the operation $*$.

Example 7.6: For each $a \in Z$, the element $-a$ is the inverse of *a* under the operation of addition, since $a+(-a) = (-a)+a = 0$, where 0 is the identity element of $\mathbb{Z}$ under addition.

7.2.4 Cancellation Property

For any *a*, *b*, *c* $\in$ S, and $a \neq 0$, the following expressions, illustrate the cancellation property.

$$a*b = a*c \Rightarrow b = c$$

$$b*a = c*a \Rightarrow b = c$$

Example 7.7: Cancellation property holds good for any $a, b, c \in Z$ under addition (+) and multiplication (×).

Consider the following examples

Example 7.8: On the set of real numbers R, $*$ is defined by $(a*b)=\sqrt{ab}\ \forall a, b \in R$. Check whether $*$ is a binary operation on R.

Proof: Given $\forall a,b \in R, (a*b) = \sqrt{ab}$ consider $a=2, b=-1$, then $a*b= \sqrt{2 \times (-1)} = \sqrt{-2} = i\sqrt{2} \notin R$
So $*$ is not a binary operation.

Example 7.9: If $S=\left\{\begin{pmatrix} 1 & 0 \\ a & 1 \end{pmatrix} / a \in R\right\}$, where R is the set of real numbers, examine whether

(i) matrix addition (ii) matrix multiplication, are binary operations on S.

Proof: Given that S is the set of all matrices of the form $\begin{pmatrix} 1 & 0 \\ a & 1 \end{pmatrix}$, where a is any real number. Let

$A=\begin{pmatrix} 1 & 0 \\ a & 1 \end{pmatrix} : B=\begin{pmatrix} 1 & 0 \\ b & 1 \end{pmatrix}$ and $A, B \in S,\ a, b \in R$

Now, $A+B=\begin{pmatrix} 1 & 0 \\ a & 1 \end{pmatrix}+\begin{pmatrix} 1 & 0 \\ b & 1 \end{pmatrix}=\begin{pmatrix} 2 & 0 \\ a+b & 2 \end{pmatrix} \notin S,\ a, b \in R$

Therefore matrix addition is not a binary operation on S.

Now $AB = AB = \begin{pmatrix} 1 & 0 \\ a & 1 \end{pmatrix}\begin{pmatrix} 1 & 0 \\ b & 1 \end{pmatrix}=\begin{pmatrix} 1+0 & 0+0 \\ a+b & 0+1 \end{pmatrix}=\begin{pmatrix} 1 & 0 \\ c & 1 \end{pmatrix} \in S$

where $c=(a+b) \in R$.

Hence, $AB=\begin{pmatrix} 1 & 0 \\ c & 1 \end{pmatrix}$ belongs to S.

Therefore, matrix multiplication is a binary operation on S. Here, we say that S is not closed under matrix addition and S is closed under matrix multiplication.

Example 7.10: Consider the set S of rational numbers, and $*$ is defined by $a*b=\frac{ab}{5}$. Show that $*$ is both commutative and associative.

Proof: The binary operation $*$ is defined by $a*b=\frac{ab}{5}$.

A. Commutative Property $= \frac{a*b}{5}=\frac{ab}{5}$; $\frac{b*a}{5}=\frac{ba}{5}$

Also $ab = ba$ (multiplication is commutative on the set of rational numbers)
$\therefore\ a*b=b*a$, so $*$ is commutative.

B. Associative property:

$\forall a, b, c \in S: \quad a*(b*c)=(a*b)*c$

(i) $a*(b*c)=a*\left(\frac{bc}{5}\right)$

$$= a*x \text{ where } x = \frac{bc}{5}$$

$$= \frac{ax}{5} = \frac{a\left(\frac{bc}{5}\right)}{5} = \frac{abc}{25} \quad \text{(I)}$$

(ii) $(a*b)*c = \left(\frac{ab}{5}\right)*c$

$$= y*c \text{ where } y = \frac{ab}{5}$$

$$= \frac{yc}{5} = \frac{\left(\frac{ab}{5}\right)c}{5} = \frac{abc}{25} \quad \text{(II)}$$

From A and B, $a*(b*c) = (a*b)*c$. Hence, operation$*$ is associative.

7.3 GROUPS

Let G be a non-empty set and let$*$ be a binary operation on G. Then G is called a group under the operation $*$ if the following properties hold good.

(i) Closure Property: $\forall\, a, b \in \mathrm{G},\; a*b \in \mathrm{G}$. Closure property holds good.

(ii) Associative Property: $\forall\, a, b, c \in \mathrm{G}$: $\quad a*(b*c) = (a*b)*c$

(iii) Identity element: $\forall\, a \in \mathrm{G},\; \exists\, e \in \mathrm{G}$, such that $a*e = e*a = a$

Here e- is called an identity element under the binary operation defined.

(iv) Inverse element: $\forall\, a \in \mathrm{G},\; \exists\, a^{-1} \in \mathrm{G}$, such that $a*a^{-1} = a^{-1}*a = e$

Here, $a^{-1}\left(\neq \frac{1}{a}\right)$ is called an inverse of a under the binary operation defined. A group G under the binary operation $*$ is denoted as $(\mathrm{G}, *)$. In addition to the foresaid four properties, if the group $(\mathrm{G}, *)$ satisfies the commutative property, then the group is called a commutative group or an Abelian group.

$\forall\, a, b \in \mathrm{G},\; a*b = b*a$.

7.3.1 Finite and Infinite Groups

A group (G , $*$) is said to be finite if the number of elements in the set G is finite, otherwise it is said to be an infinite group. The number of elements in a finite group is said to be the order of the group G and is denoted by $\circ$(G).

Example 7.11: One of the most familiar groups is the set of integers, $\mathbb{Z}$, which consists of the numbers, {… −4, −3, −2, −1, 0, 1, 2, 3, 4,…} together with the addition operation. The following properties of integer addition serve as a model for the abstract group axioms.

(i) For any two integers a and b, the sum $a+b$ is also an integer. Thus, adding two integers never yields some other type of number, such as a fraction. This property is known as closure property under addition.

(ii) For all integers a, b and c, $(a + b) + c = a + (b + c)$. Expressed in words, adding a to b first, and then adding the result to c gives the same final result as adding a to the sum of b and c. This property is known as associativity.
(iii) If a is any integer, then $0 + a = a + 0 = a$. The element 0 (also called zero) is called the identity element of addition because adding it to any integer returns the same integer.
(iv) For every integer a, there is an integer b such that $a + b = b + a = 0$. The integer b is called the inverse element of the integer a and is denoted $-a$.

The integers, together with the operation +, form a mathematical object belonging to a broad class sharing similar structural aspects. The set of integers $\mathbb{Z}$ is an Abelian group with respect to the binary operation under addition.

Example 7.12: If $Q_1 = \{Q - (1)\}$ is the set of all rational numbers other than 1, with a binary operation defined by $\forall\, a, b \in Q_1,\ a * b = a + b - ab$, Show that $(Q_1, *)$ is an Abelian group and solve $6 * x = 4$ in Q_1.

Proof:

A. Closure property:

Now $Q_1 = \{Q - (1)\}$ is the set of rational numbers and $a + b - ab$ is a rational number. Since $a \neq 1, b \neq 1, a + b \neq 1$.

If $a + b - ab = 1$, then $(a - 1)(b - 1) = 0$, which implies that $a = 1$, or $b = 1$, which is a contradiction.

$\therefore a * b = a + b - ab \in Q_1\ \forall\, a, b \in Q_1$.

The closure axiom is satisfied.

B. Associative property:

$$\forall\, a, b, c \in Q_1 : \quad a*(b*c) = (a*b)*c$$

(i) $a*(b*c) = a*(b + c - bc)$

$= a * x$ where $x = b + c - bc$

$= a + x - ax$

$= a + b + c - bc - a(b + c - bc)$

$= a + b + c - bc - ab - ac + abc$ (1)

(ii) $(a*b)*c = (a + b - ab)*c$

$= y * c$ where $y = a + b - ab$

$= y + c - yc$

$= a + b - ab + c - c(a + b - ab)$

$= a + b + c - bc - ab - ac + abc$ (2)

From (1) and (2) it is clear that the associative property holds good.

C. Identity axiom:

$\forall\, a \in Q_1,\ \exists\, e \in Q_1,\ a * e = e * a = a$

$a * e = a$

$\forall\, a, b \in Q_1,\ a * e = a + e - ae = a$

$e - ae = 0,\ e(1 - a) = 0$

This implies that $e = 0$ or $a = 1$, but $a \neq 1$, $\therefore e = 0 \in Q_1$. Therefore, an identity element exists.

D. Inverse element:

$\forall\, a \in Q_1,\ \exists\, a_1^{-1} \in Q_1,\ a * a^{-1} = a^{-1} * a = e$

$a * a^{-1} = a + a^{-1} - aa^{-1} = e$, since $e = 0$,

$a + a^{-1} - aa^{-1} = 0$

$a^{-1}(1 - a) = -a$

$a^{-1} = \dfrac{a}{a-1}\ (a \neq 1)$

Hence, the inverse of every element exists.

E. Commutative property holds good:

$a * b = b * a$

$a * b = a + b - ab = b + a - ba = b * a$

Hence $Q_1 = Q - \{1\}$ is an infinite Abelian group under the binary operation defined.

Example 7.13: Let $R_0 = R - \{0\}$, be the set of non-zero real numbers, and a binary operation is defined by $a * b = \dfrac{1}{3}ab$. Show that $(R_0, *)$ is an Abelian group.

A. Closure Property:

$\forall\, a, b \in R_0,\ a * b = \dfrac{1}{3}ab \in R_0.$

Hence, the closure property holds good.

B. Associative property:

$\forall\, a, b, c \in R_0 : \quad a * (b * c) = (a * b) * c$

(i) $a * (b * c) = a * \left(\dfrac{bc}{3}\right)$

$= a * x$ where $x = \dfrac{bc}{3}$

$= \dfrac{ax}{3} = \dfrac{abc}{9}$ (I)

(ii) $(a * b) * c = \left(\dfrac{ab}{3}\right) * c$

$= y * c\ \ y = \dfrac{ab}{3}$

$= \dfrac{yc}{3} = \dfrac{abc}{9}$ (II)

Hence, from (A) and (B), the associative property is satisfied.

C. Identity element:

$\forall\, a \in R_0,\ \exists\, e \in R_0$, such that $a * e = e * a = a$

$\frac{ae}{3} = a \quad e = 3$ is the identity element.

D. Inverse element:

$\forall\, a \in R_0,\ \exists\, a^{-1} \in R_0,\ a * a^{-1} = a^{-1} * a = e$

$\frac{aa^{-1}}{3} = 3$ or $a^{-1} = \frac{9}{a}$ is the inverse of a in R_0 under the operation *.

E. The commutative property holds good. $a * b = b * a$

$$\frac{ab}{3} = \frac{ba}{3}$$

Hence, R_0 is an Abelian group under the defined operation.

Example 7.14: Show that $G = \{----, 2^{-2}, 2^{-1}, 2^0, 2^1, 2^2, ----\}$ is an infinite Abelian group under multiplication.

Proof: Let $2^r, 2^s, 2^t \in G$, where r, s, t are integers.

A. Closure property:

$2^r 2^s = 2^{r+s} \in G$, where $r + s$ is again an integer.

$\therefore$ Closure property is satisfied.

B. Associative property:

$\forall\ 2^r, 2^s, 2^t \in G$

$2^r(2^s 2^t) = 2^r(2^{s+t}) = 2^{r+(s+t)}$

$(2^r 2^s)2^t = 2^{r+s} 2^t = 2^{(r+s)+t}$

as $r + (s + t) = (r + s) + t$

The associative law is satisfied.

C. Identity axiom:

An identity element exists. $2^0 = 1$ is the identity element.

D. Inverse axiom:

Inverse of 2^r is 2^{-r} i.e $2^r 2^{-r} = 2^0 = 1$

And $2^{-r} \in G$. Hence, an inverse of every element exists.

E. Commutative property holds good.

$2^r 2^s = 2^{r+s} = 2^{s+r} = 2^s 2^r$.

G is an infinite set. So (G, .) is an infinite Abelian group.

It is to be noted that if a is any non-zero integer, then it can be proved that $G = \{a^n / n \in I\} = \{---, a^{-2}, a^{-1}, a^0, a^1, a^2, ---\}$ is an infinite Abelian group under multiplication.

Example 7.15: Prove that the set of complex numbers is an Abelian group under the operation of addition of complex numbers.

Proof: Let C be the set of complex numbers. Then $C = \{(x+iy)/x, y \in R$ and $i = \sqrt{-1}\}$. Let α, β, γ be any three elements of C. Then take $\alpha = a + ib$, $\beta = c + id$, $\gamma = e + if$. where $a, b, c, d, e, f \in R$.

1. Closure Property:
   Now $\alpha + \beta = (a + ib) + (c + id)$
   $= (a + c) + i\,(b + d)$
   $= a^1 + i\,b^1$ where $a^1_i = a + c \in R$, $b^1 = b + d \in R$
   $\alpha + \beta \in C$. So the closure axiom is satisfied.

2. Associative Property:
   $(\alpha + \beta) + \gamma = \alpha + (\beta + \gamma)$

   (i) $(\alpha + \beta) + \gamma = [(a + ib + c + id\,] + e + if$
   $= [(a + c) + i(b + d)] + e + if$
   $= (a + c) + e + i\,[(b + d) + f\,]$ (I)

   (ii) $\alpha + (\beta + \gamma) = (\,a + ib) + [(c + id) + (e + if\,)]$
   $= (a + ib) + [(c + e\,) + i\,(d + f\,)]$
   $= a + (c + e) + i[b + (d + f\,)]$ (II)

   From (I) and (II), it is clear that associative property holds good in R. Hence, $(\alpha + \beta) + \gamma = \alpha + (\beta + \gamma)$.
   Therefore, associativity property holds good for the set of complex numbers.

3. Identity axiom: C contains $0 + i.0$, which is such that
   $\alpha + (0 + i.0) = (0 + i.0) + \alpha$.
   Hence, $(0 + i.0) = 0$ is the identity element.

4. Inverse element: Now $-\alpha = -a - ib$ and is the additive inverse of $\alpha = a + ib$.
   And $\alpha + (-\alpha) = (-\alpha) + \alpha = 0$.
   Every element in C has an inverse. The inverse property is satisfied.

5. Commutative Property:
   Now, $\alpha + \beta = (a + ib) + (c + id)$
   $= (a + c) + i(b + d)$
   $= (c + id) + (a + ib)$
   $= \beta + \alpha$
   Commutative property is valid. Hence, the set of complex numbers is an Abelian group under the addition operation.

Example 7.16: Prove that G = {(cos θ + *i* sin θ)/θ is real} is an Abelian group under the operation multiplication.

G = {(cos θ + *i* sin θ)/θ is real}
Let *x*, *y*, *z* be three elements of G. Take $x = \cos\alpha + i\sin\alpha$, $y = \cos\beta + i\sin\beta$,
$z = \cos\gamma + i\sin\gamma$, where α, β, γ are real numbers.

1. Closure property:
   $\forall\, x, y \in G,\ x * y = xy \in G.$

   $xy = (\cos\alpha + \mathrm{i}\sin\alpha)(\cos\beta + \mathrm{i}\sin\beta)$
   $= [\cos(\alpha+\beta) + \mathrm{i}\sin(\alpha+\beta)]$

$= \left[\cos\alpha^1 + i\sin\alpha^1\right]$ where $\alpha^1 = \alpha + \beta$ – is real.

The closure property is satisfied.

2. Associative property:

$$\forall\, x, y, z \in G: \qquad x*(y*z) = (x*y)*z \text{ or } (x\,y)\,z = x\,(y\,z)$$

i) $(xy)\,z = [(\cos\alpha + i\sin\alpha)(\cos\beta + i\sin\beta)](\cos\gamma + i\sin\gamma)$
$= [(\cos(\alpha+\beta) + i\sin(\alpha+\beta)](\cos\gamma + i\sin\gamma)$
$= [\cos[(\alpha+\beta)+\gamma] + i\sin[\alpha+(\beta+\gamma)]]$
$= x\,(yz)$ ($\because$ – addition is associative in R)

Associativity property is satisfied.

3. Identity axiom:
Identity element exists. $1 = \cos 0^0 + i\sin 0^0 \in G$ is the identity element.
4. Inverse axiom: Every element has an inverse.
Inverse of $(\cos\alpha + i\sin\alpha)$ is $(\cos\alpha - i\sin\alpha)$ such that
$(\cos\alpha + i\sin\alpha)(\cos\alpha - i\sin\alpha) = \cos^2\alpha + \sin^2\alpha = 1$.
$(\cos\alpha - i\sin\alpha)$ is the multiplicative inverse of $(\cos\alpha + i\sin\alpha)$ and vice versa.
Hence inverse axiom is satisfied.
5. Commutative property:
Now $xy = (\cos\ \alpha + i\sin\ \alpha)(\cos\ \beta + i\sin\beta)$
$= [\cos(\alpha+\beta) + i\sin(\alpha+\beta)]$
$= [\cos(\beta+\alpha) + i\sin(\beta+\alpha)]$
$= yx$.
The commutative property holds good. Hence, G is an Abelian group under multiplication.

Example 7.17: Show that the four matrices $\begin{pmatrix}1 & 0\\ 0 & 1\end{pmatrix}$, $\begin{pmatrix}-1 & 0\\ 0 & 1\end{pmatrix}$, $\begin{pmatrix}1 & 0\\ 0 & -1\end{pmatrix}$ and $\begin{pmatrix}-1 & 0\\ 0 & -1\end{pmatrix}$ form a group under matrix multiplication.

Proof: Let $I = \begin{pmatrix}1 & 0\\ 0 & 1\end{pmatrix}$; $A = \begin{pmatrix}-1 & 0\\ 0 & 1\end{pmatrix}$; $B = \begin{pmatrix}1 & 0\\ 0 & -1\end{pmatrix}$; $C = \begin{pmatrix}-1 & 0\\ 0 & -1\end{pmatrix}$

Then G = {I, A, B, C}
Now, IA = AI = A, IB = BI = B, IC = CI = C.

The multiplication $AB = \begin{pmatrix}-1 & 0\\ 0 & 1\end{pmatrix}\begin{pmatrix}1 & 0\\ 0 & -1\end{pmatrix} = \begin{pmatrix}-1+0 & 0+0\\ 0+0 & 0-1\end{pmatrix} = \begin{pmatrix}-1 & 0\\ 0 & -1\end{pmatrix} = C$

Similarly it can be shown that BA = C, AC = CA = B, BC = CB = A,
AA = I, BB = I, CC = I.
The composition operation is given in Table 7.1.

Table 7.1 Composition operation

.	I	A	B	C
I	I	A	B	C
A	A	I	C	B
B	B	C	I	A
C	C	B	A	I

1. The entries in the table are the same as the elements of set G. The closure property is satisfied.
2. Associative axiom.
   A(BC) = A(A) = I
   (AB)C = C(C) = I
   ∴ Associative property is satisfied.
3. Identity axiom: I is the identity element.
4. Inverse axiom:
   Every element of G has an inverse.
   Inverses of I, A, B, C are I, A, B, C respectively.
   ∴ G is a group under multiplication.

7.4 SQUARE ROOTS, CUBE ROOTS AND FOURTH ROOTS OF UNITY

1. The square roots of unity are the roots of the equation $x^2 - 1 = 0$ i.,e $x = \pm 1$
   $$G = \{+1, -1\}$$
2. The cube roots of unity are the roots of the equation $x^3 - 1 = 0$ or $(x-1)(x^2+x+1) = 0$
   $x - 1 = 0$ or $x = 1$.
   $$x^2 + x + 1 = 0 \text{ or } x = \frac{-1 \pm \sqrt{1-4}}{2} = \frac{-1 \pm i\sqrt{3}}{2}$$
   Thus the cube roots of unity are 1, $\frac{-1+i\sqrt{3}}{2}, \frac{-1-i\sqrt{3}}{2}$
   The cube roots of unity are also written as 1, ω, ω^2, where $\omega = \frac{-1+i\sqrt{3}}{2}$
   Note: (a) $1 + \omega + \omega^2 = 0$. (b) $\omega^3 = 1$. (c) $\omega^4 = \omega^3\omega = \omega$. (d) $\omega^5 = \omega^3\omega^2 = \omega$.
3. The fourth roots of unity are the roots of the equation
   $x^4 - 1 = 0$ or $(x^2-1)(x^2+1) = 0$
   $(x^2 - 1) = 0, x = \pm 1,$
   $(x^2 + 1) = 0, x = \pm i$
   Thus $G = \{1, -1, i, -i\}$.
   Note: (a) $i^2 = -1$ (b) $i^3 = i^2 i = -i$ (c) $i^4 = i^2 . i^2 = (-1)(-1) = 1$
   (d) $i^5 = i^3 . i^2 = (-i)(-1) = i$.

Example 7.18: Show that square roots of unity forms a group under multiplication.

G = {1, −1}. The operation is exemplified by Table 7.2.

Table 7.2 Operation description

.	1	−1
1	1	−1
−1	−1	1

It can be easily verified that all the axioms are satisfied.

∴ (G,*) is a group.

Example 7.19: Show that cube roots of unity form an Abelian group under multiplication.

$$G=\left\{1, \omega, \omega^2\right\}$$

.	1	ω	ω^2
1	1	ω	ω^2
ω	ω	ω^2	1
ω^2	ω^2	1	ω

∴ (a) $\omega^3 = 1$. $(b)\,\omega^4 = \omega^3\omega = \omega$.

1. All the entries in the table are the same as the elements of the set. This means that closure property is satisfied.
2. Associative axiom:

   Consider $1\,.(\omega.\omega^2) = \;1.1 = 1.$

   $(1.\omega).\omega^2 = \omega.\omega^2 = \omega^3 = 1$

   Associative property is satisfied.
3. An identity element exists. 1 is the identity element.
4. Inverse element: Every element has an inverse. The inverses of 1, ω, ω^2 are respectively 1, ω^2, ω.
5. Commutative property holds good, because the table is symmetrical about the principal diagonal.

   Hence cube roots of unity forms an Abelian group under multiplication.

Example 7.20: The fourth roots of unity form an Abelian group with respect to the multiplication operation.

G = {1, −1, i, −i}. The composition operation with respect to usual multiplication is as tabulated in Table 7.3.

Table 7.3 Composition operation

.	1	−1	i	−i
1	1	−1	i	−i
−1	−1	1	−i	I
i	I	−i	−1	1
−i	−i	i	1	−1

The entries in the table are elements of set G. Hence, closure property is satisfied.

1. Associative axiom.

   Consider $(-1)[(i)(-i)] = (-1)\,(1) = -1$

   $[(-1)(i)](-i) = (-i)\,(-i) = -1$

   Associative property holds good.

2. Identity element exists and is equal to 1, i.e., $e = 1$.

3. Inverse axiom. For every element there is an inverse element. The inverses of 1, –1, i, and $-i$, are respectively 1, –1, $-i$, i.

4. Commutative property holds good, since the elements in the table are symmetrical with respect to the principal diagonal.

   Hence (G,*) is a finite Abelian group of order 4.

Example 7.21: Prove that a group G is Abelian if and only if $(ab)^2 = a^2b^2 \;\; \forall a, b \in G$.

Proof: First let us consider that G is Abelian.

$(ab)^2 = (ab)(ab)$

$= a(ba)b$ by associative law

$= a(ab)b$ G is Abelian

$= (aa)(bb)$ by associative law

$= a^2b^2$

Conversely, suppose that

$(ab)^2 = a^2b^2 \; \forall a, b \in G.$

$(ab)(ab) = (aa)(bb)$

$a(ba)b = a(ab)b$

Now, the left and right cancellation law holds well.

$\therefore\; ba = ab$. G is Abelian.

Example 7.22: Prove that a group G is Abelian if and only if $(ab)^{-1} = a^{-1}b^{-1} \;\; \forall a, b \in G$.

Proof: Suppose G is Abelian.

Then, for any $a, b \in G$, $a^{-1}b^{-1} = b^{-1}a^{-1}$

$= (ab)^{-1} \quad (\because b^{-1}a^{-1} = (ab)^{-1})$

Conversely, suppose

$(ab)^{-1} = a^{-1}b^{-1} \; \forall a, b \in G$

Then $\forall\ x, y \in G$

$$xy = \left(x^{-1}\right)^{-1}\left(y^{-1}\right)^{-1} \quad (\because x=(x^{-1})^{-1}, y=(y^{-1})^{-1})$$

$$=\left(x^{-1}y^{-1}\right)^{-1} \quad (a^{-1}b^{-1}=(ab)^{-1})$$

$$=\left(y^{-1}\right)^{-1}\left(x^{-1}\right)^{-1} \quad (\because (ab)^{-1}=b^{-1}a^{-1})$$

$$=yx.$$

Therefore, G is Abelian.

Example 7.23: In a group G having more than one element, if $x^2 = x$ for every $x \in G$. Prove that G is Abelian.

Proof: Take any $a, b \in G$, then from the hypothesis, $a^2 = a$, $b^2 = b$.

$(ab)^2 = ab$.

$\therefore\ a(ab)b = (aa)(bb) = a^2 b^2 = ab = (ab)^2 = (ab)(ab) = a(ba)b$

or $a(ab)b = a(ba)b$. The left and right cancellation property holds good.

$ab = ba$.

$\therefore$ G is Abelian.

Example 7.24: Show that a group G in which every element is its own inverse, is Abelian.

Proof: Take any $a, b \in G$. From the given hypothesis,

$a = a^{-1}, b = b^{-1}, (ab) = (ab)^{-1}$.

But $(ab)^{-1} = b^{-1}a^{-1}$ then,

$(ab) = (ab)^{-1} = b^{-1}a^{-1} = ba$

Hence, G is Abelian.

7.5 PROPERTIES OF GROUPS

This section describes the unique property of the identity element, the inverse element, etc. in a group.

Example 7.25: Show that in a group (G,*), the identity element is unique.

Proof: Let us assume that there are two identity elements e and e^1 in a group (G,*).

Since e is an identity element, $e^1 * e = e * e^1 = e^1$ (I)

Since e^1 is also an identity element, $e * e^1 = e^1 * e = e$ (II)

From (I) and (II) it follows that $e = e^1$.

$\therefore$ The identity element in a group is unique.

Example 7.26: Show that in a group (G,*), inverses of each element are unique.

Proof: Let e be the identity element in (G,*) and let a be any element of the group.
If possible, let b and c be inverses of a in G.
Since b is an inverse of a, $a*b=b*a=e$ (I)
Since c is also an inverse of a, $a*c=c*a=e$ (II)

Now $b=b*e$ (by identity axiom)

$=b*(a*c)$ (using property II)

$=(b*a)*c$ (by associative axiom)

$=e*c$ (using property I)

$=c$

$b=c$, so the inverse of a is unique.

Example 7.27: Show that if a is any element of a group (G,*), then $(a^{-1})^{-1}=a$.

Proof: Let $a^{-1}=x$.

Consider $x*a=a^{-1}*a=e.$ (by inverse axiom)

$a*x=a*a^{-1}=e.$ (by inverse axiom)

This means that the inverse of x is a.

i.e, $x^{-1}=a$ i.e., $(a^{-1})^{-1}=a$.

Example 7.28: Show that if a and b are any two elements of a group (G,*), then $(a*b)^{-1}=b^{-1}*a^{-1}$

Proof: Since $a, b\in$ G, $\rightarrow b^{-1}*a^{-1}\in$ G.

Let us consider that,

$(a*b)*(b^{-1}*a^{-1})=a*[b*(b^{-1}*a^{-1})]$ by associative law.

$=a*[(b*b^{-1})*a^{-1})]$ by associative law.

$=a*(e*a^{-1})$ $(\because b*b^{-1}=e)$

$=a*a^{-1}$

$=e$ (I) $(\because a*a^{-1}=e)$

Similarly, we consider,

$(b^{-1}*a^{-1})*(a*b)=b^{-1}*[a^{-1}*(a*b)]$ by associative axior

$=b^{-1}*[(a^{-1}*a)*b)]$ by associative axiom

$=b^{-1}*(e*b)$ by inverse axiom

$=b^{-1}*b$ by identity axiom

$=e$ (II) by inverse axiom.

From (I) and (II) we have,

$(a*b)*(b^{-1}*a^{-1})=(b^{-1}*a^{-1})*(a*b)=e$

This means that inverse of $(a*b)$ is $b^{-1}*a^{-1}$.

$(a*b)^{-1}= b^{-1}*a^{-1}$. This is known as inverse reversal law.

It is necessary to note the following.

(i) If $a, b, c \in G$, then $(a*b*c)^{-1}=c^{-1}*b^{-1}*a^{-1}$

(ii) If (G,*) is an Abelian group then $a*b=b*a$

$$\forall a,b \in G, \text{ so } b^{-1}*a^{-1}=a^{-1}*b^{-1} \text{ and } (a*b)^{-1}=a^{-1}*b^{-1}.$$

7.6 CANCELLATION LAWS

Left and right cancellation laws hold in a group. If *a*, *b* and *c* are elements of a group (G,*), then

(i) $a*b=a*c \rightarrow b=c$ (Left cancellation law)

(ii) $a*b=c*b \rightarrow a=c$ (Right cancellation law)

Proof:

(i) $a*b=a*c$ given

Operate on left on both sides by a^{-1}.

$a^{-1}*(a*b)= a^{-1}*(a*c)$

$(a^{-1}*a)*b=(a^{-1}*a)*c$ by associative axiom

$e*b=e*c$ by inverse axiom

$b=c$ by identity axiom

$a*b=a*c$ implies $b=c$

(ii) $a*b=c*b$ given

Operate on right on both sides by b^{-1}

$(a*b)*b^{-1}=(c*b)*b^{-1}$

$a*(b*b^{-1})=c*(b*b^{-1})$ by associative axiom

$a*e=c*e$ by inverse axiom

$a=c$ by identity axiom

Example 7.29: Let G be a group and $a, b \in G$. Then, the equations $a*x=b$ and $y*a=b$ have unique solutions in G.

(i) $a*x=b$ since $a, b \in G, a^{-1}*b \in G$

Then $a*(a^{-1}*b)=(a*a^{-1})*b$ by associativity

$=b$ ($\because - e$ is the identity)

$\because x=a^{-1}*b \in G$ is the solution of $a*x=b$

To prove uniqueness, let x_1 and x_2 be two solutions of

$a*x=b$ $\because$ $a*x_1=b$, $a*x_2=b$

$\because$ $a*x_1=a*x_2$, By the left cancellation law, $x_1=x_2$

$\because a*x=b$ has unique solution $x=a^{-1}*b \in G$.

(ii) $y*a=b$ since $a, b \in G, b*a^{-1} \in G$

Then $(b*a^{-1})*a=b*(a^{-1}*a)$ by associativity

$=b*e$ ($\because a^{-1}*a=e$)

$=b$ ($\because$ e is the identity)

$\because$ $y*a=b$ has the solution $y=b*a^{-1} \in G$.

To prove uniqueness, let y_1 and y_2 be two solutions of the equation

$y*a=b$ $\because$ $y_1*a=b$, $y_2*a=b$

$y_1*a=y_2*a$ $\therefore y_1=y_2$ by right cancellation law

$\therefore y=b*a^{-1}$ is the unique solution for $y*a=b$

7.7 SEMI-GROUPS AND MONOIDS

Let S- be a non-empty set and $*$ be a binary operation on S, then the algebraic system $(S,*)$ is called a semi-group if the operation $*$ is associative. The algebraic system is called semi-group.

$$\forall a, b, c \in S, \quad (a*b)*c = a*(b*c)$$

It is to be noted that since closure property is a characteristic property of a binary operation on a set S, it is not necessary to mention it explicitly when an algebraic system is defined.

7.7.1 Definition of a Monoid

If a semi-group $(M,*)$ has an identity element with respect to the operation$*$, then $(M,*)$ is called a monoid.

$$\forall\ a, b, c \in M, \text{ the associative property holds good } (a*b)*c = a*(b*c)$$

$$\forall\, e \in M, \text{ such that } \forall\ a \in M,\ a*e=e*a=a$$

Then the algebraic system $(M,*)$ is called a monoid.

Consider the set of positive even numbers to be a semi-group with respect to the binary operations addition and multiplication.

Let S be the set of even numbers. Since the sum of any two even numbers is even and the product of any two even numbers is even, we can say

$\forall a, b \in S,\ a*b = a+b \in S\ ;\ a*b = ab \in S$

$\forall a, b, c \in S: \quad a+(b+c) = (a+b)+c\ ;\ a(bc) = (ab)c$

$\therefore$ (S, +) and (S, ×) are semigroups.

Example 7.30: The set of negative integers is not a semi-group. Hence, the operation – is closed in Z. But the operation is not associative.

Consider $a = 2$, $b = 5$, and $c = 6$

$(2 - 5) - 6 = -3 - 6 = -9$

$2 - (5 - 6) = 2 - (-1) = 3$, which is not an element of Z.

Example 7.31: The set of natural numbers N = {1, 2, 3,...} is a semi-group under the operation addition, but not a monoid, since the identity element does not exist i.e., $0 \notin N$. $\therefore$(N, +) is a semi-group not a monoid

7.8 PERMUTATIONS

Let S be a finite set of n-elements. $S = \{x_1, x_2, \ldots x_n\}$. A bijection function ϕ S onto S itself is defined as a permutation of S. We can write ϕ as

$\Phi = \begin{pmatrix} x_1 & x_2 & - & x_n \\ i_1 & i_2 & - & i_n \end{pmatrix}$ where $i_1, i_2, \ldots i_n$ are images of $x_1, x_2, \ldots x_n$.

Example 7.32: Let S = {1, 2, 3}. The possible permutations are 3! = 6.

$$S_3 = \{p_1, p_2, p_3, p_4, p_5, p_6.\}$$

$$p_1 = \begin{pmatrix} 1 & 2 & 3 \\ 1 & 2 & 3 \end{pmatrix} \quad p_2 = \begin{pmatrix} 1 & 2 & 3 \\ 1 & 3 & 2 \end{pmatrix} \quad p_3 = \begin{pmatrix} 1 & 2 & 3 \\ 2 & 3 & 1 \end{pmatrix}$$

$$p_4 = \begin{pmatrix} 1 & 2 & 3 \\ 2 & 1 & 3 \end{pmatrix} \quad p_5 = \begin{pmatrix} 1 & 2 & 3 \\ 3 & 1 & 2 \end{pmatrix} \quad p_6 = \begin{pmatrix} 1 & 2 & 3 \\ 3 & 2 & 1 \end{pmatrix}$$

7.8.1 Product or Composition of Two Permutations

The product or composition of two permutations P_2 and P_3 is denoted by $P_2 \circ P_3$ or simply (P_2P_3), and is obtained by first carrying out the operation defined by P_2 and then by P_3.

Consider $P_2 = \begin{pmatrix} 1 & 2 & 3 \\ 1 & 3 & 2 \end{pmatrix}$ and $P_3 = \begin{pmatrix} 1 & 2 & 3 \\ 2 & 3 & 1 \end{pmatrix}$

$$P_2 \circ P_3 = \begin{pmatrix} 1 & 2 & 3 \\ 1 & 3 & 2 \end{pmatrix}\begin{pmatrix} 1 & 2 & 3 \\ 2 & 3 & 1 \end{pmatrix}$$

$P_2(1) = 1, P_2(2) = 3, P_2(3) = 2$

$P_3(1) = 2, P_3(2) = 3, P_3(3) = 1$

$(P_2 \circ P_3)(1) = P_3[P_2(1)] = P_3[1] = 2$

$(P_2 \circ P_3)(2) = P_3[P_2(2)] = P_3[3] = 1$

$(P_2 \circ P_3)(3) = P_3[P_2(3)] = P_3[3] = 3$

$P_2 \circ P_3 = \begin{pmatrix} 1 & 2 & 3 \\ 2 & 1 & 3 \end{pmatrix}$. Alternatively it can also be computed as follows.

$$P_2 \circ P_3 = \begin{pmatrix} 1 & 2 & 3 \\ 1 & 3 & 2 \end{pmatrix}\begin{pmatrix} 1 & 2 & 3 \\ 2 & 3 & 1 \end{pmatrix} = \begin{pmatrix} 1 & 2 & 3 \\ 1 & 3 & 2 \end{pmatrix}\begin{pmatrix} 1 & 3 & 2 \\ 2 & 1 & 3 \end{pmatrix}$$

$$= \begin{pmatrix} 1 & 2 & 3 \\ 2 & 1 & 3 \end{pmatrix} = P_4.$$

The first row of P_3 is written so as to coincide with the second row of P_2. This is shown as follows:

$$\begin{array}{ccc} 1 & 2 & 3 \\ \downarrow & \downarrow & \downarrow \\ 2 & 3 & 1 \\ \downarrow & \downarrow & \downarrow \\ 2 & 1 & 3 \end{array}$$

$$\text{Now } P_3 \circ P_2 = \begin{pmatrix} 1 & 2 & 3 \\ 2 & 3 & 1 \end{pmatrix}\begin{pmatrix} 1 & 2 & 3 \\ 1 & 3 & 2 \end{pmatrix} = \begin{pmatrix} 1 & 2 & 3 \\ 2 & 3 & 1 \end{pmatrix}\begin{pmatrix} 2 & 3 & 1 \\ 3 & 2 & 1 \end{pmatrix} = \begin{pmatrix} 1 & 2 & 3 \\ 3 & 2 & 1 \end{pmatrix} = p_6$$

$\therefore$ $P_2 \circ P_3 \neq P_3 \circ P_2$. This is due to the fact that composition of mappings need not be commutative.

7.8.2 Identity Permutation

$$\text{Now } P_1 \circ P_2 = \begin{pmatrix} 1 & 2 & 3 \\ 1 & 2 & 3 \end{pmatrix}\begin{pmatrix} 1 & 2 & 3 \\ 1 & 3 & 2 \end{pmatrix} = \begin{pmatrix} 1 & 2 & 3 \\ 1 & 3 & 2 \end{pmatrix} = P_2$$

$P_1 = \begin{pmatrix} 1 & 2 & 3 \\ 1 & 2 & 3 \end{pmatrix}$ is the identity permutation.

7.8.3 Inverse of a Permutation

Let $P_2 = \begin{pmatrix} 1 & 2 & 3 \\ 1 & 3 & 2 \end{pmatrix}$ be a permutation. Then the inverse of P_2 is defined by

$$P_2^{-1} = \begin{pmatrix} 1 & 3 & 2 \\ 1 & 2 & 3 \end{pmatrix} = \begin{pmatrix} 1 & 2 & 3 \\ 1 & 3 & 2 \end{pmatrix} = P_2$$

$$P_2 \circ P_2^{-1} = \begin{pmatrix} 1 & 2 & 3 \\ 1 & 3 & 2 \end{pmatrix} \circ \begin{pmatrix} 1 & 2 & 3 \\ 1 & 3 & 2 \end{pmatrix} = \begin{pmatrix} 1 & 2 & 3 \\ 1 & 2 & 3 \end{pmatrix} = P_1 = \text{The identity permutation.}$$

It is important to note that the inverse of a permutation is obtained by interchanging the first row and the second row.

7.9 SYMMETRIC GROUP

The set S_3 of all permutations of a finite set containing three elements forms a non-Abelian group under composition of permutations.

$$S_3 = \{p_1, p_2, p_3, p_4, p_5, p_6\}$$

Cayley's composition table of permutations on S_3 is given in Table 7.4.

Table 7.4 Composition of permutations

*	P_1	P_2	P_3	P_4	P_5	P_6
P_1	P_1	P_2	P_3	P_4	P_5	P_6
P_2	P_2	P_1	P_4	P_3	P_6	P_5
P_3	P_3	P_6	P_5	P_2	P_1	P_4
P_4	P_4	P_5	P_6	P_1	P_2	P_3
P_5	P_5	P_4	P_1	P_6	P_3	P_2
P_6	P_6	P_3	P_2	P_5	P_4	P_1

1. Closure property: All the elements in the table are elements of S_3. Closure property is satisfied. That is,

$$\forall\, a, b \in S_3,\ \ a*b \in S_3$$

2. Associativity property holds good.

$\forall\, a, b, c \in S_3\ :\ \ a*(b*c)=(a*b)*c$

For example, $P_2 \circ (P_3 \circ P_4) = P_2 \circ P_2 = P_1$

$(P_2 \circ P_3) \circ P_4 = P_4 \circ P_4 = P_1$

1. Existence of Identity: $p_1 = \begin{pmatrix} 1 & 2 & 3 \\ 1 & 2 & 3 \end{pmatrix}$ is the identity permutation.

2. Existence of Inverse: Every permutation has an inverse.

$$p_1^{-1} = p_1,\ p_2^{-1} = p_2,\ p_3^{-1} = p_5,\ p_4^{-1} = p_4,\ p_5^{-1} = p_3 \text{ and } p_6^{-1} = p_6$$

Hence $(S_3, *)$ is a group. However the symmetric group is not Abelian.

Example 7.33: The symmetric group S_4 consists of all permutations of the set $S = \{1, 2, 3, 4\}$. If

$$f = \begin{pmatrix} 1 & 2 & 2 & 4 \\ 2 & 4 & 3 & 1 \end{pmatrix},\ g = \begin{pmatrix} 1 & 2 & 3 & 4 \\ 4 & 2 & 1 & 3 \end{pmatrix}, \text{ verify that } (fg)^{-1} = g^{-1}f^{-1}.$$

Proof: S_4 are the permutations of the set $S = \{1, 2, 3, 4\}$.
The order of S_4 is $4! = 24$.

$$\text{Now } (fg) = \begin{pmatrix} 1 & 2 & 2 & 4 \\ 2 & 4 & 3 & 1 \end{pmatrix} \begin{pmatrix} 1 & 2 & 3 & 4 \\ 4 & 2 & 1 & 3 \end{pmatrix}$$

$$= \begin{pmatrix} 1 & 2 & 3 & 4 \\ 2 & 4 & 3 & 1 \end{pmatrix} \begin{pmatrix} 2 & 4 & 3 & 1 \\ 2 & 3 & 1 & 4 \end{pmatrix} = \begin{pmatrix} 1 & 2 & 3 & 4 \\ 2 & 3 & 1 & 4 \end{pmatrix}$$

$$(fg)^{-1}=\begin{pmatrix}2 & 3 & 1 & 4\\ 1 & 2 & 3 & 4\end{pmatrix}=\begin{pmatrix}1 & 2 & 2 & 4\\ 3 & 1 & 2 & 4\end{pmatrix} \qquad \text{(I)}$$

$$g^{-1}=\begin{pmatrix}4 & 2 & 1 & 3\\ 1 & 2 & 3 & 4\end{pmatrix}=\begin{pmatrix}1 & 2 & 3 & 4\\ 3 & 2 & 4 & 1\end{pmatrix}$$

$$f^{-1}=\begin{pmatrix}2 & 4 & 3 & 1\\ 1 & 2 & 3 & 4\end{pmatrix}=\begin{pmatrix}1 & 2 & 3 & 4\\ 4 & 1 & 3 & 2\end{pmatrix}$$

$$g^{-1}f^{-1}=\begin{pmatrix}1 & 2 & 3 & 4\\ 3 & 2 & 4 & 1\end{pmatrix}\begin{pmatrix}1 & 2 & 3 & 4\\ 4 & 1 & 3 & 2\end{pmatrix}$$

$$=\begin{pmatrix}1 & 2 & 3 & 4\\ 3 & 2 & 4 & 1\end{pmatrix}\begin{pmatrix}3 & 2 & 4 & 1\\ 3 & 1 & 2 & 4\end{pmatrix}=\begin{pmatrix}1 & 2 & 3 & 4\\ 3 & 1 & 2 & 4\end{pmatrix} \qquad \text{(II)}$$

From (I) and (II), $(fg)^{-1}=g^{-1}f^{-1}$.

Example 7.34: If the permutations of the elements of S = {1, 2, 3, 4, 5} are given by

$$\alpha=\begin{pmatrix}1 & 2 & 3 & 4 & 5\\ 2 & 3 & 1 & 4 & 5\end{pmatrix} \quad \beta=\begin{pmatrix}1 & 2 & 3 & 4 & 5\\ 1 & 2 & 3 & 5 & 4\end{pmatrix}$$

$$\gamma=\begin{pmatrix}1 & 2 & 3 & 4 & 5\\ 5 & 4 & 3 & 1 & 2\end{pmatrix} \quad \delta=\begin{pmatrix}1 & 2 & 3 & 4 & 5\\ 3 & 2 & 1 & 5 & 4\end{pmatrix}$$

Find (i) $\alpha\beta$ (ii) $\beta\alpha$ (iii) α^2 (iv) $\gamma\delta$ (v) δ^{-1} (vi) $\alpha\beta\gamma$

Also, solve the equation $\alpha x = \beta$.

(i) $$\alpha\beta=\begin{pmatrix}1 & 2 & 3 & 4 & 5\\ 2 & 3 & 1 & 4 & 5\end{pmatrix}\begin{pmatrix}1 & 2 & 3 & 4 & 5\\ 1 & 2 & 3 & 5 & 4\end{pmatrix}$$

$$=\begin{pmatrix}1 & 2 & 3 & 4 & 5\\ 2 & 3 & 1 & 4 & 5\end{pmatrix}\begin{pmatrix}2 & 3 & 1 & 4 & 5\\ 2 & 3 & 1 & 5 & 4\end{pmatrix}=\begin{pmatrix}1 & 2 & 3 & 4 & 5\\ 2 & 3 & 1 & 5 & 4\end{pmatrix}$$

(ii) $$\beta\alpha=\begin{pmatrix}1 & 2 & 3 & 4 & 5\\ 1 & 2 & 3 & 5 & 4\end{pmatrix}\begin{pmatrix}1 & 2 & 3 & 4 & 5\\ 2 & 3 & 1 & 4 & 5\end{pmatrix}$$

$$=\begin{pmatrix}1 & 2 & 3 & 4 & 5\\ 1 & 2 & 3 & 5 & 4\end{pmatrix}\begin{pmatrix}1 & 2 & 3 & 5 & 4\\ 2 & 3 & 1 & 5 & 4\end{pmatrix}=\begin{pmatrix}1 & 2 & 3 & 4 & 5\\ 2 & 3 & 1 & 5 & 4\end{pmatrix}$$

(iii) $$\alpha^2=\begin{pmatrix}1 & 2 & 3 & 4 & 5\\ 2 & 3 & 1 & 4 & 5\end{pmatrix}\begin{pmatrix}1 & 2 & 3 & 4 & 5\\ 2 & 3 & 1 & 4 & 5\end{pmatrix}$$

$$=\begin{pmatrix}1 & 2 & 3 & 4 & 5\\ 2 & 3 & 1 & 4 & 5\end{pmatrix}\begin{pmatrix}2 & 3 & 1 & 4 & 5\\ 3 & 1 & 2 & 4 & 5\end{pmatrix}$$

$$= \begin{pmatrix} 1 & 2 & 3 & 4 & 5 \\ 3 & 1 & 2 & 4 & 5 \end{pmatrix}$$

(iv) $\gamma\delta = \begin{pmatrix} 1 & 2 & 3 & 4 & 5 \\ 5 & 4 & 3 & 1 & 2 \end{pmatrix} \begin{pmatrix} 1 & 2 & 3 & 4 & 5 \\ 3 & 2 & 1 & 5 & 4 \end{pmatrix}$

$$= \begin{pmatrix} 1 & 2 & 3 & 4 & 5 \\ 5 & 4 & 3 & 1 & 2 \end{pmatrix} \begin{pmatrix} 5 & 4 & 3 & 1 & 2 \\ 4 & 5 & 3 & 1 & 2 \end{pmatrix} = \begin{pmatrix} 1 & 2 & 3 & 4 & 5 \\ 4 & 5 & 1 & 3 & 2 \end{pmatrix}$$

$$\alpha^2 = \begin{pmatrix} 1 & 2 & 3 & 4 & 5 \\ 2 & 3 & 1 & 4 & 5 \end{pmatrix} \begin{pmatrix} 1 & 2 & 3 & 4 & 5 \\ 2 & 3 & 1 & 4 & 5 \end{pmatrix}$$

$$= \begin{pmatrix} 1 & 2 & 3 & 4 & 5 \\ 2 & 3 & 1 & 4 & 5 \end{pmatrix} \begin{pmatrix} 2 & 3 & 1 & 4 & 5 \\ 3 & 1 & 2 & 4 & 5 \end{pmatrix}$$

$$= \begin{pmatrix} 1 & 2 & 3 & 4 & 5 \\ 3 & 1 & 2 & 4 & 5 \end{pmatrix}$$

(vi) δ^{-1} is obtained by interchanging the two rows of δ and then rearranging the elements of the first row so as to assume the natural order.

$$\delta^{-1} = \begin{pmatrix} 3 & 2 & 1 & 5 & 4 \\ 1 & 2 & 3 & 4 & 5 \end{pmatrix} = \begin{pmatrix} 1 & 2 & 3 & 4 & 5 \\ 3 & 2 & 1 & 5 & 4 \end{pmatrix}$$

While rearranging the elements of first row, the correspondence between the elements of the two rows is maintained.

(vii) $\alpha\beta\gamma = \left[\begin{pmatrix} 1 & 2 & 3 & 4 & 5 \\ 3 & 2 & 1 & 5 & 4 \end{pmatrix} \begin{pmatrix} 1 & 2 & 3 & 4 & 5 \\ 1 & 2 & 3 & 5 & 4 \end{pmatrix}\right] \begin{pmatrix} 1 & 2 & 3 & 4 & 5 \\ 5 & 4 & 3 & 1 & 2 \end{pmatrix}$

$$= \begin{pmatrix} 1 & 2 & 3 & 4 & 5 \\ 2 & 3 & 1 & 5 & 4 \end{pmatrix} \begin{pmatrix} 1 & 2 & 3 & 4 & 5 \\ 5 & 4 & 3 & 1 & 2 \end{pmatrix}$$

$$= \begin{pmatrix} 1 & 2 & 3 & 4 & 5 \\ 2 & 3 & 1 & 5 & 4 \end{pmatrix} \begin{pmatrix} 2 & 3 & 1 & 5 & 4 \\ 4 & 3 & 5 & 2 & 1 \end{pmatrix}$$

$$= \begin{pmatrix} 1 & 2 & 3 & 4 & 5 \\ 4 & 3 & 5 & 2 & 1 \end{pmatrix}$$

(viii) To solve $\alpha x = \beta$, multiply by α^{-1} to the left on both sides.

$$\alpha^{-1}\alpha\, x = \alpha^{-1}\beta$$

$$e\, x = \alpha^{-1}\beta$$

$$x = \alpha^{-1}\beta \text{ and } \alpha^{-1} = \begin{pmatrix} 2 & 3 & 1 & 4 & 5 \\ 1 & 2 & 3 & 4 & 5 \end{pmatrix} = \begin{pmatrix} 1 & 2 & 3 & 4 & 5 \\ 3 & 1 & 2 & 4 & 5 \end{pmatrix}$$

$$x = \alpha^{-1}\beta = \begin{pmatrix} 1 & 2 & 3 & 4 & 5 \\ 3 & 1 & 2 & 4 & 5 \end{pmatrix} \begin{pmatrix} 1 & 2 & 3 & 4 & 5 \\ 1 & 2 & 3 & 5 & 4 \end{pmatrix}$$

$$= \begin{pmatrix} 1 & 2 & 3 & 4 & 5 \\ 3 & 1 & 2 & 4 & 5 \end{pmatrix} \begin{pmatrix} 3 & 1 & 2 & 4 & 5 \\ 3 & 1 & 2 & 5 & 4 \end{pmatrix}$$

$$= \begin{pmatrix} 1 & 2 & 3 & 4 & 5 \\ 3 & 1 & 2 & 5 & 4 \end{pmatrix}$$

7.10 GEOMETRICAL MEANINGS OF SIX PERMUTATIONS IN S = {1, 2, 3}

The geometrical meaning of six permutations, namely a dihedral group, is given as follows.

7.10.1 Dihedral Group

The set of transformations due to all rigid motions of a regular polygon of n-sides resulting in identical polygons but with different vertex names under the binary operation of right composition *, is a group and is called a dihedral group. Let us consider a three sided regular polygon i.e., an equilateral triangle, whose vertices are named as 1, 2, 3.

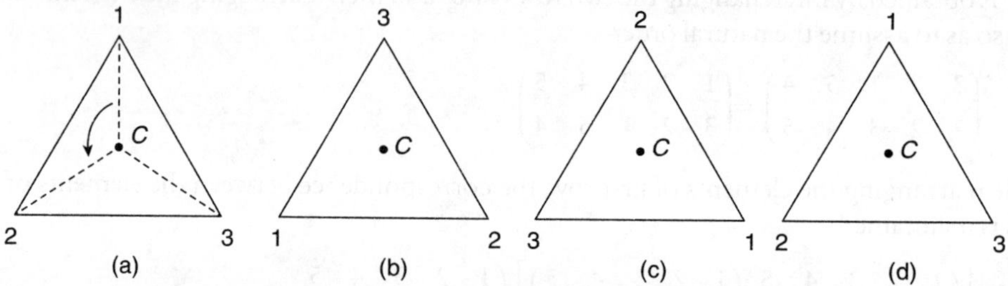

Figure 7.1 Rotations of a triangle

An equilateral triangle is made to rotate about its centroid through the angles $1 \times \frac{360}{3}$, $2 \times \frac{360}{3}$, $3 \times \frac{360}{3}$, **in the anticlockwise direction and a reflection of the triangle is made about its lines of** symmetry. (The process is known as rigid motion). The rotations are given in Fig. 7.1.

1. Rotating through $1 \times \frac{360}{3} = 120°$, in the anticlockwise direction about the centroid, we get the triangle 7.1(b). The vertices originally labelled as 1, 2, 3 have now become 3, 1, 2 respectively. The transformation is denoted as $r_5 = (p_5) = \begin{pmatrix} 1 & 2 & 3 \\ 3 & 1 & 2 \end{pmatrix}$
2. Rotation through $2 \times \frac{360}{3} = 240°$, we get the triangle 7.1(c). The transformation is denoted as $r_3 = (p_3) = \begin{pmatrix} 1 & 2 & 3 \\ 2 & 3 & 1 \end{pmatrix}$.
3. Rotation through $3 \times \frac{360}{3} = 120°$, we get the triangle 7.1(d). The transformation is denoted as $r_1 = (p_1) = \begin{pmatrix} 1 & 2 & 3 \\ 1 & 2 & 3 \end{pmatrix}$, which is an identity permutation.

In addition to these rotations, the triangle can be reflected along an axis passing through a vertex and the midpoint of the opposite side. Let us consider the reflections of the equilateral triangle about its lines of symmetry, namely 1A, 2B and 3C. Now the triangle is reflected about the line 1A, the vertex 1 remains in the original position and the other two vertices 2 and 3 interchange their positions and hence results in the triangle (b). Similarly, the reflections of original triangle about the lines 2B and 3C result in the triangles (c) and (d) respectively. The transformations are given in Fig. 7.2.

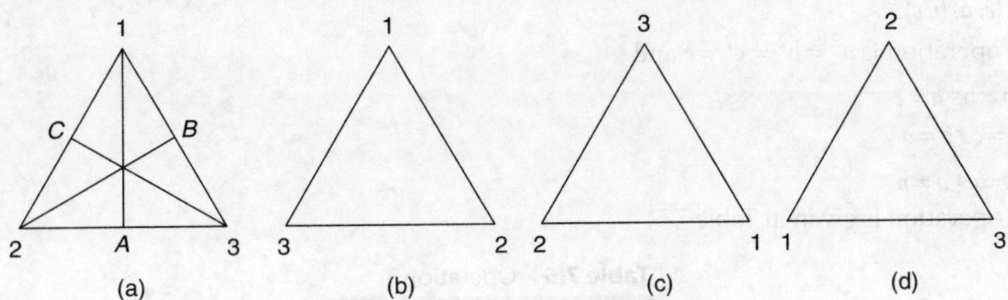

Figure 7.2 Rotations of triangle

$$r_2 = (p_2) = \begin{pmatrix} 1 & 2 & 3 \\ 1 & 3 & 2 \end{pmatrix} \quad r_6 = (p_6) = \begin{pmatrix} 1 & 2 & 3 \\ 3 & 2 & 1 \end{pmatrix} \quad r_4 = (p_4) = \begin{pmatrix} 1 & 2 & 3 \\ 2 & 1 & 3 \end{pmatrix}$$

Now the set $D_3 = \{r_1, r_2, r_3, r_4, r_5, r_6\}$ is the same as permutation set

$S_3 = \{p_1, p_2, p_3, p_4, p_5, p_6\}$. Hence, the set $D_3 = \{r_1, r_2, r_3, r_4, r_5, r_6\}$is a group with respect to the composition of the function and is called a dihedral group $(D_3, *)$. The order of the group is 6, and degree 3.

7.11 CYCLIC GROUPS

A group (G, *) is said to cyclic group if there exists an element $a \in G$, such that every element of G, can be expressed as some integral power of *a*. Every element *x* of G can be expressed as $x = a^n$ for some integer *n*. The cyclic group is said to be generated by *a*, or *a* is a generator of G, and is denoted by $G = \{a\}$.

Example 7.35: Let $G = \{1, i, -1, -i\}$, then (G, *) is a cyclic group. The generator is *i*.

$$i^4 = 1 \ , i^3 = -i \ , i^2 = -1, \ i^1 = i$$

$\therefore$ i - is a generator and $G = \{i\}$.
Similarly, $-i$, is also a generator, and $G = \{-i\}$.

Example 7.36: Let $G = \left\{1, \omega, \omega^2\right\}$ be the cube roots of unity. Then, (G, *) is a cyclic group generated by ω. ω^2 is also a generator. $\omega^1 = \omega, \ \omega^2 = \omega, \ \omega^3 = 1, \ \omega^4 = \omega^3\omega = \omega$

Example 7.37: (Z x) is a cyclic group generated by 1. For 0.1 = 1, 1 = 1.1, 2 = 2.1, ... Similarly –1 = (–1) 1, –2 = (–2) .1, –3 = (–3) .1,...... Another generator is –1.

7.11.1 Properties of Cyclic Groups

Cyclic groups exhibit the following properties:

(i) **Every cyclic group is Abelian**

Proof: Let (G, *) be a cyclic group with $a \in G$ as a generator. Let $b, c \in G$. Then $b = a^m$, and $c = a^n$, where *m* and *n* are integers.

Now, $b*c = a^m * a^n = a^{m+n} = a^{n+m} = c*b.$
Hence, (G, *) is an Abelian group.

Note: The converse of the above theorem need not be true. That is, an Abelian group need not be cyclic.
For example: let us consider Klein's 4- group.
$G = \{e, a, b, c\}$.
The operation is $a^2 = b^2 = c^2 = e$ and

$a*b = b*a = c$

$b*c = c*b = a$

$a*c = c*a = b$

The operation is given in Table 7.5.

Table 7.5 Operation *

*	e	a	b	c
e	e	a	b	c
a	a	e	c	b
b	b	c	e	a
c	c	b	a	e

(G, *) is an Abelian group, but it is not cyclic, since no element of G can generate G.

(ii) Let (G, *) be a cyclic group generated by a. Then a^{-1} is also a generator.

Proof: Let $G = \{a\}$ be a cyclic group generated by a. If $x \in G$, then $x = a^n$ for some $n \in Z$.

$\therefore x = a^n = (a^{-1})^{-n}$ for some $n \in Z$,

$\therefore a^{-1}$ is also a generator of G.

(iii) If (G, *) is a finite cyclic group generated by an element $a \in G$, and is of order n, then $a^n = e$ so that $G = \{a, a^2, a^3, \ldots\ldots, a^n (= e)\}$. Also n is the least positive integer for which $a^n = e$.

Proof: If possible, let there exist a positive integer $m < n$, such that $a^m = e$. Since G is cyclic, any element of G, can be expressed as a^k, for some $k \in Z$. Let k be divided by m, and let q be the quotient and r be the remainder, where $0 \le r < m$. Then, we can write k as,

$k = mq + r.$

$a^k = a^{mq + r} = a^{mq}\, a^r.$

$= (a^m)^q\, a^r$

$= e^q\, a^r \; (\because a^m = e)$

$= e\, a^r$

$= a^r.$

This means that every element of G can be expressed as a^r, where $0 \le r < m$. This implies that G has at most m elements, i.e., the order of $G = m < n$, which is a contradiction.

$\therefore a^m = e$ for $m < n$ is not possible. Hence $a^n = e$ where n is the least positive integer.

(iv) Let (G,*) be a finite cyclic group of order n, with a as a generator, then a^m is also a generator of (G,*), if and only if the greatest common divisor of m and n is equal to 1, or m and n are relatively prime, where $m < n$ or the G.C.D of m and n is 1, then they are said to be relatively prime; $(m, n) = 1$.

Proof: Let us assume that a^m is a generator of (G,*). For some integer r,

$a = (a^m)^r = a^{mr}$.

$= a^{mr} * e = a^{mr} * e^s$, where s is an integer.

$= a^{mr} * (a^n)^s$ ($\because a^n = e$)

$a = a^{mr} +^{ns}$

$\Rightarrow mr = ns = 1$.

G.C.D of $(m, n) = 1$.

If d is said to be G.C.D of a and b, then there exist two integers x and y such that

$$ax + by = d.$$

If a and b are relatively prime then $ax + by = 1$, i.e, $(a, b) = 1$.

Example 7.38: Find the number of generators of a cyclic group of order 5.

Proof: Let G = (a) be a cyclic group of order 5. Then $G = \{a, a^2, a^3, a^4, a^5 = e\}$.
Since $(1, 5) = 1, (2, 5) = 1, (3, 5) = 1, (4, 5) = 1$, the generators are a, a^2, a^3, a^4.

$\therefore$ The number of generators is 4.

Example 7.39: Find the generators and the number of generators of a cyclic group of order 8.

Proof: Let G = (a) be a cyclic group of order 8.

$G = \{a, a^2, a^3, a^4, a^5, a^6, a^7, a^8 = e\}$

Since $(1, 8) = 1, (3, 8) = 1, (5, 8) = 1, (7, 8) = 1$.

$\therefore$ The generators are a, a^3, a^5, a^7..

The number of generators is 4.

Example 7.40: Show that every group of order three is cyclic.

Proof: Since G is of order 3, it must have two distinct elements a, b apart from the identity element e.
G is closed under the operation $*$ i.e., $a * b \in G$.

$\therefore a*b = a$ or $a*b = b$, or $a*b = e$.

If $a*b = a$, $a*b = a*e$ and by cancellation law $b = e$

If $a*b = b$, $a*b = e*b$ and by cancellation law $a = e$ but a and b are not equal to e.

$\therefore a*b = e.$ (1)

Again by closure law, $a^2 \in G$, $\therefore a^2 = a$ or $a^2 = b$, or $a^2 = e$.

If $a^2 = a$ or $a*a = a*e$, then $a = e$, which is not true

If $a^2 = e$ then $a^2 = a*b$ by (1).

$\therefore\ a = b$, is not true. Hence, $a^2 = b$ (2)

Also, $a^3 = a * a^2 = a * b$ by equation (2). Hence $a^3 = e$ by equation(1)

Hence G = $\{a, a^2, a^3 (= e)\}$ is a cyclic group with generator ***e***.

Example 7.41: Show that the following group (G,*) is cyclic.

G = $\{a, b, c, d, e, f\}$

Proof: Let us write the group as given in Table 7.6.

Table 7.6 Group definition

*	a	b	c	d	e	f
a	a	b	c	d	e	f
b	b	c	d	e	f	a
c	c	d	e	f	a	b
d	d	e	f	a	b	c
e	e	f	a	b	c	d
f	f	a	b	c	d	e

1. From the above table we note that *a* is an identity element.
2. Also, $b^2 = b * b = c$

   $b^3 = b^2 * b = c * b = d$

   $b^4 = b^3 * b = d * b = e$

   $b^5 = b^4 * b = e * b = f$

   $b^6 = b^5 * b = f * b = a$

   Every element of G is 'an integral power' of *b*.

   $\therefore\ (G, *)$ is cyclic, with *b*- as a generator. Now $bf = fb = a$ (identity). Therefore $f = b^{-1}$, hence *f* is also a generator.

Example 7.42: The set of integers modulo *n*, $z_n = \{0, 1, 2, \ldots n{-}1\}$, is a group under addition performed modulo *n*. It is cyclic generated by 1: $z_n = (1)$.

Since $k = 1 + 1 + 1 + \ldots + 1$ € (1), for each *k*.

Example 7.43: Prove that $(Z_5, \bullet)$ is a cyclic group. Find all its generators.

Proof: $Z_5 = \{1, 2, 3, 4\}$ and operation is multiplication modulo 5 is given in Table 7.7

[2] = $[2]^1$

[3] = [8] = [2].[2].[2] = $[2]^3$

[4] = [2].[2] = $[2]^2$

[1] = [16] = [2].[2.[2].[2]=]$2]^4$.

Every element is an integral power of the element [2].

$\therefore\ (Z_5, \bullet)$ is a cyclic group with [2] as a generator. Since [2] – is a generator of $(Z_5, \bullet)$, $[2]^{-1} = [3]$ is also a generator. [1] – is identity element and it cannot be a generator. Also $[4]^n \neq 2$, [4] cannot be a generator.

∴ (Z_5, •) is cyclic group with [2] and [3] as the only generators.

Table 7.7 Operation definition

•	1	2	3	4
1	1	2	3	4
2	2	4	1	3
3	3	1	4	2
4	4	3	2	1

Subgroups: A non-empty subset H of a group (G,*) is called a subgroup of G if (h,*) is itself a group. For the set H to be subgroup of G, two conditions must be satisfied, namely

(i) H must be a subset of G i.e, H ⊂ G.

(ii) H must be a group under the composition of G.

Example 7.44: Consider the fourth roots of unity, G = {1, −1, *i*, −*i*} and the square roots of unity H = {1, −1}. Now G is a group under multiplication, and H is a group under multiplication. Further H is subset of G as given in Tables 7.8 and 7.9 respectively.
∴ H is subgroup of G. i.e., the set of square roots of unity is a subgroup of the set of fourth roots of unity under multiplication. Let G = {1, −1, *i*, −*i*}

Table 7.8 Operation definition

•	1	−1	*i*	−*i*
1	1	−1	i	−i
−1	−1	1	−i	i
i	i	−i	−1	1
−i	−i	i	1	−1

Let H = {1, −1}

TABLE 7.9 Operation definition

•	1	−1
1	1	−1
−1	−1	1

Example 7.45: Consider G = {1, −1, *i*, −*i*} and : H = {1, *i*}
G- is group under multiplication but H is not a group under multiplication because the closure axiom is not satisfied. As −1 ∉ H, even though H ⊂ G, H is not a subgroup of G. From this example, it follows that every subset need not be a subgroup of G as given in Table 7.10.

Table 7.10 Operation definition

•	1	I
1	1	I
i	I	−1

Similarly, H = {i, –i}, H = {–1, i}, H= {–1, –i} are not subgroups, as the closure axiom is not satisfied.

The necessary and sufficient conditions for a subgroup have been put down as Theorems 7.1 and 7.2, which are given as follows:

Theorem 7.1: The necessary and sufficient conditions for a non-empty subset H of a group (G, *) to be a subgroup of G are,

(i) $\forall\ a, b \in H, a * b \in H$ (ii) $\forall a \in H, a^{-1} \in H$.

Proof: Since H– is given to be a subgroup of G, (H,*) is itself a group. So, the closure and inverse axioms are satisfied in H.

By the closure axiom, $\forall\ a, b \in H, a * b \in H$

By the inverse axiom, $\forall a \in H, a^{-1} \in H$

Hence, conditions (i) and (ii) are satisfied.

Conversely, suppose H is a subset of G, it is enough if we prove (H, *) is a group.

From conditions (i) and (ii), closure and inverse axioms are satisfied. In the given condition (i) replace b by a^{-1}, we get $a * a^{-1} \in H$, i.e, $e \in H$

$\therefore$ Identity axiom is satisfied.

Let $a, b, c \in G$, then $a, b, c \in H$ $\because H \subset G$

Then $(a * b) * c = a * (b * c)$ by associative law

So (H, *) is a group.

$\therefore$ H is a subgroup of G.

Theorem 7.2: The necessary and sufficient condition for a non-empty set H of G to be subgroup of (G,*) is that $\forall\ a, b \in H, a * b^{-1} \in H$

Proof: Since H– is a subgroup of the group G, (H, *) is itself a group.

So, $\forall\ b \in H \Rightarrow b^{-1} \in H$ (by inverse axiom)

Now, $\forall\ a, b^{-1} \in H,\ a * b^{-1} \in H$ (by closure axiom)

Thus, $\forall\ a, b^{-1} \in H,\ a * b^{-1} \in H$.

Conversely, H is given to be a subset of G. Now it is enough to prove that (H, *) is a group.

In the given condition $a * b^{-1} \in H$, replacing a by b, we get $b * b^{-1} \in H$, i.e, $e \in H$. $\therefore$ identity axiom is satisfied.

If in the above condition, we replace a by e, then $e * b^{-1} \in H$, i.e, $b^{-1} \in H$. So, the inverse axiom is satisfied.

Again, in above condition, replace b by b^{-1}, then we get

$a * (b^{-1})^{-1} \in H$, i.e, $(a * b) \in H$. So closure axiom is satisfied.

Let $a, b, c \in H$, then $a, b, c \in G$ $\because H \subset G$

The associative law $(a * b) * c = a * (b * c)$ holds in G. The same also holds in *H*.

Thus all four axioms are satisfied in *H*.

$\therefore$ *(H, *) is a group. H is a subgroup of G.*

Theorem 7.3: The intersection of two subgroups of a group is again a subgroup.

Proof: Let (G, *) be a group and S and T be two subgroups of G.

Since S and T are subgroups of G, $e \in S$ and $e \in T$.

$\therefore e \in (S \cap T)$ and hence $(S \cap T)$ is a non-empty subset of G.
Let $a, b \in S \cap T \Rightarrow a, b \in S$ and $a, b \in T$.

$\Rightarrow a * b^{-1} \in S$ and $\Rightarrow a * b^{-1} \in T$ (S and T are subgroups)

$\Rightarrow a * b^{-1} \in S \cap T$

$\therefore S \cap T$ is a subgroup of G.

7.12 CONGRUENCE MODULO *n*

Let *n* be positive integer and $n > 1$. For integers *x* and *y*, *x* is said to be **congruent** to *y* **modulo** *n* if and only if $(x - y)$ is a multiple of *n*. We write $x \equiv y(\text{mod } n)$ to indicate that *x* is congruent to *y* modulo *n*.

Thus, $x \equiv y(\text{mod } n)$ if and only if *n* divides $x - y$, and this is equivalent to

$x - y = nq$, or $x = y + nq$. Another way to describe this relation is to say that *x* and *y* yield the same remainder when each is divided by *n*. To see that this true, let

$$x = nq_1 + r_1 \text{ with } 0 \le r_1 < n$$

and

$$y = nq_2 + r_2 \text{ with } 0 \le r_2 < n$$

Then

$$x - y = n(q_1 - q_2) + (r_1 - r_2) \text{ with } 0 \le |r_1 - r_2| < n.$$

Thus, $x - y$ is a multiple of *n* if and only if $r_1 - r_2 = 0$; that is, if and only if $r_1 = r_2$. In particular, any integer *x* is congruent to its remainder when divided by *n*. This means that any *x* is congruent to one of 0, 1, 2, 3, ...,(*n*–1).

7.12.1 Congruence Classes

In connection with the relation of congruence modulo *n*, we have observed that there are *n* distinct congruence classes. Let Z_n denote this set of classes.

$$Z_n = \{[1],[2],\ldots,[n-1]\}.$$

7.12.2 Addition and Multiplication of Residue Classes

Let Z_n be the set of residues. Let *a*, $b \in Z_n$, then their sum is denoted by $+_n$ and is defined as

$$[a] +_n [b] = \begin{cases} [a+b] & ; \text{if } a+b < n \\ [r] & ; \text{if } a+b \ge n \end{cases}$$

Where *r* is the least non-negative remainder when $a + b$ is divided by *n* and $0 \le r < n$.

Example 7.46: Consider $Z_5 = \{0, 1, 2, 3, 4\}$

$[1] +_5 [2] = [1+2] = [3] \because 3 < 5$

$[3] +_5 [10] = 3$, $(\because 3 +_5 10 = 3 < 5)$ and $13 = 2 * 5 + 3$

$[3] +_5 [2] = 0.$

Consider the multiplication of residue classes. Let Z_n be the set of residues. Let *a*, $b \in Z_n$, then their product is denoted by X_n and is defined as

$$[a] \ X_n \ [b] = \begin{cases} [a \text{ x } b] & ; \text{if } ab < n \\ [r] & ; \text{if } ab \ge n \end{cases}$$

Where r is the least non-negative remainder when ab is divided by n and $(0 \le r < n)$.
Consider $Z_5 = \{0, 1, 2, 3, 4\}$ $2 \times_5 2 = 4$; $4 < 5$, $3 \times_5 2 = 1$;

7.13 COSETS

Consider the subgroup of integers divisible by 3. This forms a subgroup of the additive group of integers. Its elements are $\{\ldots -9, -6, -3, 0, 3, 6, 9, \ldots\}$. By adding 1 to any multiple of 3, we get another subset of a group of integers, $\{\ldots -8, -5, -2, 1, 4, 7, 10, \ldots\}$. Adding 2 to the elements of the multiples-of-3 subgroup gives us the subset $\{\ldots -7, -4, -1, 2, 5, 8, 11, \ldots\}$. These three subsets exhaust the integers. Any integer will be in one and only one of these subsets. One interesting point about the three subsets is that if we take any two consecutive elements a and b out of one of the subsets (both from the same subset) then their difference, $a - b$, will be a multiple of three. Their difference will be in the subgroup that originally generated them.

Let (H,*) be a subgroup of a group (G,*). The set aH, where $a \in G$, defined by

$aH = \{a * h \mid h \in H\}$, is called the left coset of H in G generated by the element $a \in G$, a is called the representative (element) of the left coset aH.

Similarly, the set Ha is defined by the following:

$Ha = \{h * a \mid h \in H\}$, and is called the right coset of H in G, generated by $a \in G$. It is important to note the following observations.

1. The left and right cosets of H need not be the same.
2. If G is Abelian, then every left coset of H is a right coset also.
3. $e * H = H * e = H$, whether or not G is Abelian.

7.13.1 Definition

If H is a subgroup of a group G then for any element a of the group, the set of products of the form $h * a$ where h is in H, is a **right coset** of H denoted by the symbol Ha. The set of all products of the form $a * h$ where h is in H, is a **left coset** of the subgroup H denoted by the symbol aH.

In fact, if we take any subset U of a group G (which may or may not be a subgroup) we can multiply every element in it from the right by some group element v to get another subset V of **G.** If the subset H happens to be a subgroup of G, then the subset it is *transformed* into, is called a **right coset** of the subgroup H.

In set-builder notation we can define Uv as

$Uv = \{u * v \mid u \in U\}$ where U is a subgroup that is a right coset of U.

The operation of 'multiplication on the right by a group element' which transforms subsets into subgroups (and subgroups into right cosets) will also transform any right coset of a subgroup H into a right coset of H. (This could be a different right coset than the one we started with or may be the same). This follows from the definition since (H u) v is all elements of the form $(h * u) * v$ which by the associative property is the same as the set of elements of the form $h * (u * v)$ which is none other than H(u,v). This is also true in the case of left cosets. We are able to get more insight into the properties of right and left cosets from Theorem 7.4.

Theorem 7.4: If $\boldsymbol{x}$ and $\boldsymbol{y}$ are in the same right coset of a subgroup H then $x * y^{-1}$ is in H.

Proof: If x and y are in the same right coset of H then there are h_1 and h_2 in H and some element a of **G** such that

$h_1 * a = x$ and $h_2 * a = y$

By the properties of groups, the inverse of $\boldsymbol{y}$ is $(h_2 * a)^{-1} = a^{-1} * h_2^{-1}$.

Thus $x * y^{-1} = (h_1 * a) * (a^{-1} * h_2^{-1}) = h_1 * (a * a^{-1}) * h_2^{-1} = h_1 * h_2^{-1}$

By the closure property, $\boldsymbol{h_1} \bullet \boldsymbol{h_2}^{-1}$ is in H and our original property is preserved by the new definition.

The fundamental idea of a coset is that of taking a subgroup and 'translating' it to fill up G. A coset is just one such 'translate'. No coset, but H itself is a subgroup of G, since they do not contain 1, but there is a lot of ground that can be covered by treating them like elements of a group or set instead. Since the cardinality of the cosets is constant, they represent ways of dividing up G, and formulas like |G|=|H|[H:G] are obtained, where [H:G] is called the index of H in G and is equal to the number of cosets.

The biggest use of cosets is to set up the idea of a quotient group. One may have seen the general idea — formally dividing one mathematical structure by a substructure — in other fields, like topology. Cosets arise when we want to model the idea that certain elements of a group are effectively equal. To see this, rather that looking at one coset at a time it is best to think of all possible cosets of a subgroup. Then we will find that cosets partition a group G into equivalence classes, such that two elements of a class differ by an element of the subgroup H.

Example 7.47: In number theory, we do not want to distinguish between two numbers that differ by a multiple of a given number *n*. Let $n = 4$. Then let G be the set of integers under addition and H be the set of multiples of 4. The cosets of H are as given here.

$H\cdot 0 = \{\ldots, -4, 0, 4, 8, \ldots\}$, $H\cdot 1 = \{\ldots, -3, 1, 5, 9, \ldots\}$, $H\cdot 2 = \{\ldots, -2, 2, 6, 10, \ldots\}$

$H\cdot 3 = \{\ldots, -1, 3, 7, 11, \ldots\}$, $H\cdot 4 = \{\ldots, 0, 4, 8, 12, \ldots\}$, $H\cdot 5 = \{\ldots, 1, 5, 9, 13, \ldots\}$

Here, we note that $H\cdot 0 = H\cdot 4$ and $H\cdot 1 = H\cdot 5$. In fact there are only 4 distinct cosets, each corresponding to one congruence class of integers modulo 4. If we care only about the remainder of an integer after division by 4, then all the elements in a coset are equivalent and we can think of them as a single entity.

Example 7.48: Let G be the group of possible rotations of a point on a unit circle, and we care only about where a point ends up on the circle after the rotation. Then let H be rotations that are a multiple of 2π. The cosets of H will now be rotations which differ by multiples of 2π and which therefore have the same effect on the final position of the point rotated.

Example 7.49: In general, two elements g_1 and g_2 of G are defined to be equivalent ($g_1 \equiv g_2$) if $g_1 * g_2^{-1} \in H$.

For all g in G, $g \equiv g$. Since the identity element belongs to H as it is a subgroup.

If $g_1 \equiv g_2$ then $g_2 \equiv g_1$, since if an element belongs to H then so does its inverse.

If $g_1 \equiv g_2$ and $g_2 \equiv g_3$, then $g_1 \equiv g_3$ since H is closed under multiplication.

This shows that ≡ is a genuine equivalence relation.

We define the equivalence class of an element *x* of G as

$[x] = \{y \in G \,/ x \equiv y\}$. For any $x, y \in G$, either $[x] \cap [y] = \Phi$ or $[x] = [y]$.

So the equivalence classes are either equal or disjoint and they cover G. To relate to cosets, the equivalence classes defined above are the same as the right cosets.

$y \equiv x$ then $yx^{-1} = h \quad h \in H$

or $y = hx$. So $[x] = Hx$. By defining equivalence relation by the $x^{-1}y \in H$ condition we get the left cosets.

Theorem 7.5: (Lagrange's Theorem) If(G, *) is a finite group and (H, *) is a subgroup, then the order of H is a divisor of the order of G.

Proof: Since G is finite, $H \subset G$, and H is also finite. Let $\circ(G) = n$ and $\circ(H) = m$. Now G is partitioned into a finite number of distinct right cosets of H in G, say *k*, namely $H, Ha_1, Ha_2, \ldots Ha_{k-1}$. Also, the order of each coset is the order of H.

$$\circ(Ha_1) = \circ(Ha_2) = \ldots = \circ(Ha_{k-1}) = \circ(H) = m$$

$$\circ(G) = \circ(H) + \circ(Ha_1) + \ldots + \circ(Ha_{k-1})$$

$$n = m + m + \ldots + m(k \text{ times})$$

$$n = km$$

Therefore, *m* divides *n*, i.e., the order of H divides the order of G.

7.14 HOMOMORPHISMS

Homomorphism, derived from the Greek word 'homoiosorphe' (meaning similar form), is defined as a special correspondence between the members (elements) of two algebraic systems, such as two groups, two rings, or two fields. Two homomorphic systems have the same basic structure, and, while their elements and operations may appear entirely different, results on one system often apply as well to the other system. Thus, if a new system can be shown to be homomorphic to a known system, certain known features of one can be applied to the other, thereby simplifying the analysis of the new system.

In a homomorphism, corresponding elements of two systems behave very similarly in combination with other corresponding elements. For example, let G and H be two groups. The elements of G are denoted by $g_1, g_2, \ldots g_n$ and they are subject to some operation•. (Although the symbol may be thought of as some operation like multiplication, the symbol can just as well indicate rotation or some other non-arithmetic operation.) Similarly, the elements of H are denoted by $h_1, h_2, \ldots h_n$ and they are subject to some operation∗. A homomorphism from G to H is a correspondence $g \to h$ between all elements of G and some elements of H, and that has the following property:

If $g \to h$ and $g_1 \to h_1$, then $g_1 \circ g_2 \to h_1 * h_2$. In other words, the element of H corresponding to a product of elements in G is the product, in the same order, of the elements of H corresponding to the two elements in G. Expressed more compactly, the 'image' of the product is the product of the images, or the correspondence preserves the operation.

A correspondence between members of two algebraic systems may be written as a function (*f*) from G to H, and one speaks of (*f*) as 'mapping' G to H. The condition that (*f*) be a homomorphism of the group G to the group H, may be expressed as the requirement that $f(g_1 \circ g_2) = f(g_1) * f(g_2)$. Homomorphisms impose conditions on a mapping *f*: whereby, if *e* is the identity of G, then $g \circ e = g$, so $f(g \circ e) = f(g)$. Furthermore, since f is a homomorphism, $f(g_1 \circ e) = f(g_1) * f(e)$, so $f(g) = f(g) * f(e)$. By the cancellation laws for groups, this implies that f(*e*) is equal to the identity in H. Thus, homomorphism maps the unique identity element of one group to the unique identity element of the other group. Similarly, homomorphism maps the inverse of an element *g* in one group to the inverse of the element f(*g*). This is why homomorphism is called a structure-preserving map.

Special types of homomorphisms have their own names. A one-to-one homomorphism from G to H is called a monomorphism, and a homomorphism that is 'onto,' or covers every element of H, is called an epimorphism. An especially important homomorphism is an isomorphism, in which the homomorphism from G to H is both one-to-one and onto. In this last case, G and H are essentially the same system and differ only in the names of their elements. Thus, homomorphism is useful in classifying and enumerating algebraic systems since they allow one to identify how closely different systems are related.

7.14.1 Definition

If (G, ∘) and (H, ∗) then a function f: G→H is a homomorphism, when the following relation holds.

$$f(x \circ y) = f(x) * f(y) \ \forall\, x, y \in G$$

- A homomorphism is an **isomorphism** if it is bijective — equivalently, if it has an inverse.
- If G and H are groups, G and H are **isomorphic** if there is an isomorphism. f:G→H. Isomorphic groups are the same as groups.
- If f:G→H is a group map, then $f(1)=1$ and $f(a^{-1})=f(a)^{-1}$.
- Groups G and H are not isomorphic if they have different orders, or if one satisfies the group property that the other doesn't. For example, two groups are not isomorphic if one is Abelian and the other is not; two groups are not isomorphic if the order of elements of one is not the same as the order of elements of the other. The operation tables for two groups of order 3 are shown in Table 7.11 and Table 7.12:

Table 7.11 Operations

·	1	a	a^2
1	1	a	a^2
a	a	a^2	1
a^2	a^2	1	a

Table 7.12 Operations

+	0	1	2
0	0	1	2
1	1	2	0
2	2	0	1

There is an obvious sense in which these two groups are 'the same'. You can get the second table from the first by replacing 0 with 1, 1 with a, and 2 with a^2. The following observations are interesting. i). 'The derivative of a sum is the sum of the derivatives'.

ii). Group homomorphisms are to groups as linear transformations are to vector spaces.

Consider the following definitions:

A group has a single binary operation $a*b$. A group homomorphism preserves the operation

$$f(a*b) = f(a)*f(b).$$

A vector space has two operations, vector addition (V + W) and scalar multiplication αV.

A linear transformation preserves the operators.

$$L\{V+W\} = L\{V\} + L\{W\}$$

$$L\{kV\} = kL\{V\}$$

Example 7.50: Consider the group Z under addition, and look at $f: Z \to Z$ given by $f(n) = 3$ for all n. Then, $f(1+1) = f(2) = 3$, but $f(1)+f(1) = 3+3 = 6$. Therefore, constant maps are usually not group maps. In general, constant maps are not homomorphism.

Example 7.51: Consider logs and exponentials, and that exp:(R, +)→(R$^+$,·), is a homomorphism from the real set under addition to the positive real set under multiplication. The operation on the domain R is addition, while the operation on the range R$^+$ is multiplication. Therefore, we can show that exp is a homomorphism,

$$\exp(x+y) = \exp(x).\ \exp(y)\quad \forall\ x,y \in R$$

But $\exp(x+y) = e^{x+y}$ and $\exp(x).\exp(y) = e^x e^y$, hence from identity $e^{x+y} = e^x e^y$ Thus, exp is a homomorphism. The following are the observations:

(i) The exponential function exp:(R, +)→(R$^+$,+) is not a homomorphism. In this case, the addition is an operation on both the domain and range. So, the homomorphism property that says '$\exp(x+y) = \exp(x) + \exp(y)\ \forall\ x,y \in R$' is fulfilled. In other words, $e^{x+y} = e^x + e^y$. This is not an identity, for example, if $(x = 1)$ and $(y = 1)$, $e^{x+y} = e^2$, while $e^x + e^y = 2e$, and $e^2 \neq 2e$.
(ii) Similarly, log:(R, ·)→(R$^+$,+) is a homomorphism from the set of real numbers under multiplication to the positive real set under addition. Since, log $(mnp) = \log m + \log n + \log p$.
(iii) log:(R, +)→(R$^+$,·) is not a homomorphism. In this case,

$$\log(m+n+p) \neq \log m + \log n + \log p$$

This example gives a function that is not just a rule. It is a rule together with a domain and a range. In many basic mathematical situations, the domain and range do not play a large role. In this situation, they do.

Example 7.52: Define f:Z→Z by f(x) = 5x. Check whether a function is a group map. In order to show f is a homomorphism, consider $x,y \in$ Z and the addition operation on Z.
Then $f(x+y) = f(x) + f(y)$
$f(x+y) = 5(x+y) = 5x + 5y = f(x) + f(y)$.Therefore, f is a homomorphism.

Example 7.53: Show that a function is not a homomorphism. Consider g:Z→Z by $g(x) = x^2$. To show that g is not a homomorphism, consider $x,y \in Z$, such that $g(x+y) \neq g(x) + g(y)$. Consider two values at random, say $x = 2$, and $y = 3$.
$g(2+3) = g(5) = 5^2 = 25$. But $g(2) + g(3) = 2^2 + 3^2 = 4 + 9 = 13$

Since, $g(2+3) \neq g(2) + g(3)$. g is not a homomorphism.

Example 7.54: Consider a group map on a matrix group. Let M(2, R)be the group of 2 × 2 real set matrices under matrix addition. Let tr:M(2, R)→R denote the **trace map.** Trace of a matrix is defined as the sum of the elements along the principal diagonal.

$$\mathrm{tr}\left(\begin{bmatrix} a & b \\ c & d \end{bmatrix}\right) = a + d$$

Now $\mathrm{tr}\left(\begin{bmatrix} a & b \\ c & d \end{bmatrix} + \left(\begin{bmatrix} a^1 & b^1 \\ c^1 & d^1 \end{bmatrix}\right)\right) = \mathrm{tr}\left(\begin{bmatrix} a+a^1 & b+b^1 \\ c+c^1 & d+d^1 \end{bmatrix}\right) = (a+a^1) + (d+d^1)$

and $\mathrm{tr}\left(\begin{bmatrix} a & b \\ c & d \end{bmatrix}\right) + \mathrm{tr}\left(\begin{bmatrix} a^1 & b^1 \\ c^1 & d^1 \end{bmatrix}\right) = (a+d) + (a^1 + d^1)$

Since, $(a+a^1)+(d+d^1)=(a+d)+(a^1+d^1)$,

it follows that

$$\operatorname{tr}\left(\begin{bmatrix} a & b \\ c & d \end{bmatrix}+\left(\begin{bmatrix} a^1 & b^1 \\ c^1 & d^1 \end{bmatrix}\right)\right) = \operatorname{tr}\left(\begin{bmatrix} a & b \\ c & d \end{bmatrix}\right) + \operatorname{tr}\left(\begin{bmatrix} a^1 & b^1 \\ c^1 & d^1 \end{bmatrix}\right)$$

Therefore, **tr** is a homomorphism.

Example 7.55: Consider group maps and linear transformation. It has already been stated that group homomorphism are analogous to linear transformations. In fact, a vector space is an Abelian group under vector addition. Thinking of a vector space in this manner, one can say that a linear transformation is a group homomorphism.

Recall that the definition of a linear transformation requires that

$$T\left(\vec{u}+\vec{v}\right)=T\left(\vec{u}\right)+T\left(\vec{v}\right)$$

for all vectors $\vec{u}$ and $\vec{v}$. This means that T is a group homomorphism.

Example 7.56: Consider the example. R^2: is a 2-dimensional vector space. Recall the axioms for a vector space. One can see that R^2 is an Abelian group under vector addition. Consider the linear transformation $T:R^2 \to R^2$ defined by

$T(x, y)=(4x+3y,\ 2x+5y)$.

In matrix form, this is $T\left(\begin{bmatrix} x \\ y \end{bmatrix}\right) = \begin{bmatrix} 4 & 3 \\ 2 & 5 \end{bmatrix}\begin{bmatrix} x \\ y \end{bmatrix}$

Since, matrix multiplication distributes over vector addition,

$$T\left(\begin{bmatrix} x \\ y \end{bmatrix}+\begin{bmatrix} x^1 \\ y^1 \end{bmatrix}\right)=\begin{bmatrix} 4 & 3 \\ 2 & 5 \end{bmatrix}\left(\begin{bmatrix} x \\ y \end{bmatrix}+\begin{bmatrix} x^1 \\ y^1 \end{bmatrix}\right)=\begin{bmatrix} 4 & 3 \\ 2 & 5 \end{bmatrix}\begin{bmatrix} x \\ y \end{bmatrix}+\begin{bmatrix} 4 & 3 \\ 2 & 5 \end{bmatrix}\begin{bmatrix} x^1 \\ y^1 \end{bmatrix}=T\left(\begin{bmatrix} x \\ y \end{bmatrix}\right)+\left(\begin{bmatrix} x^1 \\ y^1 \end{bmatrix}\right)$$

This gives a direct proof that T is a group homomorphism.

Example 7.57: A group map involving multiplication and addition. Let Q be the group of rational numbers under addition; let Q^+ be the group of positive rational numbers under multiplication. Define $f:Q\to Q^+$ by $f(x)=3^x$.

Let us check whether f is a homomorphism. The operation on Q is addition (+) and the operation on Q^+ is multiplication (.). It can be shown that,

$f(x+y) = f(x)\cdot f(y)\ \ \forall x,y\in Q$. Hence $f(x+y)=3^{x+y}=3^x\cdot 3^y=f(x)\cdot f(y)$.

Therefore, f is a homomorphism.

Theorem 7.6: Let $\phi:G\to H$ be a group homomorphism. Then,

(a) $\phi(e_1)=e_2$, where e_1 is the identity in G and e_2 is the identity in H.

(b) $\phi(g^{-1})=\phi(g)^{-1}$ for all $g\in G$.

Proof: $\phi(e_1)=\phi(e_1\cdot e_1)=\phi(e_1)\cdot\phi(e_2)$ because $e_1e_1=e_1$ in G.

Cancel $\phi(e_1)$ off both sides, we obtain $\phi(e_1)=e_2$.

Now let $g\in G$. $\phi(g)\cdot\phi(g^{-1})=\phi(g\cdot g^{-1})=\phi(e_1)=e_2$

$$\phi(g^{-1})\cdot\phi(g)=\phi(g^{-1}\cdot g)=\phi(e_1)=e_2$$

This shows that $\phi(g^{-1})$ is the inverse of $=\phi(g)$ i.e $=\phi(g)^{-1}=\phi(g^{-1})$

This shows that $\phi(g^{-1})$ is the inverse of $\phi(g)$, i.e., $\phi(g)^{-1}=\phi(g^{-1})$.

It is observed that the properties shown in Theorem 7.6 are not part of the definition of a homomorphism. To show the function f is a homomorphism, we need to show the expression $f(a\cdot b) = f(a)\cdot f(b)$ for all a and b. The properties are automatically true for any homomorphism. On the other hand, to show that a function is not a homomorphism, we need to check whether it maps identity to the identity. If not, then the theorem shows that it's not a homomorphism.

Example 7.58: Group maps must take the identity to the identity. Let $\mathbb{Z}$ denote the group of integers with addition. The function $f:\mathbb{Z}\rightarrow\mathbb{Z}$ given by $f(x)= x+2$ has $f(0)=2$. Since the identity $0\in\mathbb{Z}$ is not mapped to the identity $0\in Z$, f cannot be a group homomorphism.

On the other hand, consider $g:\mathbb{Z}\rightarrow\mathbb{Z}$ given by $g(x)=x^3$.

$g(0)=0$, but this doesn't mean that g is a homomorphism. In fact,

$g(2+2)\neq g(2)+g(2)$, hence g is not a homomorphism.

The point is clear that a simple looking function need not be homomorphism. There are several important subsets associated to a group homomorphism, $\phi:G\rightarrow H$.

7.14.2 Definition Let $\phi:G\rightarrow H$ be a Group Homomorphism

(a) The **kernel** of ϕ is $\ker\phi=\{g\in G ! \ \phi(g)=1\}$

(b) The **image** of ϕ is (as usual) $\mathrm{im}\phi=\{\phi(g) \ ! g\in G\}$

(c) Let $H^1 < H$. The **inverse image** of H^1 is (as usual)

The notation $\phi^{-1}(H^1)$ does not imply that the **inverse** of ϕ exists. The function $\phi^{-1}(H^1)$ is simply the set of inputs, which ϕ maps onto H^1; ϕ^{-1} is applied to the set H^1 if there is a ϕ^{-1}. But there need not be.

Theorem 7.7: Let $\phi:G\rightarrow H$ be a group map.

(a) $\ker\phi$ is a subgroup of G.

(b) $\mathrm{im}\phi$ is a subgroup of H.

(c) If H^1 is a subgroup of H, then $\phi^{-1}(H^1)$ is a subgroup of G.

Proof: (a) Let $\phi(1)=1$. So $1\in\ker\phi$

Suppose $g_1.g_2\in\ker\phi$. Then $\phi(g_1g_2)=\phi(g_1)\phi(g_2)=1.1=1$.

Hence $g_1g_2\in\ker\phi$.

Finally, suppose $g \in \ker\phi$. Then $\phi(g^{-1}) = \phi(g)^{-1} = 1^{-1} = 1$.

Hence $g^{-1} \in \ker\phi$. Therefore, $\ker\phi$ is a subgroup of G.

(b) Since $\phi(1) = 1$. Then $1 \in \mathrm{im}\phi$

Suppose $\phi(g_1).\phi(g_2) \in \mathrm{im}\phi$. Then $\phi(g_1).\phi(g_2) = \phi(g_1 g_2) \in \mathrm{im}\phi$

Finally, suppose $\phi(g) \in \mathrm{im}\phi$. Then $\phi(g^{-1}) = \phi(g)^{-1} \in \mathrm{im}\phi$.

Therefore, $\mathrm{im}\phi$ is a subgroup of H.

(c) Let H' be a subgroup of H. To show that ϕ^{-1} is a subgroup of G.

(Note: The criterion for membership in $\phi^{-1}(H^1)$ is that ϕ takes the element into H'.)

Since $1 \in H'$ and $\phi(1) = 1$, it follows that $1 \in \phi^{-1}(H')$.

Suppose $g_1, g_2 \in \phi^{-1}(H')$. This means that $\phi(g_1)$ and $\phi(g_2)$ are in H'. Since H' is a subgroup, $\phi(g_1)\phi(g_2)$ is in H' as well. But $\phi(g_1)\phi(g_2) = \phi(g_1 g_2)$.

Therefore, $\phi(g_1 g_2)$ is in H', which means that $g_1 g_2 \in \phi^{-1}(H')$.

Finally, assume that $g \in \phi^{-1}(H')$. So $\phi(g) \in H'$. Since H' is a subgroup, $\phi(g)^{-1} \in H'$.

But $\phi(g)^{-1} = \phi(g^{-1})$, so $\phi(g^{-1}) \in H'$. This means that $g^{-1} \in H'$.

Hence, $\phi^{-1}(H')$ is a subgroup of G.

Example 7.59: (Finding the kernel and image) Let $S' = \{z \in C \mid |z| = 1\}$. S' can be shown as a group under multiplication of complex numbers. Each element $z \in S'$ can be uniquely written in the form

$$z = e^{2\pi i t} = \cos 2\pi t + i \sin 2\pi t. \quad 0 \le t < 1.$$

The identity element is 1. The inverse of $e^{2\pi i t}$ is $e^{-2\pi i t}$

Let us define $\phi : R \to S'$ by $\phi(t) = e^{2\pi i t}$.

By the above statements, $\mathrm{im}\phi = S'$. To see that ϕ is a homomorphism, note that

$\phi(s+t) = e^{2\pi i (s+t)} = e^{2\pi i s} e^{2\pi i t} = \phi(s)\phi(t)$.

The kernel of ϕ is $\ker\phi = \{t \in R \mid e^{2\pi i t} = 1\}$.

Using $e^{it} = \cos t + i \sin t$, we can say that $\ker\phi = \mathbb{Z}$.

7.14.3 Definition of Isomorphism

Let G and H be groups. An **isomorphism** from G to H is a bijective homomorphism $\phi : G \to H$. If there is an isomorphism, $\phi : G \to H$. Then G and H are **isomorphic** and is denoted by $G \approx H$.

To say that two groups are isomorphic, it is necessary to say that they are the same groups. The elements of the two groups and the group operations may be different, but the two groups need to have the same structure. This means that if one has a certain group-theoretic property, the other will as well. In other words, **a group-theoretic property is a property preserved by isomorphism.** For this to be a useful concept, we will have to provide specific examples of properties that one can check.

Some old books define an isomorphism from G to H to be an injective homomorphism $\phi:G\rightarrow H$. That is, ϕ need not map G onto H. One then says that G and H are isomorphic if there is an isomorphism from G onto H. Unfortunately, this results in the odd situation that there may be an isomorphism from G to H, yet G and H may not be isomorphic. Here, the word isomorphism means a bijective map. Theorem 7.8 suggests an easy way to tell that a group map is an isomorphism.

Theorem 7.8: A group map $\phi:G\rightarrow H$ is an isomorphism if and only if it is invertible. In this case, ϕ^{-1} is also a homomorphism, hence an isomorphism.

Proof: The first statement is trivial, since a map of sets is bijective if and only if it has an inverse. Now suppose $\phi:G\rightarrow H$. is an isomorphism. We must show the inverse $\phi^{-1}:H\rightarrow G$ is a homomorphism. Let $h_1h_2 \in H$. We have to show that

$$\phi^{-1}(h_1.h_2)=\phi^{-1}(h_1)\phi^{-1}(h_2).$$

Since $\phi:G\rightarrow H$. is onto, there exist $g_1g_2 \in G$ such that $\phi(g_1)=h_1$ and $\phi(g_2)=h_2$. Then

$\phi^{-1}(h_1h_2)=\phi^{-1}(\phi(g_1)\phi(g_2))=\phi^{-1}(\phi(g_1g_2))=g_1g_2=\phi^{-1}(h_1)\phi^{-1}(h_2)$

Therefore, ϕ^{-1} is a homomorphism.

Since ϕ^{-1} is invertible – its inverse is ϕ – it is an isomorphism.

Example 7.60: (A group isomorphism) Consider the exponential map $\exp:(R,+)\rightarrow(R^+,\cdot)$ given by $\exp(x)=e^x$. By an earlier example, **exp** is a homomorphism from the real under addition to the positive real under multiplication.

The natural logarithm $\ln:(R^+,\cdot)\rightarrow(R,+)$ inverts **exp**:

$$\ln e^x = x,\ e^{\ln y}=y \quad \text{for } x\in R, y\in R^+.$$

By the theorem, **exp** is an isomorphism. The groups $(R, +)$ and $(R^+,\cdot)$ are isomorphic.

Example 7.61: (A group isomorphism on the integers mod 2) Consider the set $G=\{-1,1\}$. Make G into a group using multiplication as the group operation.

Define a map $\phi:Z_2\rightarrow G$ by $\phi(0)=1,\ \phi(1)=-1$

Clearly, ϕ is invertible: Its inverse is $\phi^{-1}(1)=0,\ \phi^{-1}(-1)=1$.

To show ϕ is a homomorphism, hence an isomorphism, by simply checking cases, consider Table 7.13

Table 7.13 Homomorphism

a	b	$\phi(a+b)$	$\phi(a)\phi(b)$
0	0	1	1 . 1 = 1
0	1	−1	1.(−1) = −1
1	0	−1	(−1).1 = −1
1	1	1	(−1).(−1) = 1

The above approach can be used to construct an isomorphism from Z_2 to any group of order 2. There is only one group of order 2, leading to isomorphism. Isomorphic groups have the same group-theoretic properties.

Theorem 7.9: Suppose G and H are isomorphic groups. If G is Abelian, so is H.

Proof: Let $h_1, h_2 \in H$. We have to prove that $h_1h_2 = h_2h_1$. Since ϕ is onto, there exist $g_1g_2 \in G$ such that $\phi(g_1) = h_1$ and $\phi(g_2) = h_2$. Then

$$h_1h_2 = \phi(g_1)\phi(g_2) = \phi(g_1g_2) = \phi(g_2g_1) = \phi(g_2g_1) = \phi(g_2)\phi(g_1) = h_2h_1.$$

Therefore, *H* is Abelian.

Example 7.62: (Non-isomorphic groups) S_3 and Z_6 are both groups of order 6. However, Z_6 is Abelian while S_3 is non-Abelian. Therefore, S_3 and Z_6 are not isomorphic. Suppose G and H are isomorphic groups. If G is finite, so is H. If G is infinite, so is H. More specifically, isomorphic groups have the same cardinality.

Example 7.63: (Groups of different cardinality aren't isomorphic) Z and R cannot be isomorphic, since the integers are countable, while the real are uncountable. Note that two groups with the same order are not necessarily isomorphic. In the previous example, we saw that S_3 and Z_6 are not isomorphic, even though both of them have an order 6.

Theorem 7.10: Suppose G and H are isomorphic groups. If *G* has a subgroup *K* of order 42, so does *H*.

Proof: If $K < G$ and $|K| = 45$, then $\phi(K) < H$ and (since ϕ maps K bijectively onto

$\phi(K)$). Hence, $\phi(K) = 45$.

Obviously, there is nothing special about '45'. If G has a subgroup of order 117, so does H. If G has a subgroup of order 91, so does H. And so on. This proposition is just to show that the property is shared by isomorphic groups. There are infinitely more properties that will be shared by isomorphic groups. In fact, the **group-theoretic property** is one that is necessarily shared by isomorphic groups. However, the earlier examples show that some properties are not shared by them. For example, the elements of one group may be letters, while the elements of the other are numbers. 'Having the same kind of elements' is not a group-theoretic property. Likewise, the operation in one group may be addition of numbers, while the operation in the other could be composition of functions. 'Having the same kind of binary operation' is not a group-theoretic property.

Example 7.64: (Showing groups are not isomorphic, by considering orders of elements) $Z_2 \times Z_2$ and Z_4 are *not* isomorphic. Both groups have 4 elements. However, every element of $Z_2 \times Z_2$ has order 1 or 2, while Z_4 has two elements of order 4, namely 1 and 3. Having different numbers of elements of a given order is a group property. Since these groups differ in this respect, they are not isomorphic.

Example 7.65: If $f: G \to H$ and $g: H \to K$ are homomorphisms, prove that $g \circ f: G \to K$ defined by $(g \circ f)(x) = g\{f(x)\}$ is homomorphism.

Take any $a, b \in G$. Then $ab \in G$ and

$$(g \circ f)(ab) = g\{f(ab)\} = g\{f(a)f(b)\} = g\{f(a)\}g\{f(b)\} = (g \circ f)(a)(g \circ f)(b)$$

This proves that $(g \circ f)$ is homomorphism.

7.15 APPLICATIONS OF GROUP THEORY

Application of group theory is widespread. Almost all structures in abstract algebra are special cases of groups. Rings, for example, can be viewed as Abelian groups corresponding to the addition operation together with a second operation, corresponding to multiplication. Therefore, group theoretic arguments underlie large parts of the theory of those entities.

Galois theory uses groups to describe the symmetries of the roots of a polynomial (or more precisely the automorphism of the algebras generated by these roots). The fundamental theorem of Galois theory provides a link between algebraic field extensions and group theory. It gives an effective criterion for the solvability of polynomial equations in terms of the solvability of the corresponding Galois group. For example, S_5, the symmetric group in 5 elements, is not solvable, which implies that the general equation of degree ($n \geq 5$) cannot be solved by radicals as done in the case of lower orders. The theory, being one of the historical roots of group theory, is still fruitfully applied to yield new results in areas such as class field theory.

Algebraic topology is another domain which prominently associates groups to the objects the theory is interested in. There, groups are used to describe certain invariants of topological spaces. They are called 'invariants' because they are defined in such a way that they do not change if the space is subjected to some deformation. Algebraic geometry and cryptography likewise uses group theory in many ways. The presence of the group operation yields additional information which makes these varieties particularly accessible. They also often serve as a test for new conjectures. The one-dimensional case, namely, elliptic curves is studied in particular detail. They are both theoretically and practically intriguing. Very large groups of prime order constructed in Elliptic-Curve Cryptography serve for public key cryptography. Cryptography methods of this kind benefit from the flexibility of the geometric objects, hence their group structures, together with the complicated structure of these groups, make the discrete logarithm very hard to calculate. One of the earliest encryption protocols, Caesar's cipher, may also be interpreted as a very easy group operation.

The concept of the Lie group, named after mathematician Sophus Lie, is important in the study of differential equations and manifolds. These describe the symmetries of continuous geometric and analytical structures. Analysis on these and other groups is called harmonic analysis. Haar measures, that is, integrals invariant under the translation in a Lie group, are used for pattern recognition and other image processing techniques. In combinatorics, the notion of permutation groups and the concept of group action are often used to simplify the counting of a set of objects.

In physics, groups are important because they describe the symmetries which the laws of physics seem to obey. According to Noethers's theorem, every symmetry of a physical system corresponds to a conservation law of the system. Physicists are very interested in group representations, especially of Lie groups, since these representations often point the way to the 'possible' physical theories. Examples of the use of groups in physics include the Standard Model, gauge theory, the Lorentz group, and the Poincar'e group.

In chemistry and materials science, groups are used to classify crystal structures, regular polyhedrals, and the symmetries of molecules. The assigned point groups can then be used to determine physical properties (such as polarity and chirality), spectroscopic properties (particularly useful for Raman spectroscopy and infrared spectroscopy), and to construct molecular orbitals.

ADDITIONAL SOLVED EXAMPLES

Example W7.1: Determine whether the set $G=\{(a, b): a, b \in R, a \neq 0\}$ under the operation defined as $(a, b) . (c, d) = (ac, bc + d)$ for all $(a, b)(c, d) \in G$ is an Abelian group.

Solution:

1. Closure property is satisfied.
2. Associative property:
   Let (a, b), (c, d) and (e, f) be any three elements of G. Then
   $[(a, b).(c, d)].[(e, f) = (ac, bc + d).(e, f)$
   $= [(ac).e, (bc + d).e + f]$
   $= [ace, bce + de + f]$ (I)
   $(a, b).[(c, d).(e, f)] = (a, b).[ce, de + f]$
   $= [a.(ce), b.(ce) + (de + f)]$
   $= [ace, bce + de + f]$ (II)
   From I and II we conclude that the operation defined is associative.
3. Identity element:
   Now consider (x, y) as an element of G.
   Then $(a, b).(x, y) = (a, b) \in G$
   $(xa, ya + b) = (a, b)$
   This gives $xa = a$ and $ya + b = b$
   $\therefore x = 1, y = 0.$
   $\therefore$ (1, 0) is the identity element.
4. Inverse element:
   For each $(a, b) \in G$, there exists $(x, y) \in G$ such that
   $(x, y) .(a, b) = (1, 0)$
   $(xa, ya + b) = (1, 0)$
   This gives $xa = 1, ya + b = 0$
   $x=\frac{1}{a}$ and $y=-\frac{b}{a}$
   Since $a \neq 0$, x, y are real numbers and $(x, y) = \left(\frac{1}{a},-\frac{b}{a}\right)$ is the inverse of (a, b).
5. Commutative property:
   Now consider $(a, b).(c, d) = (ac, bc + d)$
   $(c, d).(a, b) = (ca, da + b)$
   $(a, b).(c, d) \neq (c, d).(a, b)$
   The commutative property does not hold good.
   $\therefore$ The set G defined under the given operation is a non-Abelian group.

Example W7.2: Show that $G = \{0, 1, 2, 3, 4, 5\}$ is an Abelian group under addition modulo 6.

$+_6$	0	1	2	3	4	5
0	0	1	2	3	4	5
1	1	2	3	4	5	0
2	2	3	4	5	0	1
3	3	4	5	0	1	2
4	4	5	0	1	2	3
5	5	0	1	2	3	4

1. Closure axiom: All the entries in the table are the elements of the set G.
   Closure axiom is satisfied.
2. Associative axiom:
   Consider the elements 3, 4, 5 $\in$ G
   $3 +_6 (4 +_6 5) = 3 +_6 3 = 0$
   $(3 +_6 4) +_6 5 = 1 +_6 5 = 0$
   $\therefore$ Associative axiom is satisfied.
3. Identity element: Identity element exists.
   Row heading 0 is the same as the topmost row.
   $\therefore$ 0 is the identity element.
4. Inverse element:
   $0^{-1} = 0, 1^{-1} = 5, 2^{-1} = 4, 3^{-1} = 3, 4^{-1} = 2, 5^{-1} = 1.$
   Every element has an inverse. $1 +_6 5 = 0$, $2 +_6 4 = 0$,--- etc.
5. The table is symmetrical about the principal diagonal. Hence the commutative property holds good.
   Therefore the set G under addition modulo 6 is an Abelian group.

ExampleW7.3: Examine whether G = {1, 2, 3, 4, 5} is a group under multiplication modulo 6.

X_6	1	2	3	4	5
1	1	2	3	4	5
2	2	4	0	2	4
3	3	0	3	0	3
4	4	2	0	4	2
5	5	4	3	2	1

1. Closure property: All the entries in the table are not elements of the set G. 0 is an entry and $0 \notin$ G.
   $\therefore$ Closure law is not satisfied.
   $\therefore$ (G, X_6) is not a group.

Example W7.4: Prove that G = {1, 2, 3, 4} is an Abelian group under multiplication, modulo 5.

X_5	1	2	3	4
1	1	2	3	4
2	2	4	1	3
3	3	1	4	2
4	4	3	2	1

1. Closure axiom: All the entries in the table are elements of the set G.
   Closure axiom is satisfied.

2. Associative axiom.
   Consider the elements 2, 3, 4 $\in$ G
   $2 X_5 (3 X_5 4) = 2 X_5 2 = 4$
   $(2 X_5 3) X_5 4 = 1 X_5 4 = 4$
   $\therefore$ Associative axiom is satisfied.

3. Identity element: Identity element exists.
   Row heading 1 is the same as the topmost row.
   ∴ 1 is the identity element.

4. Inverse element:
   $1^{-1} = 1, 2^{-1} = 3, 3^{-1} = 2, 4^{-1} = 4$.
   Every element has an inverse. $2\, X_5\, 3 = 1, 4\, X_5\, 4 = 1$, and so on.

5. The table is symmetrical about the principal diagonal. Hence commutative property holds good.
   $3\, X_5\, 4 = 4\, X_5\, 3 = 2$.
   Therefore the set G under multiplication modulo 5, is an Abelian group.

Example W7.5: Prove that G = {2, 4, 6, 8} is an Abelian group under multiplication modulo 10.

X_{10}	2	4	6	8
2	4	8	2	6
4	8	6	4	2
6	2	4	6	8
8	6	2	8	4

1. Closure axiom: All the entries in the table are elements of the set G.
   Closure axiom is satisfied.

2. Associative axiom:
   Consider the elements 2, 4, 8 ∈ G
   $(2\, X_{10}\, 4)\, X_{10}\, 8 = 8\, X_{10}\, 8 = 4$
   $2\, X_{10}\, (4\, X_{10}\, 8) = 2\, X_{10}\, 2 = 4$
   ∴ Associative axiom is satisfied.

3. Identity element: Identity element exists.
   Row heading 6 is the same as the topmost row.
   ∴ 6 is the identity element.

4. Inverse element:
   $2^{-1} = 8, 4^{-1} = 4, 6^{-1} = 6, 8^{-1} = 2$.
   Every element has an inverse. $2\, X_{10}\, 8 = 6, 4\, X_{10}\, 4 = 6$, and so on.

5. The table is symmetrical about the principal diagonal. Hence commutative property holds good. $6\, X_{10}\, 8 = 8\, X_{10}\, 6 = 6$.
   Therefore the set G under multiplication modulo 10 is an Abelian group.

Example W7.6: Let the operation * be defined on the set of real numbers R, as follows. $a*b = |a-b|\ \forall\ a,b \in R$. Prove that * is a commutative binary operation on R but (R, *) is not a semi-group.

Solution:

Since $|a-b|$ is always a positive real number and $a*b \in R\ \forall\ a,b \in R$

Then * is a binary operation on R.

Now $|b-a| = |a-b|\ \forall\ a,b \in R$

So $a*b = b*a$. Thus * is a commutative operation.

Let us check for associative property.
Consider 1, 2, 3 ∈ R.

$$1*(2*3) = 1*|2-3| = 1*1 = |1-1| = 0$$
$$(1*2)*3 = |1-2|*3 = 1*3 = |1-3| = 2$$

Thus $1*(2*3) \neq (1*2)*3$. Associative property does not hold good.
Hence (R,*) is not a semigroup.

Example W7.7: Let I be the set of positive integers and * be an operation on S = I * I, defined by

(i) $(a, b) * (c, d) = (ac, bd)$
(ii) $(a, b) * (c, d) = (a + c, b + d)$

Then show that S is a semigroup.

(i) $(a, b) * (c, d) = (ac, bd)$

1. Since closure property holds good.
2. Associative property:
   Let $x = (a, b), y = (c, d)$ and $z = (e, f)$

$$(x*y)*z = [(a,b)*(c,d)]*(e,f)$$
$$= [ac, bd]*(e,f)$$
$$= [(ac)e, (bd)f] \quad \text{(I)}$$
$$x*(y*z) = (a,b)*[(c,d)*(e,f)]$$
$$= (a,b)*[ce, df]$$
$$(a(ce), b(df)) \quad \text{(II)}$$

From (I) and (II), $((x*y)*z = x*(y*z)$
∴ it is associative. S is a semi-group.

(i) Closure property holds good
(ii) Secondly, we have

$$(xy)z = [(a,b)*(c,d)]*(e,f)$$
$$= [(a+c, b+d)]*(e,f)$$
$$= [(a+c)+e, (b+d)+f] \quad \text{(I)}$$
$$x(yz) = (a*b)*[(c,d)*(e,f)] = (a,b)*[(c+e),(d+f)]$$
$$= [a+(c+e), b+(d+f)] \quad \text{(II)}$$

From I and II, $(xy)z = x(yz)$
Associative property is satisfied. This proves that S is a semi group.

Example W7.8: In the set of Integers $\mathbb{Z}$, $(\mathbb{Z}, +)$ is a monoid. The identity element is 0, with respect to addition.

Example W7.9: Let $\mathbb{N}$ be the set of natural numbers. Then $(\mathbb{N},*)$ is a monoid. 1 is the identity element with respect to addition. The set of natural numbers $\mathbb{N}$ is not a monoid under addition. The identity element 0 does not exist in the set of natural numbers.

Example W7.10: Let G = {−1, 1, i, −i) where $i = \sqrt{-1}$. Then (G,*) is a cyclic monoid. The elements *i* and −*i* are its generators.

Example W7.11:

Let G = {------, −3, −2, −1, 0, 1, 2, 3--------} be the set of integers. Let H be a subgroup of G obtained on multiplying each element of G by 3. Then

H = {-------, −9, −6, −3, 0, 3, 6, 9, -----}.The Group G is Abelian, and any right coset will be equal to the left.

The right cosets of H in G are as follows.

H + 0 = {-, -, −9, −6, −3, 0, 3, 6, 9,-----------) for 0 ∈ G

H + 1 = {-, -, -, −8, −5, −2, 1, 4, 7, 10, --------} for 1 ∈ G

H + 2 = {-, -, -, -, −7, −4, −1, 2, 5, 8, 11, -------) for 2 ∈ G,

This shows that the right cosets of H, H + 1, and H + 2 are disjoint and distinct.

H + 3 = {-----, −6, −3, 0, 3, 6, 9, ------------} for 3 ∈ G as well as 3 ∈ H.

Again H + 4 + {-----, −5, −2, 1, 4, 7, 10, 13, ---------} for 4 ∈ G is equal to H + 1. Also, 4 ∈ H + 1 and so on.

Similarly the right coset H + 5 coincides with H + 2, H + 6 with H, H + (−1) with H = 2, H + (−2) and H + 1 and so on.

Hence there exist only three distinct right cosets H, H + 1, H + 2 and

G = H ∪ (H + 1) ∪ (H + 2)

Example W7.12: Let H = {1, −1) G = {1, −1, *i*, −I} then (H,*) is a subgroup of (G,*).

Solution:

The various left cosets and right cosets of H in G are given in the following table:

Left cosets of H in G	Right cosets of H in G
1 × H = {1, −1} = H	H × 1 = {1, −1} = H
−1 × H = { −1, 1 } = H	H × −1 = { −1, 1 } = H
i × H = { *i*, −*i* } = H	H × *i* = { *i*, −*i* } = H
−*i* × H = { −*i*, *i* } = H	H × −*i* = { −*i*, *i* } = H

There are only two distinct left cosets of H in G and two distinct right cosets of H in G.

Also G = H ∪ (H × *i*)

= H ∪ (*i* × H)

= H ∪ (H × (−*i*))

= H ∪ ((−*i*) × H).

SUMMARY

Modern group theory is a research area in mathematics. We have covered fairly well, the concepts of the algebraic systems. Groups have been dealt with in more detail. Groups are algebraic systems that find many applications in computer science, especially in coding theory. Cryptography is an active research area in computer science, more so in the Internet era, where secured transmission of information is a challenge. Haar measures, that is, integrals invariant under the translation in a Lie group, are used for pattern recognition and other image processing techniques. In chemistry and materials science, groups are used to classify crystal structures, regular polyhedrals, and the symmetries of molecules.

EXERCISES

1. What does the statement, 'a set is closed with respect to a binary operation', mean?
2. Prove that {2, 4, 6, 8} is an Abelian group under multiplication mod 10.
3. In a group of integers I, an operation * is defined by $a*b = a+b-1 \ \forall a,b \in I$. Find the identity element.
4. Prove that G = {1, 5, 7, 11} is an Abelian group under multiplication mod 12.
5. In a group of integers, an operation * is defined by $a * b = a + b - 3$. Find the identity and inverse of 5.
6. In any group G = {1, 2, 3, 4, : *x* mod 5}, find inverses of 2 and 5.
7. If Q_{-1} is the set of all rational numbers except −1, and * is a binary operation defined by Q_{-1} $a*b = a+b+ab$, show that $(Q_{-1},*)$ is a group. Solve $2*x = 3^{-1}$ in Q_{-1}.
8. Show that set of all real numbers of the form $a+b\sqrt{2}$, where $a,b \in Q$, the set of rational numbers, is a group under addition.
9. Show that $H_1 = \{0, 2, 4\}$, $H_2 = \{0, 3\}$ are subgroups of the group $G = \{0, 1, 2, 3, 4, 5, + \text{mod } 6\}$
10. Prove that H = {0, 2, 4} is a subgroup of $(Z_6, +)$.
11. If G is an Abelian group with identity e, prove that all elements *x* of G satisfying the equation $x^2 = e$ form a subgroup H of G.
12. Prove that the additive group of even integers is cyclic.
13. Prove that (Z_5, °) is a cyclic group. Find all its generators.
14. Show that (H, *) is a subgroup of group (S_3, *) of degree 3, where $H = \{p_1, p_2, p_3\}$. Find also the right cosets of H in G.
15. If G is the additive group of integers and H is a subgroup of G defined by $H = \{5x \mid x \in G\}$, find the distinct left cosets of H in G.
16. In the group S_3, let $\alpha = \begin{pmatrix} 1 & 2 & 3 & 4 & 5 \\ 2 & 3 & 1 & 4 & 5 \end{pmatrix}$ and $\beta = \begin{pmatrix} 1 & 2 & 3 & 4 & 5 \\ 2 & 1 & 5 & 3 & 4 \end{pmatrix}$. Determine $\alpha\beta, \alpha^3, (\alpha\beta)^{-1}$, $\beta^{-1}\alpha^{-1}$.
17. For a group G, prove that the function $f: G \to G$ defined by $f(a) = a^{-1}$ is an Isomorphism if and only if G is Abelian.
18. Prove that the function $f: \mathbb{Z} \to \mathbb{Z}$ defined by $f(x) = 5x$ is homomorphism.
19. Prove that S × T is a commutative semi-group, where (S, *) and (T, *) is a commutative semi-group.
20. Determine whether the relation R on the semi-group S is a congruence relation.
    Also describe the quotient semi-group for S and R.
21. Write the multiplication table of the group $Z_3 \times Z_4$.
22. Consider a group G with identity e. If $x^2 = x$ for some *x* in G, then show that $x = e$. If $a^2 = a$ for all *a* in G, then prove that G is Abelian.
23. Consider a finite group G with identity *e* and let *a* be an arbitrary element of G.
    Prove that there exists a positive integer *n* such that $a_n = e$.
24. Why are groups said to have more structure than semi-groups?
25. Prove by mathematical induction that for a group G if $ab = ba$ then $(ab)^n = a^n b^n$ for all *n* that belongs to $\mathbb{Z}^+$.

8

Coding Theory

In Chapter 7, we have discussed algebraic systems or structures. One of the applications of algebraic systems is found in Coding theory. Codes are used in data compression, cryptography, error coding and more recently, in network coding. Codes are studied by various scientific disciplines such as information theory, electrical engineering, mathematics, and computer science for the purpose of designing efficient and reliable data transmission methods. One needs to study the properties of codes and their fitness for a specific application in coding theory.

8.1 INTRODUCTION

The American mathematician Claude E. Shannon (1916–2001) worked at Bell Laboratories and proposed the mathematical theory of communication. He is considered the father of the field of information theory. His friend Richard Hamming and his colleagues at Bell laboratories founded the field of coding theory. Coding theory deals with channel encoding and channel decoding activities. Coding theory typically involves the removal of redundancy and the correction (or detection) of errors in the transmitted data. There are essentially two aspects to coding theory, namely, i) Data compression (also called, source coding) and ii) Error correction (also called channel coding). The source encoding consists of compression of the data in order to transmit more efficiently. This is the established practice in Internet transactions. The Zip data compression is used to reduce the network load and make files smaller. Channel encoding adds extra data bits to make the transmission of data more robust and secure. The ordinary user is not aware of channel coding. A typical music CD uses the Reed-Solomon code to correct for scratches and dust. In this application, the transmission channel is the CD itself. Cell phones also use coding techniques to correct for the fading and noisy high frequency radio transmission. Data modems, telephone transmissions and NASA communications, all employ channel coding techniques to get the bits through, as in the case of the turbo code and LDPC codes. In general, a typical number of stages in data processing are shown in Fig. 8.1.

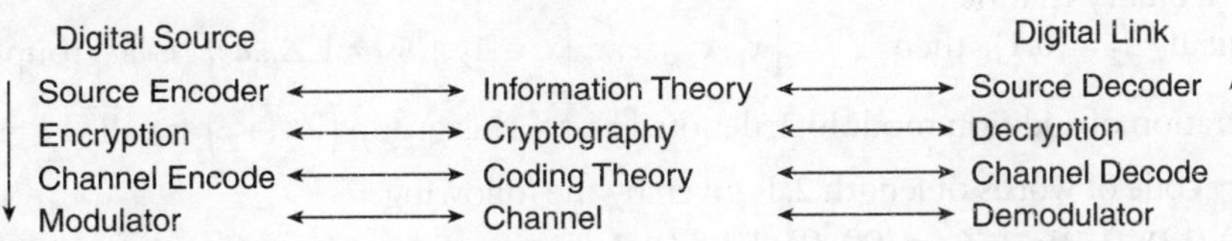

Figure 8.1 A flow representing the various stages of data processing

Source encoding and decoding is the main concern of the domain called information theory. The functional area that deals with encryption and decryption of data is called cryptography. During transmission of data through a channel, the systematic arrangement of such data from the source to the destination point is governed by a branch called coding theory. There are several coding techniques available in the present times.

8.1.1 Some Basic Coding Techniques

Transmitted messages, like data from a satellite, invariably suffer from noise. It is important to encode a message in such a way that despite the noise, one is able to decode a message to its original form. This is sometimes carried out by repeating the message two or more times; this is something very common in human speech. However, copying data stored on a compact disk, or a floppy disk more than once, requires extra space for storing. The coded data along with compression is one solution. The basic unit of information is called a message and encoding of this message is called **coding**. The detection of errors in a scrambled message is called **error detection**. The process of correcting the errors is called **error correction**. It is considered harder to generate error correcting codes than error-detecting ones.

Most messages are sent as digital-sequences of 0's and 1's, such as 10101 or 1010011. Let us assume that we want to send the message 1011 in binary for a real word, such as 'buy', or a sentence such as 'buy stocks'. One way of encoding 1011 would be to attach a binary 'tail' to it so that if the message gets distorted to, say, 0011, we can detect the error. One such tail could be a 1 or 0, depending on whether we have an odd or an even number of 1's in the word. This way all encoded words will have an even number of 1's. So 1011 will be encoded as 10111. This is called even parity. Now if this is distorted to 00111, we know that an error has occurred, because we have received an odd number of 1's. Similarly, if we look for odd number of 1's in the word at the time of encoding then it is called odd parity. In general, this error-detecting code is called a **parity check** and is considered too simple though very useful. For example, if two digits are changed, this scheme does not detect the error. Another approach would be to encode the message by repeating it twice, such as 10111011. This code, if received as 00111011, would indicate that one of the two equal halves is distorted. If only one error occurs, then it is clearly at position 1. This coding scheme also gives poor results and is not often used. We get better results by repeating the message several times, but that takes space and time in the communication process.

8.2 COMMUNICATION PROCESS

The process of communication involves transmitting information between the source and destination. The three essential parts in an ideal communication system are the (1) Transmitter (2) Channel and (3) Receivers. The communication channel is subjected to a variety of disturbances called noise. It may be due to lightning, folds in a magnetic tape or in any other reasons. Coding theory deals with minimizing the disturbances due to noise. Consider a typical communication system with noise. The information is defined by sequences of 0's and 1's, each is called a digit. It is denoted by the alphabet $Z_2 = \{0,1\}$. A word is a sequence of digits. The number of digits in the word is called the length of word. For example, the word 1011 is of length of 4 and 00101 is of length 5. A word is transmitted by sending its digits, one after the other across a binary channel.

- **Definition:** If $Z_2 = \{0,1\}$, then $Z_2^n = \left\{x_1, x_2, \ldots\ldots, x_n \,\middle|\, x_i \in Z_2, \text{for } i = 1,2,\ldots n\right\}$ is a group under the binary operation of addition modulo 2, denoted as $+_2$. This group $\left(Z_2^n, +_2\right)$ is called a group code.

The set of binary code of words of length 2, is given by the following
$Z_2^2 = \{(0,0),(0,1),(1,0),(1,1)\}$ $Z_2^2 = \{00, 01, 10, 11\}$. Let us consider $(Z_2^2, +_2)$ as given in Table 8.1.

Table 8.1 Operation table

$+_2$	00	01	10	11
00	00	01	10	11
01	01	00	11	10
10	10	11	00	01
11	11	10	01	00

It is necessary to note that $1 +_2 1 = 0$, $0 = 00\ldots0$ is the identity and every element is its own inverse. Hence $(Z_2^n, +_2)$ is an Abelian group. In general, any code which is a group under the operation $+_2$ is called a group code.

8.3 ENCODERS AND DECODERS

The following are some encoders and decoders being discussed for the purpose of illustrating the concept, along with few examples.

8.3.1 Hamming Metric

The number of 1's in the binary string $x \in Z_2^n$ is called the weight of x and is denoted as wt(x) or $|x|$.

Example 8.1: If $x = 11001 \in Z_2^n$, $\text{wt}(x) = 3$. $x = 000 \in Z_2^n$, $\text{wt}(x) = 0$ then Hamming distance is defined as follows.

Let $x = x_1, x_2, \ldots x_n$ and $y = y_1, y_2, \ldots y_n$ be binary strings. The hamming distance between x and y, denoted by $d(x, y)$, is the number of bits such that $x_i \neq y_i$ for $1 \leq i \leq n$.

$$\therefore \; d(x, y) = \text{wt of } (x +_2 y) \text{ or } |x +_2 y|$$

$$= \sum_{i=1}^{n} (x_i +_2 y_i)$$

Example 8.2: Let $x = 110010, y = 010110$, where $x, y \in Z_2^6$

Then $$x +_2 y = 100100 \text{ and } \text{wt}(x +_2 y) = 2$$

$$\therefore \; d(x, y) = \text{wt of } (x +_2 y) = 2$$

It is necessary to take note of the following observations in the context of Hamming distance.

(i) If $x, y \in Z_2^n$, $d(x, y) = \text{wt}(x +_2 y)$ and d be a real valued function on

$$Z_2^n \text{ X } Z_2^n, \text{ satisfying } \forall\, x, y, z \in Z_2^n,$$

then (i) $d(x, y) \geq 0$ (ii) $d(x, y) = 0 \;\; x \Leftrightarrow y$

(iii) $d(x, y) = d(y, x)$ (iv) $d(x, z) \leq d(x, y) + d(y, z)$

(Z_2^n, d) is a metric space and d is a Hamming metric.

(ii) Let $M \subset Z_2^n$ be a code. The minimum distance between the code words of M is min{d(x, y): x, y ∈ M. For example M = {1001, 0100, 1000},

$$x = 1001, y = 0100, z = 1000$$
$$x +_2 y = 1101 \quad d(x, y) = 3$$
$$x +_2 z = 0001 \quad d(x, z) = 1$$
$$y +_2 z = 1100 \quad d(y, z) = 3$$
$$\therefore \text{Minimum distance} = 1$$

8.3.2 Encoding Function

Due to noise, the receiver may make some mistakes that go undetected, like an interchange of **0** or **1**. For example, the code 000 may be changed to 100 or 010 or 001 due to noise. The codes obtained by introducing additional digits called parity digits to the digits in the original message are called Hamming codes. If the original message is a binary string of length *m*, the Hamming encoded message consisting of a string of length *n*, where $(n > m)$. Out of *n* digits, *m* digits are being used to represent the information part of the message and the remaining *(n-m)* digits are used for the detection and correction of errors in the received message.

8.3.2.1 *Even parity checks*

In Hamming's single-error detection code of length *n*, the first (*n*-1) digits contain the information part of the message and last digit is made either **0** or **1**. If the digit introduced in the last position gives an even number of 1's in the encoded word of length *n*, then the entire message is said to have even parity.

Let $E: Z_2^m \to Z_2^n$ be an encoding function such that $x = x_1x_2 \ldots x_m \in Z_2^n$ and $E(x) = x_1x_2 \ldots x_n$ and

$$x_n = \begin{cases} 0 & \text{if } wt(x) \text{ - is even} \\ 1 & \text{if } wt(x) \text{ - is odd} \end{cases}$$

This encoding procedure is called an even parity check.

Example 8.3: Let us consider $E: Z_2^3 \to Z_2^4$ and define $E(x_1x_2x_3) = x_1x_2x_3x_4$.

$$x_4 = \begin{cases} 0 & \text{if } wt(x) \text{ -is even} \\ 1 & \text{if } wt(x) \text{ - is odd} \end{cases}$$

Table 8.2 Code and code with redundant digit

Original code	Code with redundant digit
000	0000
001	0011
010	0101
011	0110
100	1001
101	1010
110	1100
111	1111

Let us consider the word 000 and E (000) = 0000. If a single error occurs in 0000, it must be 1000, or 0100 or 0010 or 0001. Hence, with the use of redundant digits, we are able to detect and correct a single error occurring in the code, as shown in Table 8.2.

8.3.2.2 *Odd parity check*

In the information part of the message, the last digit is appended as either **0** or **1**. If the digit introduced in the last position gives an odd number of 1's in the encoded word of length *n*, then the entire message is said to have an odd parity.

Let $E{:}Z_2^m \to Z_2^n$ be an encoding function such that $x = x_1x_2 \ldots x_m \in Z_2^n$ and

$$E(x) = x_1x_2 \ldots x_n \text{ and } x_n = \begin{cases} 1 \text{ if } \mathrm{wt}(x) \text{ - is even} \\ 0 \text{ if } \mathrm{wt}(x) \text{ - is odd} \end{cases}$$

This type of encoding procedure is called an odd parity check.

Example 8.4: Let us consider $E{:}Z_2^3 \to Z_2^4$ and define $E(x_1x_2x_3) = x_1x_2x_3x_4$.

$$x_4 = \begin{cases} 1 \text{ if } \mathrm{wt}(x) \text{ - is even} \\ 0 \text{ if } \mathrm{wt}(x) \text{ - is odd} \end{cases}$$

Table 8.3 Code and code with redundant digit

Original code	Code with redundant digit
000	0001
001	0010
010	0100
011	0111
100	1000
101	1011
110	1101
111	1110

It is necessary to note the following observations, as given in Table 8.3.

(i) A code detects at the most *k* errors if and only if the minimum distance between any two words is at least $(k + 1)$. Hence, if the code word *x* is transformed to the word *y* due to at least $(k + 1)$ errors almost *k* errors can be detected.

(ii) A code can correct a set of at the most *k* errors if and only if the minimum distance between any two code words is at least $(2k + 1)$.

8.4 DECODING AND ERROR CORRECTION

Let W denote the set of words to be transmitted and $W \in Z_2^m$. In order to reduce the number of errors in the transmission of the words, the following procedure is adopted. Choose an integer $(n > m)$, such that for each word $w \in W$, a unique word $c \in Z_2^n$ is assigned. This is performed by appending $(n - m)$ extra components to *w* by means of a one-to-one function, $E{:}W \to Z_2^n$. Hence, $c = E(w)$. The appending of digits helps to detect or correct errors that occur during transmission. The process of appending extra 0's and 1's to w through the function E is called **the encoding process.**

The word $c = E(w)$ is transmitted through a channel. Let $r = T(c) \in Z_2^n$ be the word received. The extra components contained are removed with the help of an onto function, $D: Z_2^n \to Z_2^m$, to get the word $D(r) \in Z_2^m$. The process of obtaining the original word w from D(r) is called **decoding**. It is necessary to take note of the following observations.

(i) The word $w \in W$ is called a message and $c = E(w) \in Z_2^n$ is called the code word assigned to w.
(ii) The function E is one-one. This means that different words in W have different code words.
(iii) The set $r = T(c)$ is called the received word and the word D(r) is called the decoded word.
(iv) The function *E* is called the encoding function and the function *D* is called the decoding function.

Example 8.5: Consider the encoding function $E: Z_2^m \to Z_2^n$ $(n > m)$ defined as follows:

$$E(x) = E(x_1, x_2, \ldots x_m) = x_1 x_2 \ldots x_m, x_{m+1}$$

$$x_{m+1} = \begin{cases} 0 & \text{if } wt(x) \text{ - even} \\ 1 & \text{if } wt(x) \text{ - odd} \end{cases}$$

and the corresponding decoding function, $D: Z_2^{m+1} \to Z_2^m$ $D(r) = D(r_1, r_2, \ldots r_{m+1}) = r_1 r_2 \ldots r_m$, which is an example of parity-check code.

(a) Find the code words assigned by E to the following messages in Z_2^3,

000, 001, 011, 100, 110, 101, 111, 010.

(b) Find the decoded words assigned by D, to the following received words in Z_2^4

0000, 0001, 0101, 1111, 1010, 1100, 1101, 1001.

Solution:

By the definition of E, we have the following

(a) $E(000) = 0000$, $E(001) = 0011$, $E(011) = 0110$, $E(100) = 1001$
$E(110) = 1100$, $E(101) = 1010$, $E(111) = 1111$, $E(010) = 0101$.
(b) By the definition of D,
$D(0000) = 000$, $D(0001) = 000$, $D(0101) = 010$, $D(1111) = 111$
$D(1010) = 101$, $D(1100) = 110$, $D(1101) = 110$, $D(1001) = 100$

8.5 PARITY AND GENERATOR MATRICES

The encoding and decoding functions can be expressed as matrices over Z_2. Given $m, n \in Z^+, m < n$, the encoding function $E: Z_2^m \to Z_2^n$ is given by an $m \times n$ matrix. The matrix obtained is called the generator matrix for the code and is of the form $[I_m | A] = G$, where I_m is a unit matrix of order *m* and A is an $m \times (n - m)$ matrix to be chosen appropriately.

Let $G = \begin{pmatrix} 1 & 0 & 0 & 1 & 1 & 0 \\ 0 & 1 & 0 & 0 & 1 & 1 \\ 0 & 0 & 1 & 1 & 0 & 1 \end{pmatrix}_{3\times 6}$ be a 3×6 matrix over Z_2. Then $G = [I_m | A]$, where $I_3 = \begin{pmatrix} 1 & 0 & 0 \\ 0 & 1 & 0 \\ 0 & 0 & 1 \end{pmatrix}$: $A = \begin{pmatrix} 1 & 1 & 0 \\ 0 & 1 & 1 \\ 1 & 0 & 1 \end{pmatrix}$. The augmented matrix G is called a generator matrix.

Example 8.6: Let $E:Z_2^3 \to Z_2^6$ be an encoding function defined by $E(w) = wG \in Z_2^6$.

Now $Z_2^3 = \{000, 100, 010, 001, 110, 101, 011, 111\}$

Now, $E(000) = (000)\begin{pmatrix} 1 & 0 & 0 & 1 & 1 & 0 \\ 0 & 1 & 0 & 0 & 1 & 1 \\ 0 & 0 & 1 & 1 & 0 & 1 \end{pmatrix} = (000000)$

$$E(100) = (100)\begin{pmatrix} 1 & 0 & 0 & 1 & 1 & 0 \\ 0 & 1 & 0 & 0 & 1 & 1 \\ 0 & 0 & 1 & 1 & 0 & 1 \end{pmatrix} = (100110)$$

$$E(010) = (010)\begin{pmatrix} 1 & 0 & 0 & 1 & 1 & 0 \\ 0 & 1 & 0 & 0 & 1 & 1 \\ 0 & 0 & 1 & 1 & 0 & 1 \end{pmatrix} = (010011)$$

$$E(001) = (001)\begin{pmatrix} 1 & 0 & 0 & 1 & 1 & 0 \\ 0 & 1 & 0 & 0 & 1 & 1 \\ 0 & 0 & 1 & 1 & 0 & 1 \end{pmatrix} = (001101)$$

and so on. The set of code words are given by

$$D = \{000000, 100110, 010011, 001101, 110101, 101011, 011110, 111000\} \in Z_2^6 \qquad \text{(I)}$$

We calculate the distances of words in D and the minimum distance here is 3. Hence, we can detect 2 or less errors, and correct a single error.

Let $\quad W = w_1 w_2 w_3 \in Z_2^3$ and $E(w) = w_1 w_2 w_3 w_4 w_5 w_6 \in Z_2^6$

$$E(w) = (w_1 w_2 w_3)\begin{pmatrix} 1 & 0 & 0 & 1 & 1 & 0 \\ 0 & 1 & 0 & 0 & 1 & 1 \\ 0 & 0 & 1 & 1 & 0 & 1 \end{pmatrix}$$

$$= [w_1 w_2 w_3 (w_1 +_2 w_3)(w_1 +_2 w_2)(w_2 +_2 w_3)]$$

$$w_1 w_2 w_3 w_4 w_5 w_6 = [w_1 w_2 w_3 (w_1 +_2 w_3)(w_1 +_2 w_2)(w_2 +_2 w_3)]$$

Comparing both sides, we get

$$w_4 = (w_1 +_2 w_3), \quad w_5 = (w_1 +_2 w_2), \quad w_6 = (w_2 +_2 w_3).$$

These equations are called parity-check equations. $w_4 w_5 w_6$ are either 0's or 1's.
The equations are rewritten as

$$w_1 +_2 w_3 +_2 w_4 = 0, \qquad w_1 +_2 w_2 +_2 w_5 = 0, \qquad w_2 +_2 w_3 +_2 w_6 = 0$$

Let us consider the matrix H defined as $H = \left[A^T | I_3\right]$, where A is a matrix of size $m \times (n - m)$ (i.e., 3 × 3). The matrix, A^T is also a 3 × 3 matrix. Consequently, H is a matrix of size $(n - m) \times n$, i.e., (3 × 6). This matrix is uniquely determined by the matrix A, and is called the parity-check matrix.

We can prove that $H.(E(w))^T = \begin{pmatrix} 0 \\ 0 \\ 0 \end{pmatrix}$, where $H = \left[A^T | I_3\right]$ and $(E(w))^T$ is a transpose of $(E(w))$. Now $r = r_1, r_2 \ldots r_6 \in Z_2^6$ is a code word if and only if $H\,r^T = \begin{pmatrix} 0 \\ 0 \\ 0 \end{pmatrix}$.

It is to be noted that the parity-check matrix H provides a decoding scheme that corrects single errors in transmission. If the parity-check matrix H has no column of 0's and no two columns of H are the same, then the following decoding technique helps in correcting a single error.

Example 8.7: Consider the parity-check matrix (H) for an encoding function E: $Z_2^3 \to Z_2^6$ given by

$$H = \begin{bmatrix} 1 & 0 & 1 & 1 & 0 & 0 \\ 1 & 1 & 0 & 0 & 1 & 0 \\ 1 & 0 & 1 & 0 & 0 & 1 \end{bmatrix}$$

and the receive a word $r = 110110$, which is not a code word.

$$\text{The column matrix } H.r^1 = \begin{pmatrix} 1 & 0 & 1 & 1 & 0 & 0 \\ 1 & 1 & 0 & 0 & 1 & 0 \\ 0 & 1 & 1 & 0 & 0 & 1 \end{pmatrix} \begin{pmatrix} 1 \\ 1 \\ 0 \\ 1 \\ 1 \\ 0 \end{pmatrix} = \begin{pmatrix} 0 \\ 1 \\ 1 \end{pmatrix}$$

and is the same as the second column of H. We get to know that an error occurred in the second component of r. By changing the second component, we get 100110 and dropping the last three digits, the message sent is 100.

Now $r = 110110$, $c = 100110$ $e = 010000$ which is a transmission error of weight 1.

Hence, $$r = c +_2 e \text{ or } c = r +_2 e$$

If $H\,r^T$ is not one of the columns of H, but can be obtained as the sum of the two columns of H, multiple errors are generated and cannot be corrected.

(i) If $H\,r^T = 0$, the transmission is correct and r is the code word that is transmitted. The decoded message consists of the first m components of r.

(ii) If $H\,r^T$ equals the i-th column of H, then there is a single error in the i- th component of r and changing it, we get the correct code word C. The first m components of C give the original message.

It is to be noted that decoding of 111111 is not possible by using H. $r = 111111$

$$H.r^T = H(1\,1\,1\,1\,1\,1)^T = \begin{pmatrix} 1 & 0 & 1 & 1 & 0 & 0 \\ 1 & 1 & 0 & 0 & 1 & 0 \\ 0 & 1 & 1 & 0 & 0 & 1 \end{pmatrix} \begin{pmatrix} 1 \\ 1 \\ 1 \\ 1 \\ 1 \\ 1 \end{pmatrix} = \begin{pmatrix} 1 \\ 1 \\ 1 \end{pmatrix}$$

We observe that this matrix is not a zero matrix and is not equal to any of the columns of H. Hence, we cannot decode $r = 111111$ by using H.

Example 8.8: For the message $w \in Z_2^2$ and encoding function $E: Z_2^2 \to Z_2^5$, the generator matrix is given by $G = \begin{pmatrix} 1 & 0 & 1 & 1 & 0 \\ 0 & 1 & 0 & 1 & 1 \end{pmatrix}$

Find all possible code words. Obtain parity-check matrix H.

Solution:
The words that belong to Z_2^2 are $Z_2^2 = (00,10,01,11)$. The code words corresponding to the above message words are given as under.

$$E(00) = (0\,0)\begin{pmatrix} 1 & 0 & 1 & 1 & 0 \\ 0 & 1 & 0 & 1 & 1 \end{pmatrix} = (00000)$$

$$E(10) = (1\,0)\begin{pmatrix} 1 & 0 & 1 & 1 & 0 \\ 0 & 1 & 0 & 1 & 1 \end{pmatrix} = (10110)$$

$$E(00) = (0\,1)\begin{pmatrix} 1 & 0 & 1 & 1 & 0 \\ 0 & 1 & 0 & 1 & 1 \end{pmatrix} = (01011)$$

$$E(11) = (1\,1)\begin{pmatrix} 1 & 0 & 1 & 1 & 0 \\ 0 & 1 & 0 & 1 & 1 \end{pmatrix} = (11101)$$

From the above matrices, we get back the message word from the corresponding code word by dropping last 3 $(n-m)$ digits.

For all $w_1 w_2 \in Z_2^2$, the code words are C = {00000, 10110, 01011, 11101}.

$$E(w) = w_1 w_2 w_3 w_4 w_5 \in Z_2^5 \qquad \text{(I)}$$

The matrix G is of the form $G = (I_2 | A)$

where $I_2 = \begin{pmatrix} 1 & 0 \\ 0 & 1 \end{pmatrix}$ $A = \begin{pmatrix} 1 & 1 & 0 \\ 0 & 1 & 1 \end{pmatrix}$

$$E(w) = wG = (w_1 w_2)\begin{pmatrix} 1 & 0 & 1 & 1 & 0 \\ 0 & 1 & 0 & 1 & 1 \end{pmatrix}$$

$$= (w_1, w_2, w_1, w_1 + w_2, w_2) \qquad \text{(II)}$$

From (I) and (II) we have $w_1 = w_3$, $w_1 + w_2 = w_4$ and $w_2 = w_5$ (III)
The above set of equations becomes

$$\begin{aligned} x_1 + x_3 \quad &= 0 \\ x_1 + x_2 + x_4 \quad &= 0 \\ x_2 + x_5 &= 0 \end{aligned} \qquad \text{(IV)}$$

We can easily show that

$$\begin{pmatrix}1&0&1&0&0\\1&1&0&1&0\\0&1&0&0&1\end{pmatrix}\begin{pmatrix}w_1\\w_2\\w_3\\w_4\\w_5\end{pmatrix}=\begin{pmatrix}0\\0\\0\end{pmatrix}$$

$$\mathrm{HE}\left[(w)\right]^{\mathrm{T}}=\begin{bmatrix}0\\0\\0\end{bmatrix} \text{ where } H=\left(A^{\mathrm{T}}|I_3\right) \quad A^{\mathrm{T}}=\begin{pmatrix}1&0\\1&1\\0&1\end{pmatrix} \quad I_{\text{n-m}}=I_3=\begin{pmatrix}1&0&0\\0&1&0\\0&0&1\end{pmatrix}$$

H does not contain a column of only 0's and no two columns of H are the same. The parity-check matrix H provides a decoding scheme that corrects a single error. Therefore,

$$\begin{aligned}&d\big(\mathrm{E}(00),\mathrm{E}(01)\big)=3, \quad d\big(\mathrm{E}(00),\mathrm{E}(10)\big)=3\\&d\big(\mathrm{E}(00),\mathrm{E}(11)\big)=4, \quad d\big(\mathrm{E}(01),\mathrm{E}(10)\big)=4\\&d\big(\mathrm{E}(01),\mathrm{E}(11)\big)=3, \quad d\big(\mathrm{E}(10),\mathrm{E}(11)\big)=3\end{aligned} \tag{II}$$

min(E) =3. The code can detect all errors of weight less than or equal to 2 and can correct all the single errors. It is to be noted that if $r=[11101]$, then H is written as follows.

$$H\,r^{\mathrm{T}}=\begin{pmatrix}1&0&1&0&0\\1&1&0&1&0\\0&1&0&0&1\end{pmatrix}\begin{pmatrix}1\\1\\1\\0\\1\end{pmatrix}=\begin{pmatrix}0\\0\\0\end{pmatrix}$$

Hence, r itself is the code word transmitted and the decoded message is 11. The decoded message is obtained by taking first two ($m=2$) components of r.

If $H\,r^{\mathrm{T}}=i^{\text{th}}$ column of H, then we conclude that a single error has occurred during the transmission and it has occurred in the ith component of r. Changing the ith component of r, we get the code word C, which is transmitted. As stated before, the first m components of C give the original message.

Let $r=(11011)$, then

$$H\,r^{\mathrm{T}}=\begin{pmatrix}1&0&1&0&0\\1&1&0&1&0\\0&1&0&0&1\end{pmatrix}\begin{pmatrix}1\\1\\0\\1\\1\end{pmatrix}=\begin{pmatrix}1\\1\\0\end{pmatrix}$$

which is the first column of H. A single error has occurred in the first component of r. We get the transmitted code word as 01011. Taking the first two components of the code, we get 01 as original message. When neither (1) nor (2) occur, more than one transmission error would have occurred. The detection of error is possible in this case, but correction is not possible.

$r = (11010)$ then

$$Hr^T = \begin{pmatrix} 1 & 0 & 1 & 0 & 0 \\ 1 & 1 & 0 & 1 & 0 \\ 0 & 1 & 0 & 0 & 1 \end{pmatrix}\begin{pmatrix} 1 \\ 1 \\ 0 \\ 1 \\ 0 \end{pmatrix} = \begin{pmatrix} 1 \\ 1 \\ 1 \end{pmatrix}$$

Since $Hr^T \neq$ any column of H, hence more than one transmission error has occurred.

Since $\begin{pmatrix} 1 \\ 1 \\ 1 \end{pmatrix} = \begin{pmatrix} 1 \\ 1 \\ 0 \end{pmatrix} + \begin{pmatrix} 0 \\ 0 \\ 1 \end{pmatrix} =$ first column of H + fifth column of H.

two errors have occurred in transmission, one in the first component and the other in the fifth component of *r*. By changing these components, the code word transmitted can be assumed as 01011 and hence the original message can be taken as 01.

On the other hand, the right hand side matrix of Hr^T can also be taken as

$\begin{pmatrix} 1 \\ 1 \\ 1 \end{pmatrix} = \begin{pmatrix} 0 \\ 1 \\ 1 \end{pmatrix} + \begin{pmatrix} 1 \\ 0 \\ 0 \end{pmatrix} =$ second column of H + third column of H.

Here, two errors may have occurred, one in the second component and the other in third component of *r*. By changing these components in *r*, the code word transmitted is 10110, and hence the original message can be taken as 10. There is an ambiguity as to which message has been encoded and transmitted. Hence, the correction is not possible, even though errors have been detected; the minimum distance between any pair of words as given by equation (II) is three. Hence, two errors can be detected and at the most, one error can be corrected.

Example 8.9: Find the code words generated by the encoding function $E: Z_2^2 \to Z_2^5$, with respect to the parity-check matrix

$$H = \begin{pmatrix} 0 & 0 & 1 & 0 & 0 \\ 1 & 1 & 0 & 1 & 0 \\ 1 & 1 & 0 & 0 & 1 \end{pmatrix} = \left(A^T | I_{n\text{-}m}\right)$$

The generator matrix G is given by

$$G = (I_2 | A) = \begin{pmatrix} 1 & 0 & 0 & 1 & 1 \\ 0 & 1 & 0 & 1 & 1 \end{pmatrix}$$

Now $Z_2^2 = \{00, 01, 10, 11\}$ E (w) = WG.

$$E(00) = (0\,0)\begin{pmatrix} 1 & 0 & 0 & 1 & 1 \\ 0 & 1 & 0 & 1 & 1 \end{pmatrix} = (00000)$$

$$E(01) = (0\,1)\begin{pmatrix} 1 & 0 & 0 & 1 & 1 \\ 0 & 1 & 0 & 1 & 1 \end{pmatrix} = (01\,011)$$

$$E(10)=(10)\begin{pmatrix}1&0&0&1&1\\0&1&0&1&1\end{pmatrix}=(10011)$$

$$E(11)=(11)\begin{pmatrix}1&0&0&1&1\\0&1&0&1&1\end{pmatrix}=(11000)$$

The code words generated are as follows.

$$C=\{00000, 01011, 10011, 11000\}.$$

Example 8.10: Finds the code words generated by the encoding function $E:Z_2^3 \to Z_2^6$ with respect to the parity-check matrix

$$H=\begin{pmatrix}1&1&0&1&0&0\\1&0&1&0&1&0\\1&1&1&0&0&1\end{pmatrix}=\left(A^T|I_3\right)$$

As per our notation, the generator matrix G is given by

$$G=\left(I_3|A\right)=\begin{pmatrix}1&0&0&1&1&1\\0&1&0&1&0&1\\0&0&1&0&1&1\end{pmatrix}$$

$$Z_2^3=(000,001,010,100,011,101,110,111)$$

$$E(000)=(000)\begin{pmatrix}1&0&0&1&1&1\\0&1&0&1&0&1\\0&0&1&0&1&1\end{pmatrix}=(000000)$$

$$E(001)=(001)\begin{pmatrix}1&0&0&1&1&1\\0&1&0&1&0&1\\0&0&1&0&1&1\end{pmatrix}=(001011)$$

$$E(010)=(010)\begin{pmatrix}1&0&0&1&1&1\\0&1&0&1&0&1\\0&0&1&0&1&1\end{pmatrix}=(010101)$$

$$E(100)=(100)\begin{pmatrix}1&0&0&1&1&1\\0&1&0&1&0&1\\0&0&1&0&1&1\end{pmatrix}=(100111)$$

$$E(011)=(011)\begin{pmatrix}1&0&0&1&1&1\\0&1&0&1&0&1\\0&0&1&0&1&1\end{pmatrix}=(011110)$$

$$E(101)=(101)\begin{pmatrix}1&0&0&1&1&1\\0&1&0&1&0&1\\0&0&1&0&1&1\end{pmatrix}=(101100)$$

$$E(110)=(110)\begin{pmatrix}1&0&0&1&1&1\\0&1&0&1&0&1\\0&0&1&0&1&1\end{pmatrix}=(110010)$$

$$E(111)=(111)\begin{pmatrix}1&0&0&1&1&1\\0&1&0&1&0&1\\0&0&1&0&1&1\end{pmatrix}=(111001)$$

Thus, the code words generated are as follows.

$$C=\{000000, 001011, 010101, 100111, 011110, 101100, 110010, 111001\}.$$

8.6 GROUP CODES

We have already discussed a group code, that is, any code which is a group under the binary operation of addition modulo 2. In general when the code words form a group, it is easier to find the minimum distance between code words.

Definition: Let $E:Z_2^m \to Z_2^n$ be an encoding function. The code $C=E(Z_2^m)$ is called a group code if C is a subgroup of Z_2^n.

Example 8.11: Consider the encoding function $E:Z_2^2 \to Z_2^6$, where $Z_2^2=\{00,10,01,11\}$.

Solution:

We have $E(00)=000000$, $E(10)=101010$, $E(01)=010101$, $E(11)=111111$.

Now $C=\{000000, 101010, 010101, 111111\} \subset Z_2^6$, and is a subgroup of Z_2^6 under component-wise addition modulo 2 i.e.,

$$101010+101010=000000,\ 010101+010101=000000,\ 111111+111111=000000$$
$$101010+010101=111111,\ 101010+111111=010101,\ 010101+111111=101010.$$

This shows that C is closed under component-wise addition modulo 2. Consequently, C is a group code. It is to be noted that in a group code, the minimum distance between distinct code words is the minimum of the weights of the non-zero elements of the code. If $E:Z_2^m \to Z_2^n$ is an encoding function given by a generator matrix G or the associated parity-check matrix H, then $C=E(Z_2^m)$ will be a group code. Let us explain the group structure of C together with its cosets in Z_2^n which helps in decoding techniques.

Example 8.12: Let $E:Z_2^3 \to Z_2^6$ be an encoding function given by

$$E(000)=000000,\ E(100)=100110,\ E(010)=010011,\ E(001)=001101$$
$$E(110)=110101,\ E(101)=101011,\ E(011)=011110,\ E(111)=111000$$

Let $C=\{000000, 100110, 010011, 001101, 110101, 101011, 011110, 111000\} \subset Z_2^6$

Is a subgroup with respect to $+_2$.

Let us prepare the decoding table for this code with the aid of coset leaders.

Step 1: Write the elements of C in a row starting with identity.

$$000000, 100110, 010011, 001101, 110101, 101011, 011110, 111000.$$

The order of C, that is, $O(C) = 8 = 2^3$ and $O(Z_2^6) = 2^6$

Hence, the number of distinct cosets of C in Z_2^6 is given by $\frac{2^6}{2^3} = 2^3 = 8$. The decoding Table 8.4 for C consists of eight rows.

Step 2: Let $x \in Z_2^6$, which does not appear in the above row and wt(x) be least.

If $x = 100000 \in Z_2^6$, wt(x) = 1. Find ($x +_2$ elements of C), and the second row, with the first row defined in Step 1.

$$000000, 100110, 010011, 001101, 110101, 101011, 011110, 111000.$$
$$10000, 000110, 110011, 101101, 010101, 001011, 111110, 011000$$

Step 3: Let $x \in Z_2^6$, which does not appear in the above two rows, and wt(x) be least.
If $x = 010000 \in Z_2^6$, wt(x) = 1. Find ($x +_2$ elements of C), and write the third row and repeat the process until the cosets are exhausted.

Table 8.4 Coset leaders

000000	100110	010011	001101	110101	101011	011110	111000
100000	000110	110011	101101	010101	001011	111110	011000
010000	110110	000011	011101	100101	111011	001110	101000
001000	101110	011011	000101	111101	100011	010110	110000
000100	100010	010111	001001	110001	101111	011010	111100
000010	100100	010001	001111	110111	101001	011100	111010
000001	100111	010010	001100	110100	101010	011111	111001
010100	110010	000111	011001	100001	111111	001010	101100

The entries in the first column of Table 8.4 are called coset leaders. It is to be noted that in the last row, the weights of 010100, 1000001 and 001010 are equal to two, instead of taking 010100 we can take even 1000001 or 001010 and construct the last row. Table 8.4 need not be unique. Consequently, if two errors occur during transmission, they can be detected but not corrected. Consider the following received words.

$$r_1 = 101001,\ r_2 = 111010,\ r_3 = 001001,\ r_4 = 111011.$$

Let the corresponding code words be c_1, c_2, c_3, c_4. Let the respective messages be w_1, w_2, w_3, w_4.

Now r_1= 101001 appears in the sixth row and sixth column. The corresponding code is from the first and sixth columns, i.e., c_1= 101011, and dropping the last three components in this word, we get the message w, which is $C = E(w) = w_1 = 101$.

Similarly for $r_2 = 111010$, $c_2 = 111000$, $w_2 = 111$

$$r_3 = 001001, c_3 = 001101, w_3 = 001$$

$$r_4 = 111011, c_4 = 101011, w_4 = 101$$

∴ The message is given by $w_1 = 101, w_2 = 111, w_3 = 001, w_4 = 101$

Example 8.13: Let the encoding function $E: Z_2^2 \to Z_2^5$ and the code C be defined by the generator matrix

$$G = \begin{pmatrix} 1 & 0 & 1 & 1 & 0 \\ 0 & 1 & 0 & 1 & 1 \end{pmatrix}$$

Develop the encoding table. Also denote the following received words using cosets of C.

$$r_1 = 11110,\ r_2 = 11011.$$

Solution: Now $[E(w)] = wG.\ \forall w \in Z_2^2$

We have $E(00) = (00)\begin{pmatrix} 1 & 0 & 1 & 1 & 0 \\ 0 & 1 & 0 & 1 & 1 \end{pmatrix} = (00000)$

$E(01) = (01)\begin{pmatrix} 1 & 0 & 1 & 1 & 0 \\ 0 & 1 & 0 & 1 & 1 \end{pmatrix} = (01011)$

$E(10) = (10)\begin{pmatrix} 1 & 0 & 1 & 1 & 0 \\ 0 & 1 & 0 & 1 & 1 \end{pmatrix} = (10110)$

$E(11) = (11)\begin{pmatrix} 1 & 0 & 1 & 1 & 0 \\ 0 & 1 & 0 & 1 & 1 \end{pmatrix} = (11101)$

Since $O(C) = 4 = 2^2$, $O(Z_2^5) = 2^5 = 32$, the number of distinct cosets of C in Z_2^5

is, by Lagrange's theorem, $\dfrac{O(Z_2^5)}{O(C)} = 2^3 = 8.$

Now $Z_2^2 \subset Z_2^5$. Select $x \in Z_2^5$ such that $x \notin C$ and wt(x) is minimum, and construct the coset $x +_2$ elements of C.

Let us take $x = 10000$, so that $x +_2$ elements of C = {10000, 11011, 00110, 01101}

Let us take $x = 01000$, so that $x +_2$ elements of C = {0 1000, 00011, 11110, 10101} and continue the process. The decoding for the code is given by Table 8.5.

Table 8.5 Decoding values

00000	01011	10110	11101
10000	11011	00110	01100
01000	00011	11110	10101
00100	01111	10010	11001
00010	01001	10100	11111
00001	01010	10111	11100
11000	10011	01110	00101
10100	10111	00010	01001

The entries in the first column of this table are called coset leaders. The table enables us to decode any received $r \in Z_2^5$, by the following technique.

(i) Find the column containing r and identify the code word c in C.
(ii) Decode this c by deleting the last three components of c, resulting in the word (w) that is the message sent.

Now consider $r_1 = 11110$, which appears in the third column and third row, and the corresponding code word is $c = 10110$. Hence, the message $w = 10$. Now consider $r_2 = 11011$, which appears in the second column, the corresponding code word for which is $c = 01011$. Hence, the message $w = 01$.

8.7 APPLICATIONS OF CODING THEORY

The concern of coding theory is to design codes that help synchronization. A code may be designed so that a phase shift can be easily detected and corrected and that multiple signals can be sent on the same channel. Another application of codes, used in some mobile phone systems, is code division multiple access (CDMA). Each phone is assigned a code sequence that is approximately uncorrelated with the codes of other phones. When transmitting, the code word is used to modulate the data bits representing the voice message. At the receiver, a demodulation process is performed to recover the data. The properties of this class of codes allow many users (with different codes) to use the same radio channel at the same time. To the receiver, the signals of other users will appear to the demodulator only as low-level noise.

Another general class of codes is the automatic repeat request (ARQ) codes. In these codes, the sender adds redundancy to each message for error checking, usually by adding the check bits. If the check bits are not consistent with the rest of the message, when it arrives, the receiver will ask the sender to retransmit the message. All but the simplest wide area network protocols use ARQ. Common protocols include SDLC (IBM), TCP (Internet), X.25 (International) and many others. The following are some codes used.

8.7.1 Hamming Codes

In telecommunication, **Hamming codes** are a family of linear error-correcting codes that generalise the Hamming (7, 4) code invented by Richard Hamming in 1950. Hamming codes can detect up to two-bit errors or correct one-bit errors without detection of uncorrected errors. By contrast, the simple parity code cannot correct errors, and can detect only an odd number of bits in error. Hamming codes are perfect codes, that is, they achieve the highest possible rate for codes with their block length and minimum distance.

In mathematical terms, Hamming codes are a class of binary linear codes. For each integer $r \geq 2$ there is a code with block length $n = 2^r - 1$ and message length $k = 2^r - r - 1$. Hence the rate of Hamming codes is $R = \frac{k}{n} = 1 - \frac{r}{2^r - 1}$, which is the highest possible for codes with distance 3 and block length $2^r - 1$. The parity-check matrix of a Hamming code is constructed by listing all columns of length r that are non-zero. This means that the dual code of the Hamming code is the punctured Hadmard code. The parity-check matrix has the property that any two columns are pair-wise linearly independent.

Due to the limited redundancy that Hamming codes add to the data, these can only detect and correct errors when the error rate is low. This is the case in computer memory (ECC memory), where bit errors are extremely rare and Hamming codes are widely used. In this context, an extended Hamming code having one extra parity bit is often used. Extended Hamming codes achieve a Hamming distance of 4, which allows the decoder to distinguish between one-bit and two-bit errors. In this sense, the extended Hamming codes are single-error correcting and double-error detecting codes, and are abbreviated as **SECDED**.

8.7.2 Hamming Codes with Additional Parity (SECDED)

Hamming codes have a minimum distance of three, which means that the decoder can detect and correct a single error, but it cannot distinguish between a double-bit error of some codeword and a single-bit error of a different codeword. Thus, they can detect double-bit errors only if correction is not attempted. To remedy this shortcoming, Hamming codes can be extended by an extra parity bit. This way, it is possible to increase the minimum distance of the Hamming code to 4, which allows the decoder to distinguish between single-bit errors and two-bit errors. Thus, the decoder can detect and correct a single error and at the same time detect (but not correct) a double error. If the decoder does not attempt to correct errors, it can detect up to 3 errors.

This extended Hamming code is popular in computer memory systems, where it is known as SECDED ('single error correction, double error detection'). Particularly popular is the (72, 64) code, a truncated (127, 120) Hamming code plus an additional parity bit, which has the same space overhead as that of a (9, 8) parity code. In 1950, Hamming introduced the [7, 4] Hamming code. It encodes 4 data bits into 7 bits by adding three parity bits. It can detect and correct single-bit errors. With the addition of an overall parity bit, it can also detect (but not correct) double-bit errors.

8.7.3 Construction of *G* and *H*

The matrix $G = (I_k \mid -A^T)$ is called a generator matrix of a linear (*n*, *k*) code, and $H = (A \mid I_{n-k})$ is called a parity-check matrix. This is the construction of **G** and **H** in standard form. Regardless of the form, **G** and **H** form linear block codes that satisfy $HG^T = 0$. Since $[7, 4, 3] = [n, k, d] = [2^m - 1, 2^m - 1 - m, m]$. The parity-check matrix **H** of a Hamming code is constructed by listing all columns of length *m* that are pair-wise independent. Thus, **H** is a matrix whose left side is all of the nonzero *n*-tuples, where the order of the *n*-tuples in the columns of the matrix does not matter. The right hand side is just the $(n-k)$-identity matrix. So **G** can be obtained from **H** by taking the transpose of the left hand side of **H** with the *k*-identity matrix on the left hand side of **G**.

The code generator matrix G and the parity-check matrix H are:

$$G = \begin{pmatrix} 1 & 0 & 0 & 0 & 1 & 1 & 0 \\ 0 & 1 & 0 & 0 & 1 & 0 & 1 \\ 0 & 0 & 1 & 0 & 0 & 1 & 1 \\ 0 & 0 & 0 & 1 & 1 & 1 & 1 \end{pmatrix}_{4 \times 7}$$

And

$$H = \begin{pmatrix} 1 & 1 & 0 & 1 & 1 & 0 & 0 \\ 1 & 0 & 1 & 1 & 0 & 1 & 0 \\ 0 & 1 & 1 & 1 & 0 & 0 & 1 \end{pmatrix}_{3 \times 7}$$

Finally, these matrices can be mutated into equivalent non-systematic codes by the following operations:

(i) Column permutations (swapping columns)
(ii) Elementary row operations (replacing a row with a linear combination of rows)

8.7.4 Encoding

Consider Example 8.11. From the matrix, we have $2^k = 2^4 = 16$ code words. The code words *x* of this binary code can be obtained from $x = wG$. $w = w_1 w_2 w_3 w_4$. Thus the code words are all 4-tuples (*k*-tuples). Therefore (1,0,1,1) gets encoded as (1,0,1,1,0,1,0).

8.7.5 [7, 4] Hamming Code with an Additional Parity Bit

The [7, 4] Hamming code can easily be extended to an [8, 4] code by adding an extra parity bit on top of the (7, 4) encoded word. This can be summed up with the revised matrices:

$$G=\begin{pmatrix}1&1&1&0&0&0&0&1\\1&0&0&1&1&0&0&1\\0&1&0&1&0&1&0&1\\1&1&0&1&0&0&1&0\end{pmatrix}_{4\times 8}$$

and

$$H=\begin{pmatrix}1&0&1&0&1&0&1&0\\0&1&1&0&0&1&1&0\\0&0&0&1&1&1&1&0\\1&1&1&1&1&1&1&1\end{pmatrix}_{4\times 8}$$

Note that H is not in standard form. To obtain G, elementary row operations can be used to obtain an equivalent matrix to H in systematic form:

$$H=\left(\begin{array}{cccc|cccc}0&1&1&1&1&0&0&0\\1&0&1&1&0&1&0&0\\1&1&0&1&0&0&1&0\\1&1&1&0&0&0&0&1\end{array}\right)_{4\times 8}$$

For example, the first row in this matrix is the sum of the second and third rows of H in non-systematic form. Using the systematic construction for Hamming codes from above, the matrix A is apparent and the systematic form of G is written as

$$G=\left(\begin{array}{cccc|cccc}1&0&0&0&0&1&1&1\\0&1&0&0&1&0&1&1\\0&0&1&0&1&1&0&1\\0&0&0&1&1&1&1&0\end{array}\right)_{4\times 8}$$

The non-systematic form of G can be row reduced (using elementary row operations) to match this matrix. The addition of the fourth row effectively computes the sum of all the codeword bits (data and parity) as the fourth parity bit.

Example 8.14: Let $r=1011$ be the code, encoded into 01100110 where the third , fifth ,sixth and seventh digits are data, first , second and fourth digits are parity from a [7, 4] Hamming code and the last digit is parity, added by an [8,4] code . The last digit makes the parity of the [7, 4] code even. Finally, it can be shown that the minimum distance has increased from 3, as with the [7, 4] code, to 4 with the [8, 4] code. Therefore, the code can be defined as an [8, 4] Hamming codes.

ADDITIONAL SOLVED EXAMPLES

Example W8.1: The parity-check matrix for an encoding $E{:}Z^3 \to Z^6$ is as follows,

$$H=\left(A^T/I_{n-m}\right)=\begin{pmatrix}1&0&1&1&0&0\\1&1&0&0&1&0\\1&0&1&0&0&1\end{pmatrix}$$

(i) Determine the associated generator matrix.
(ii) Can the given code correct all single errors in transmission?

Solution:

We have $$H=\left(A^T/I_{n-m}\right)=\begin{pmatrix}1&0&1&1&0&0\\1&1&0&0&1&0\\1&0&1&0&0&1\end{pmatrix}$$

Hence $$A_T=\begin{pmatrix}1&0&1\\1&1&0\\1&0&1\end{pmatrix} \text{ and } A=\begin{pmatrix}1&1&1\\0&1&0\\1&0&1\end{pmatrix}$$

The associated generator matrix is

$$G=(I_3/A)=\begin{pmatrix}1&0&0&1&1&1\\0&1&0&0&1&0\\0&0&1&1&0&1\end{pmatrix}$$

(ii) Two columns of H (second and fifth) are identical. Hence H does not provide a decoding matrix that corrects single errors in transmission.

It is to be noted that H does not contain a column having all zeroes and no two columns of H should be identical. The careful choice of A provides unique parity matrix H and will be a good decoding scheme that corrects a single error in transmission.

Example W8.2:

Let $H=\begin{pmatrix}1&0&0\\0&1&1\\1&1&1\\1&0&0\\0&1&0\\0&0&1\end{pmatrix}$ be a parity-check matrix. Determine the group codes, given that the encoding function is $E:Z^3\to Z^6$.

Solution:

Taking $$H=\begin{pmatrix}1&0&1&1&0&0\\0&1&1&0&1&0\\0&1&1&0&0&1\end{pmatrix}=\left(A^T/I_{n-m}\right)$$

The generator matrix G is given by

$$G=(I_m/A)=\begin{pmatrix}1&0&0&1&0&0\\0&1&0&0&1&1\\0&0&1&1&1&1\end{pmatrix}$$

Now $Z^3=\{000, 001, 010, 100, 011, 101, 110, 111\}$

Now $$e(000)=(0\,0\,0).G=(000)\begin{pmatrix}1&0&0&1&0&0\\0&1&0&0&1&1\\0&0&1&1&1&1\end{pmatrix}=[0\,0\,0\,0\,0\,0]$$

$$e(001) = [0\,0\,1].G=[001]\begin{pmatrix}1&0&0&1&0&0\\0&1&0&0&1&1\\0&0&1&1&1&1\end{pmatrix}=[0\,0\,1\,1\,1\,1]$$

$$e(010) = [0\,1\,0].G=[010]\begin{pmatrix}1&0&0&1&0&0\\0&1&0&0&1&1\\0&0&1&1&1&1\end{pmatrix}=[0\,1\,0\,0\,1\,1]$$

$$e(100) = [1\,0\,0].G=[10\,0]\begin{pmatrix}1&0&0&1&0&0\\0&1&0&0&1&1\\0&0&1&1&1&1\end{pmatrix}=[1\,0\,0\,1\,0\,0]$$

$$e(011) = (0\,1\,1).G=(011)\begin{pmatrix}1&0&0&1&0&0\\0&1&0&0&1&1\\0&0&1&1&1&1\end{pmatrix}=(0\,1\,1\,1\,0\,0)$$

$$e(101) = (1\,0\,1).G=(101)\begin{pmatrix}1&0&0&1&0&0\\0&1&0&0&1&1\\0&0&1&1&1&1\end{pmatrix}=(1\,0\,1\,0\,1\,0)$$

$$e(110) = (1\,1\,0).G=(110)\begin{pmatrix}1&0&0&1&0&0\\0&1&0&0&1&1\\0&0&1&1&1&1\end{pmatrix}=(1\,1\,0\,1\,1\,1)$$

$$e(111) = (111).G=(111)\begin{pmatrix}1&0&0&1&0&0\\0&1&0&0&1&1\\0&0&1&1&1&1\end{pmatrix}=(1\,1\,1\,0\,0\,0)$$

The code words generated are {000 000}, {001 111}, {010 011}, {100 100}, {011 100}, {101 010}, {110 111}, {111 000}.

Example W8.3: Let $H=\begin{pmatrix}1&0&1\\1&1&0\\0&1&1\\1&0&0\\0&1&0\\0&0&1\end{pmatrix}$ be a parity-check matrix. Decode the following words relative to maximum likelihood decoding functions.

(i) 0 1 1 0 0 1 (ii) 1 0 1 0 1 1 (iii) 1 1 1 0 1 0 (iv) 1 1 0 1 1 0

Solution:

Now

$$H=\left(A^T/I_{n-m}\right)=\begin{pmatrix}1&1&0&1&0&0\\0&1&1&0&1&0\\1&0&1&0&0&1\end{pmatrix}$$. The decoding function is $E{:}Z^3 \rightarrow Z^6$

(i) $H.(r)^T = \begin{pmatrix} 1 & 1 & 0 & 1 & 0 & 0 \\ 0 & 1 & 1 & 0 & 1 & 0 \\ 1 & 0 & 1 & 0 & 0 & 1 \end{pmatrix} \begin{pmatrix} 0 \\ 1 \\ 1 \\ 0 \\ 0 \\ 1 \end{pmatrix} = (100)$

Since $H.[e/w]^T = [1\,0\,0]^T$ = the fourth column of H. A single error has occurred in the fourth component of r. Changing the fourth component of r we get the code 0 1 1 1 0 1.

The code word transmitted is 0 1 1 1 0 1. Taking the first three components of the code word we get 0 1 1 as the original message.

(ii) $H.(r)^T = \begin{pmatrix} 1 & 1 & 0 & 1 & 0 & 0 \\ 0 & 1 & 1 & 0 & 1 & 0 \\ 1 & 0 & 1 & 0 & 0 & 1 \end{pmatrix} \begin{pmatrix} 1 \\ 0 \\ 1 \\ 0 \\ 1 \\ 1 \end{pmatrix} = (101)$

Since $H.(e/w)^T = (1\,0\,1)^T$ = the first column of H, a single error has occurred in the first component of r. Changing the first component of r we get the code. The code word transmitted is 0 0 1 0 1 1. Taking the first three components of the code we get 0 0 1 as the original message.

(iii) $H.(r)^T = \begin{pmatrix} 1 & 1 & 0 & 1 & 0 & 0 \\ 0 & 1 & 1 & 0 & 1 & 0 \\ 1 & 0 & 1 & 0 & 0 & 1 \end{pmatrix} \begin{pmatrix} 1 \\ 1 \\ 1 \\ 0 \\ 1 \\ 0 \end{pmatrix} = (010)$

Since $H.(e/w)^T = (0\,1\,0)^T$ = the fifth column of H, a single error has occurred in the fifth component of r. Changing the fifth component of r we derive the code. The code word was transmitted as 1 1 1 0 0 0. Taking the first three components of the code we get 111 as the original message.

(iv) $H.(r)^T = \begin{pmatrix} 1 & 1 & 0 & 1 & 0 & 0 \\ 0 & 1 & 1 & 0 & 1 & 0 \\ 1 & 0 & 1 & 0 & 0 & 1 \end{pmatrix} \begin{pmatrix} 1 \\ 1 \\ 0 \\ 1 \\ 1 \\ 0 \end{pmatrix} = (101)$

Now the syndrome is $[10\,1]^T$ which is the same as the first column of H. By changing the first position of r, the decoded word is determined to be 0 1 0 1 1 0 and the original message is 0 1 0.

Example W8.4: Given the generator matrix

$$G=\begin{pmatrix}1&0&0&1&1&0\\0&1&0&0&1&1\\0&0&1&1&0&1\end{pmatrix}$$

Corresponding to the encoding function $E:Z^3 \rightarrow Z^6$

(i) Find the corresponding parity-check matrix.

(ii) Decode the following received words and find the original message.

(i) 110101 (ii) 001111 (iii) 110001 (iv) 111111.

Solution:

Now $G = [I_3/A]$, then

$$H=\begin{pmatrix}1&0&1&1&0&0\\1&1&0&0&1&0\\0&1&1&0&0&1\end{pmatrix}$$. Let us find the nature of the received word by using $H.[r]^T$.

(i) $$H.(r)^T=\begin{pmatrix}1&0&1&1&0&0\\1&1&0&0&1&0\\0&1&1&0&0&1\end{pmatrix}\begin{pmatrix}1\\1\\0\\1\\0\\1\end{pmatrix}=\begin{pmatrix}0\\0\\0\end{pmatrix}$$

The received word in this case is the transmitted (encoded) word itself. Hence, the original message is 1 1 0 (by taking first three components).

(ii) $$H.(r)^T=\begin{pmatrix}1&0&1&1&0&0\\1&1&0&0&1&0\\0&1&1&0&0&1\end{pmatrix}\begin{pmatrix}0\\0\\1\\1\\1\\1\end{pmatrix}=\begin{pmatrix}0\\1\\0\end{pmatrix}$$

Now the syndrome is $[0\ 1\ 0]^T$ which is the same as the fifth column of H. By changing the fifth position of *r*, the decoded word is 0 0 1 1 0 1 and the original message is 0 0 1.

(iii) $$H.(r)^T=\begin{pmatrix}1&0&1&1&0&0\\1&1&0&0&1&0\\0&1&1&0&0&1\end{pmatrix}\begin{pmatrix}1\\1\\0\\0\\0\\1\end{pmatrix}=\begin{pmatrix}1\\0\\0\end{pmatrix}$$

Now the syndrome is $[1\ 0\ 0]^T$ is the same as the fourth column of H. By changing the fourth position of *r*, the decoded word is 1 1 0 1 0 1 and the original message is 1 1 0.

(iv) $$H.(r)^T = \begin{pmatrix} 1 & 0 & 1 & 1 & 0 & 0 \\ 1 & 1 & 0 & 0 & 1 & 0 \\ 0 & 1 & 1 & 0 & 0 & 1 \end{pmatrix} \begin{pmatrix} 1 \\ 1 \\ 1 \\ 1 \\ 1 \\ 1 \end{pmatrix} = \begin{pmatrix} 1 \\ 1 \\ 1 \end{pmatrix}$$

Now the syndrome $[1\ 1\ 1]^T$ is not identical with any column of H. The received word cannot be decoded uniquely.

Example W8.5: Let $H = \begin{pmatrix} 1 & 1 & 0 & 1 & 1 & 0 & 0 \\ 1 & 0 & 1 & 1 & 0 & 1 & 0 \\ 0 & 1 & 1 & 1 & 0 & 0 & 1 \end{pmatrix}$ be the parity-check matrix for a Hamming (7,4) code. Encode the following messages.

(i) 1000 (ii) 1100 (iii) 1011 (iv) 1110
(v) 1001 (vi) 1111.

Solution:
Now

$$H = \begin{pmatrix} 1 & 1 & 0 & 1 & 1 & 0 & 0 \\ 1 & 0 & 1 & 1 & 0 & 1 & 0 \\ 0 & 1 & 1 & 1 & 0 & 0 & 1 \end{pmatrix} = (B_{3\times4} : I_{3\times3})$$

The generator matrix $G = (I_{4\times4} : A_{4\times3})$ and $A_{4\times3}$ is the transpose of $B_{4\times3}$.

$$G = \begin{pmatrix} 1 & 0 & 0 & 0 & \vdots & 1 & 1 & 0 \\ 0 & 1 & 0 & 0 & \vdots & 1 & 0 & 1 \\ 0 & 0 & 1 & 0 & \vdots & 0 & 1 & 1 \\ 0 & 0 & 0 & 1 & \vdots & 1 & 1 & 1 \end{pmatrix}$$

Encoding

(i) $$(1000)G = (1000)\begin{pmatrix} 1 & 0 & 0 & 0 & \vdots & 1 & 1 & 0 \\ 0 & 1 & 0 & 0 & \vdots & 1 & 0 & 1 \\ 0 & 0 & 1 & 0 & \vdots & 0 & 1 & 1 \\ 0 & 0 & 0 & 1 & \vdots & 1 & 1 & 1 \end{pmatrix} = 1000110$$

And is same as the first row of G.

(ii) $$(1100)G = (1100)\begin{pmatrix} 1 & 0 & 0 & 0 & \vdots & 1 & 1 & 0 \\ 0 & 1 & 0 & 0 & \vdots & 1 & 0 & 1 \\ 0 & 0 & 1 & 0 & \vdots & 0 & 1 & 1 \\ 0 & 0 & 0 & 1 & \vdots & 1 & 1 & 1 \end{pmatrix} = 1100011$$

and is equal to the sum of the first and second rows of G.

(iii) $(1011)G=(1011)\begin{pmatrix}1&0&0&0&\vdots&1&1&0\\0&1&0&0&\vdots&1&0&1\\0&0&1&0&\vdots&0&1&1\\0&0&0&1&\vdots&1&1&1\end{pmatrix}=1011010$

and is equal to the sum of the first, third and fourth rows of G

(iv) $(1110)G=(1110)\begin{pmatrix}1&0&0&0&\vdots&1&1&0\\0&1&0&0&\vdots&1&0&1\\0&0&1&0&\vdots&0&1&1\\0&0&0&1&\vdots&1&1&1\end{pmatrix}=1110000$

(v) $(1001)G=(1001)\begin{pmatrix}1&0&0&0&\vdots&1&1&0\\0&1&0&0&\vdots&1&0&1\\0&0&1&0&\vdots&0&1&1\\0&0&0&1&\vdots&1&1&1\end{pmatrix}=1001001$

(vi) $(1111)G=(1111)\begin{pmatrix}1&0&0&0&\vdots&1&1&0\\0&1&0&0&\vdots&1&0&1\\0&0&1&0&\vdots&0&1&1\\0&0&0&1&\vdots&1&1&1\end{pmatrix}=1111111$

SUMMARY

This chapter gives the hierarchy of data processing, illustrating where coding theory fits in. We have seen the codes as groups. The concept of code groups is introduced. Encoding and decoding techniques with a mathematical perspective are discussed. Both encoding and decoding are viewed as functions. Parity checking, both odd and even, is explained. Generator matrices are introduced. Hamming codes with their variations are seen. The applications of coding theory are also discussed.

EXERCISES

1. Find the weights of the following words.
   (a) 1011 (b) 0110 (c) 1110 (d) 011101 (e) 11111 (f) 010101
2. Consider the (3, 4) parity-check code. For each of the received words (a) 01000, and (b) 1100, determine whether an error will be detected.
3. Find the distance between x and y given as under.
   (a) $x=1100010$ and $y=1010001$
   (b) $x=0100110$ and $y=0110010$
4. The encoding function E is given by $E:Z_2^3 \to Z_2^6$. Using the parity-check matrix

$$H=\begin{pmatrix}1&0&1&1&0&0\\1&1&0&0&1&0\\1&0&1&0&0&1\end{pmatrix}$$

   (a) Determine all code words.
   (b) Does this code correct all single errors in transmission?

5. Construct a decoding table for the group code given by the generator matrix

$$g=\begin{pmatrix}1 & 0 & 1 & 1 & 0\\ 0 & 1 & 0 & 1 & 1\end{pmatrix}$$

Use the table to decode the following received words
11110 11101 11011 10100
10011 10101 11111 01100

6. A word, c, transmitted through a binary symmetric channel is received as $r=0000111$. If $e=0101111$ is the error pattern, determine c.

7. Find the distance between x and y in each of the following cases.
   (a) $x=110110,\ y=000101$ (b) $x=001100, y=010110$
   (c) $x=01010101,\ y=10101010$

8. For an encoding function $E{:}Z_2^4 \to Z_2^6$, the parity-check matrix is given by

$$H=\begin{pmatrix}1 & 0 & 1 & 0 & 1 & 0\\ 1 & 1 & 0 & 1 & 0 & 1\end{pmatrix}$$

   (a) Find the associated generator matrix
   (b) Decode the received words : 010101, 111010, 111110

9. Given the generator matrix $G=\begin{pmatrix}1 & 0 & 0 & 1 & 1 & 0\\ 0 & 1 & 0 & 0 & 1 & 1\\ 0 & 0 & 1 & 1 & 0 & 1\end{pmatrix}$ the encoding function $E{:}Z_2^3 \to Z_2^6$ find the corresponding parity-check matrix and use it to decode the following received words and hence find the original message.
   (a) 110101 (b) 001111 (c) 110001 (d) 111111

10. Find the code words generated by the parity-check matrix

$$H=\begin{pmatrix}1 & 1 & 0 & 1 & 0 & 0\\ 1 & 0 & 1 & 0 & 1 & 0\\ 0 & 1 & 1 & 0 & 0 & 1\end{pmatrix}.$$

   The encoding function is given by $E{:}Z_2^3 \to Z_2^6$

11. Consider the (2, 4) group encoding function e defined as under.
   e (00) = 0000 e (10) = 1001
   e (01) = 0111 e (11) = 1111
   Decode the words, 0011, 1011 and 1111

12. Consider the encoding function $E{:}.Z_2^3. \to Z_2^6$ and the parity-check matrix H as given under.

$$H=\begin{pmatrix}1 & 0 & 1 & 1 & 0 & 0\\ 1 & 1 & 0 & 0 & 1 & 0\\ 1 & 0 & 1 & 0 & 0 & 1\end{pmatrix}$$

   (a) Determine all code words.
   (b) Does this code correct all single errors in transmission?

13. Consider the encoding function $E{:}\ Z_2^2 \to Z_2^5$ given by the generator matrix G, which is given as under.

$$G=\begin{pmatrix}1 & 0 & 1 & 1 & 0\\ 0 & 1 & 0 & 1 & 1\end{pmatrix}$$

(a) Determine all the code words. What can we say about the error-detection capability of this code? What about its error-correction capability?

(b) Find the associated parity-check matrix H.

14. Consider the $(m+1, m)$ block code for $m = 8$. Let $W = Z_2^8$. For each $w = w_1w_2w_3\dots w_8 \in W$. Define $E: Z_2^8 \to Z_2^9$ by $E(w) = w_1w_2w_3\dots w_9$, where $w_9 = \sum_{i=1}^{8} w_i$ with addition performed modulo 2. Find the generator and parity-check matrices for the (9,8) single parity-check coding scheme.

15. What are the dimensions of the generator matrix for the Hamming (63, 57) code? What are the dimensions of for the associated parity-check matrix H? What is the rate of this matrix?

9 Fundamentals of Counting Principles

In Chapter 8, we discussed the concept of coding theory. Enumeration, the counting of objects with certain properties, is an important mathematical concept used in our life. Counting problems arise throughout mathematics and computer science. Consider for example, the instances of counting the successful outcomes of an experiment or determining the probabilities of discrete events, or counting the number of operations in an algorithm to find the efficiency of a given method. In this chapter, we deal with principles of counting.

9.1 FUNDAMENTALS OF COUNTING

In most cases, listing the possibilities and counting them is the least desirable way of finding the answer to the given tasks. Consider an automobile license plate number used in India. It consists of three letters and six digits. An example is 'KA25-M-7045', where 'KA' stands for the state of Karnataka. In this license plate number, if the first two letters and two digits are fixed (Indicating the area) then by using a letter and four digits, how many different license plates are available? The procedure is as follows:

(i) 26 choices for each letter.
(ii) 10 choices for each digit.

There are a total of $26 \times 10 \times 10 \times 10 \times 10 = 260{,}000$ possible license plates for a particular place.

9.1.1 Rules of Sum and Product

There are two basic principles of counting: the rules of sum and of product.

Sum Rule

If a task can be done in m ways, while a second can be performed in n ways, and the two tasks cannot be performed simultaneously, then performing either of tasks can be done in any one of the $(m + n)$ ways.

Example 9.1: A student has selected five courses and needs one more course for the next term. There are ten courses in English, nine courses in Hindi, and five courses in Sanskrit, that the student is eligible to take. In how many ways can the student choose the sixth course?

Solution:
Let E be the task of selecting a course in English, H be that of selecting a course in Hindi, and S that of selecting a course in Sanskrit. These can be done in 10, 9 and 5 ways respectively. These selections are mutually exclusive. Then by the addition principle, the sixth course can be selected in $|E|+|H|+|S|=10+9+6=25$ ways, $(E\cap H\cap S=\Phi)$. They are pair wise mutually exclusive). It is to be noted that the addition principle can be extended to any finite number of pair wise mutually exclusive tasks.

Example 9.2: Let T_1, T_2,----T_n be n pair-wise mutually exclusive tasks. **i).** Suppose T_1 can be done in m_1 ways, T_2 can be done in m_2 ways---- and T_n can be done in m_n ways.
ii). The addition principle is also used to define the sum rule. $|A\cup B|=|A|+|B|$ since $A\cap B=\Phi$, where A and B are mutually exclusive.

Product Rule

Before considering a formal definition of the product rule, let us consider the common experience with an illustration given in Fig. 9.1.

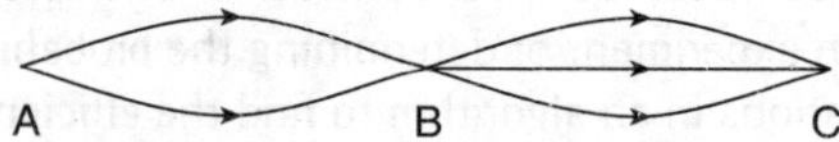

Figure 9.1 Product rule illustration

Consider that A, B and C are three different cities. A person can go from city A to city B by two different modes of transport and from B to C by three different mode of transport. Then the person can go in $2\times 3=6$ modes of transport from the city A to city C. This can be generalised as if a person can go from A to B in m different ways, B to C in n different ways, then the person can go in $(m\times n)$ ways from A to C. Thus, consider the following definition.
Definition: The procedure is made into two stages, and there are m possible outcomes for the first stage, and for each of these outcomes there are n possible outcomes for the second stage, then the total tasks can be carried out in $(m\times n)$ ways.

Example 9.3: Consider the example of manufacturing a license plate consisting of one letter followed by four digits. While referring to the earlier illustration, we can write as follows:

(a) If no letter or digit can be repeated, then there are $26\times 10\times 9\times 8\times 7=131{,}040$ different possible plates.
(b) With repetition of digits there are $26\times 10\times 10\times 10\times 10=260{,}000$ different license plates possible.
(c) A license plate contains one letter followed by four digits. If repetitions are allowed, how many plates have only vowels (A, E, I, O, U) and even digits? The first position can be filled in five ways (with vowels) while the remaining four places can be filled with five digits (0, 2, 4, 6, 8). Since an even number ends with one of the digits, the number of possible license plates of the given type is

$$5\times(5\times 5\times 5\times 5)=5^5=3125.$$

Example 9.4: How many 3-digit numbers can be formed by using the digits 2, 3, 5, 6, 7, 9, without repetition?

(i) How many of these are less than 400?
(ii) How many of these are even?
(iii) How many of these are multiples of 5?

Solution:

There are six digits and we require 3-digit numbers, using the six digits without repetition. The process is explained as follows.

As we are familiar, in counting the numbers we have to start counting from the units place, tens place, hundreds place. The units place can be filled up in six ways, as any one of the six digits can be used. After this, five digits are left and the tens place can be filled up with any one of the five digits, and similarly the hundreds place can be filled up in four ways, as given in Fig. 9.2.

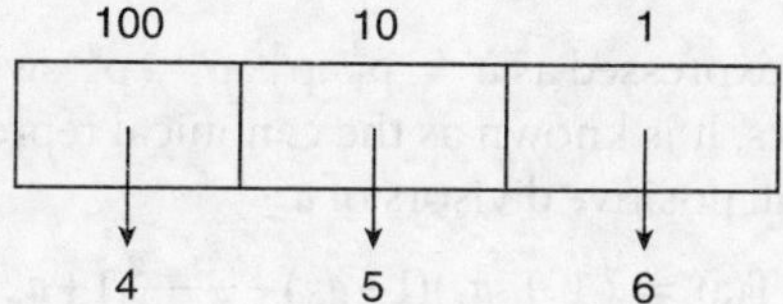

Figure 9.2 Illustration of Example 9.4

∴ The number of ways for filling up three places (formation of 3-digit numbers)

$$= 6 \times 5 \times 4 = 120.$$

(i) In this process, the numbers should be less than 400. So, we can put the digits 2 or 3 (among the given digits) in the hundreds place, which means that this place can be filled in two ways. The other two places can be filled in five and four ways respectively, as given in Fig. 9.3.

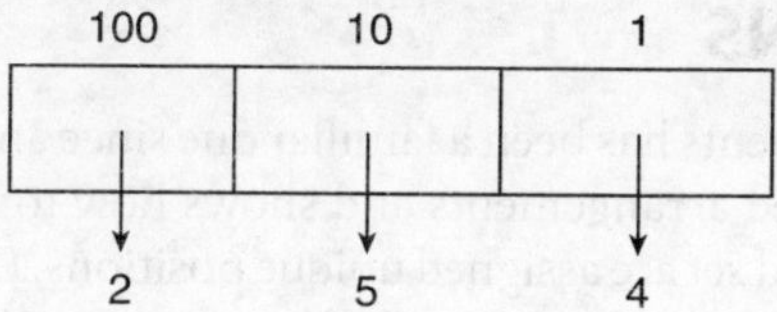

Figure 9.3 Illustration of Example 9.4 (i)

∴ The number of ways for forming 3-digit numbers which are less than $400 = 2 \times 5 \times 4 = 40$.

(ii) To form an even number using the given digits, the units place can have either 2 or 6. So, the units place can be filled up in two ways, and the other places can be filled in five and four ways respectively, as given in Fig. 9.4.

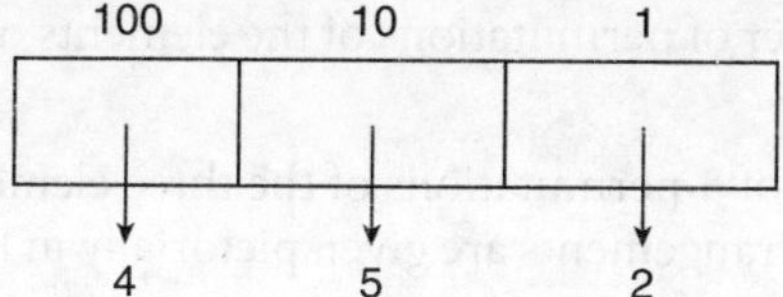

Figure 9.4 Illustration of Example 9.4 (ii)

∴ Total number of even numbers $= 2 \times 5 \times 4 = 40$. It is to be noted that the total number of odd numbers $= 120 - 40 = 80$.

(iii) A number which is a multiple of 5 ends with either 0 or 5. As 0 is not present in the given digits, so we can only place the digit 5 in the units place. So the units place can be filled in 1 way, and the other two places can be filled in 5×4 ways.

∴ forming numbers which are multiples of $5 = 1 \times 5 \times 4 = 20$ ways.

Example 9.5: Consider the technique of expressing every composite number as the product of prime factors. We shall proceed with examples not simply forcing the formulae.

(i) $408 = 2 \times 204$
$= 2 \times 102 \times 2$
$= 2 \times 2 \times 2 \times 17 \times 3$
$= 2^3 \times 3^1 \times 17^1$
In which 2, 3 and 17 are prime numbers.

(ii) $864 = 2 \times 432$
$= 2 \times 2 \times 216$
$= 2^5 \times 3^3$

In general, any integer (a) can be expressed as $a = p_1^{a_1}.p_2^{a_2}.p_3^{a_3}...p_n^{a_n}$, where $p_1p_2....p_n$ are prime numbers, and $a_1a_2....a_n$ are positive integers. It is known as the canonical representation on the number a. From this, we can find the number of all positive divisors of a.

$$T(a) = (1 + a_1)(1+a_2)----(1+a_n).$$

Consider for simplicity $a = 24$.

$$24 = 2 \times 3 \times 4 = 2^3 \times 3^1$$
$$p_1 = 2,\ p_2 = 3,\ a_1 = 3,\ a_2 = 1$$
$$T(a) = (1 + 3)\ (1 + 1) = 8$$

The positive divisors of 24 = {1, 2, 3, 4, 6, 8, 12, 24}, which are 8 in number and support the canonical representation of any composite number.

9.2 PERMUTATIONS

The concept of ordered arrangements has been a familiar one since ancient times. The following section introduces the concept of ordered arrangements and shows how to find the number of such arrangements. The elements of an ordered set are assigned unique positions. Let us denote the ordered set {*p*, *q*, *r*} as the arrangement *pqr*. The words *prq*, *pqr* are two different arrangements of the very same letters. Each of these arrangements is a permutation of these letters taken all at a time.

Definition: A permutation of a set of *n* distinct elements taken *r* at a time ($0 \le r \le n$) is an arrangement of *r* elements of the set. It is also called *r* permutations. If $r = n$, then the *r* permutations is called simply a permutation. The number of *r* permutations of a set containing *n* elements is denoted by $P(n, r)$ or ^np_r

Example 9.6: Find the number of permutations of the elements of the set {*p*, *q*, *r*}

Solution:
By the product rule the number of 3-permutations of the three elements without repetition = $3 \times 2 \times 1 = 6$ i.e., $P(3, 3) = 6$. The various arrangements are given pictorially in Fig. 9.5.

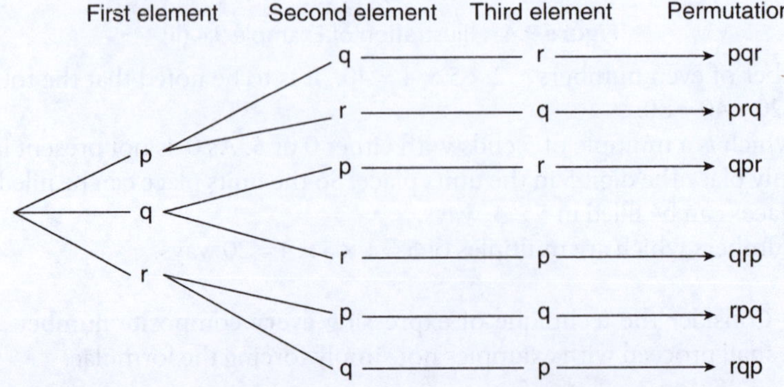

Figure 9.5 Illustration of permutations

The various permutations are *pqr, prq, qpr, qrp, rpq, rqp*. Before proceeding to find the general formula to find permutations of *n* different things taken *r* at a time, let us consider an illustration which we come across in our routine life.

Suppose we are waiting for a bus at the bus-stand which is the starting point. The bus contains 40 seats. When the bus arrives, the first person who enters the bus has 40 chances to sit in a bus. Ultimately he will select one seat (has to select one seat!). Now there are 39 seats left and the second person has 39 choices, and so on. And finally the 40th person has no choice but to occupy the remaining seat. By considering this, we can define the general permutation formula using the multiplication principle.

Let us consider *n* letters $a_1, a_2, \text{---}\, a_n$ and *r* blank spaces. The first place can be filled up in *n* ways as we can put any one of the *n* letters in the first place. After this is done, there are $(n-1)$ letters and we have to fill $(r-1)$ places. The second place can be filled up in $(n-1)$ ways by using any one of the remaining $(n-1)$ letters. Continue like this until the rth element is ready to be chosen. At this point there are $(n-r+1)$ elements left. Consequently, the rth element can be chosen in $n-r+1$ ways. So all the *r* places can be filled up in

$$n\,(n-1)\,(n-2)\,(n-3)\ldots(n-r+1) \text{ ways.}$$

Thus, the permutation of *n* elements taking *r* at a time is given by

$$P(n, r) = n\,(n-1)\,(n-2)\,(n-3)\ldots(n-r+1)$$

$$= \frac{n(n-1)(n-2)\text{----}(n-r+1)(n-r)\text{----}3.2.1}{(n-r)\text{---}3.2.1}$$

$$p(n,r) = \frac{n!}{(n-r)!}$$

The pictorial representation is shown in Fig. 9.6.

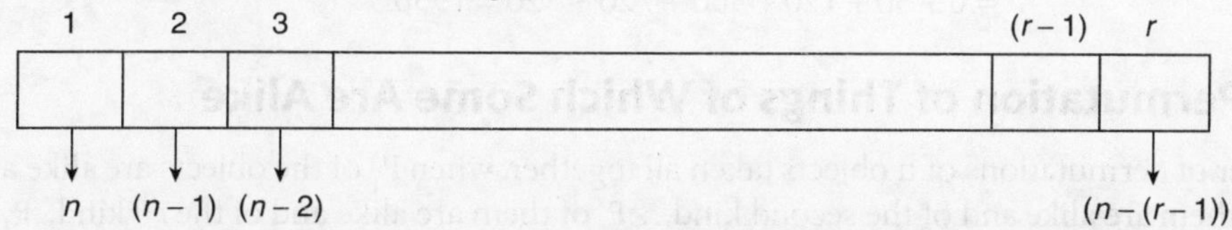

Figure 9.6 Pictorial representation of permutation

The following cases arise:

(i) If $r = 0$, $p(n, 0) = \dfrac{n!}{(n-0)!} = 1$ so, $p(n, r) = \dfrac{n!}{(n-r)!}$ holds for all $0 \le r \le n$.

(ii) If $r = n$, $p(n, n) = \dfrac{n!}{(n-n)!} = \dfrac{n!}{(0)!} = n!$. $(0! = 1)$

(iii) P (n, r) counts linear arrangements in which objects cannot be repeated. If repetitions are allowed, then by the product rule there are n^r possible arrangements with $(r \ge 0)$.

Example 9.7: Find the number of words that can be formed with the letters of the word 'COMPUTER'. Remember that a word is just an arrangement of symbols. It need not make any sense.

Solution:

The word "COMPUTER" contains eight distinct letters. The number of words that can be formed equals the number of arrangements, that is, P(8, 8) = 8!

$P(8, 8) = 8! = 8 \times 7 \times 6 \times 5 \times 4 \times 3 \times 2 \times 1 = 40{,}320.$

If only five of the letters are used, the number of permutations (of size 5) is $P(8, 5) = p(n,r) = \frac{8!}{(8-5)!} = \frac{8!}{3!}$ $= 6720$

If repetitions are allowed, the number of possible 10-letter sequence is 8^{10}.

Example 9.8: Eight chairs are numbered 1 to 8. Two women and three men are to occupy one chair each. First the women choose the chairs from amongst the chairs 1 to 4 and then men select from the remaining chairs. Find the number of possible arrangements.

Solution:
The two women can be arranged in the four chairs in P(4, 2) ways. Further, the men can be arranged in the remaining six chairs in P(6, 3) ways.

Hence, the number of ways in which two women and four men can be arranged is

$$= P(4, 2) \times P(6, 3)$$

Example 9.9: Find the number of signals that can be generated by using six different coloured flags, when any number of them may be hoisted at any time.

Solution:
With one flag, the number of signals that can be generated = P (6, 1) and so on.

Hence, the number of different signals that can be generated

$$= P(6, 1) + P(6, 2) + P(6, 3) + P(6, 4) + P(6, 5) + P(6, 6)$$
$$= 6 + 30 + 120 + 360 + 720 + 720 = 1956.$$

9.2.1 Permutation of Things of Which Some Are Alike

The number of permutations of n objects taken all together, when P_1 of the objects are alike and of one kind, P_2 of them are alike and of the second kind,…P_r of them are alike and of the rth kind, $P_1 + P_2 + \ldots + P_r = n$.

$$P(n, r) = \frac{n!}{p_1!p_2!\ldots p_r!}$$

Example 9.10: Find the number of permutations of the letters of the word INSTITUTION.

(i) How many of these begin with I?
(ii) How many of these begin with I and end with N?
(iii) In how many words are the three T's are together?

Solution:
The word contains 11 letters of which there are 3 Is, 2 Ns, 3 Ts , 1 S, 1 U, 1 O that are alike.

$\therefore$ The number of permutations $= \frac{11!}{3!\,2!\,3!}$.

(i) The numbers of permutations that begin with I: Now put I in the first place and arrange the rest of letters. Now there are ten letters of which 3 Ts, 2 Is, 2 Ns are alike and the rest are different.

Number of permutations which begin with I $= \frac{10!}{3!\,2!\,2!}$.

(ii) Put I in the first place and N in the last place and arrange the remaining letters.

There are nine letters of which 3 Ts, 2 Is are alike, and the rest of them are different.

Number of permutations for words which begin with I and end with N are $=\frac{9!}{3!\,2!}$.

(iii) Regard the 3 Ts as one letter say X. Then we have nine letters, of which the 3 Is, 2 Ns are alike, while the rest of them are different.

Number of permutations in which the 3 Ts are together $=\frac{9!}{3!\,2!}$.

Example 9.11: Find n if p(n, 4) = 12 p(n, 2)

$$\frac{n!}{(n-4)!}=12\frac{n!}{(n-2)!}.$$

Solution:

$$\frac{n(n-1)(n-2)(n-3)(n-4)!}{(n-4)!}=12\frac{n(n-1)(n-2)!}{(n-2)!}$$

$$(n-2)(n-3)=12.$$

$$n^2-5n-6=0$$

$$(n+1)(n-6)=0.\ n\neq-1\quad \therefore n=6$$

Example 9.12: If p(6, r) = 360, then find r.

$$\frac{6!}{(6-r)!}=360$$

Solution:

$$\frac{6!}{360}=(6-r)!\Rightarrow\frac{6\times5\times4\times4\times3\times2\times1}{360}=(6-r)!$$

$(6-r)!=2!$. Since $r\leq6$, by inspection we have $r=4$.

9.2.2 Circular Permutation

So far we have discussed linear permutations. Permutation of elements arranged in a circle is somewhat different from linear permutations. The arrangements of n things around a circle are known as circular permutations. In order to find the number of circular permutations of n things taken all at a time, let us consider four letters A, B, C and D and their circular permutations. Let x be the circular permutations of these four letters. If we shift A to the place B, B to C, C to D and D to A, we get the same circular permutations. The four circular permutations shown in Fig. 9.7 are the same. But the corresponding linear permutations are different. From one circular permutation we get four linear permutations. Hence, from x permutations we get (4x) linear permutations which are equal to 4!

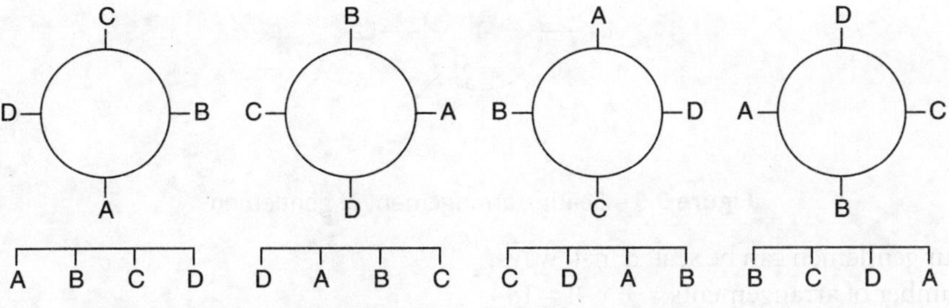

Figure 9.7 Circular permutations

$$\therefore 4x = 4!, \; x = \frac{4!}{4} = (4-1)! = 3!$$

The number of circular permutations of four letters A, B, C, D = (4 – 1)! = 3!, as shown in Fig. 9.7. The circular permutation is generalised. The number of cyclic permutations on n (distinct) items is $(n-1)!$. This can be explained as follows.

Let us assign a fixed position to the first item a_1 around the circle. Now $(n-1)$ positions are left. The second item a_2 can be placed in any one of the $(n-1)$ positions. Now $(n-2)$ positions are left. Continue this until all items have been placed. Then by the product rule, the number of cyclic permutations is= $(n-1).(n-2)....2.1 = (n-1)!$, as shown in Fig. 9.8.

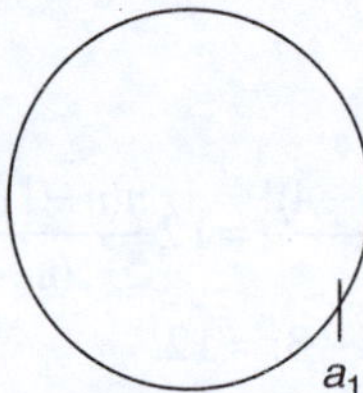

Figure 9.8 Number of cyclic permutations

It is observed that when we arrange n persons around a circular table, the clockwise and anticlockwise arrangements are different. But when we arrange n beads or n flowers to form a garland, the anticlockwise and clockwise arrangements are the same. If no distinction is to be made between clockwise and anticlockwise arrangements, then the number of arrangements is equal to

$$\frac{1}{2}(n-1)!$$

Example 9.13: In how many ways can a party of four ladies and four gentlemen be seated at a round table, with the ladies and gentlemen being seated alternatively?

Solution:

Let the four ladies be seated around a table. This can be done in (4 – 1)! =3! ways. Consider one of the arrangements of gentlemen. There are four possible places for the four gentlemen to sit. The arrangement of gentlemen is linear.

This is because if we shift G_1 to G_2, G_2 to G_3 and G_3 to G_4 and G_4 to G_1 then we get a different arrangement as the relative positions are no longer the same. The arrangement is shown in Fig. 9.9.

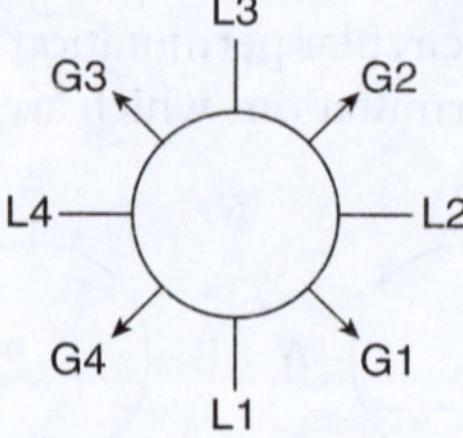

Figure 9.9 Seating arrangement of gentlemen

Hence the four gentlemen can be seated in 4! ways.

$\therefore$ Total number of arrangements = $3! \times 4! = 144$

9.3 COMBINATIONS

A permutation is an ordered arrangement of elements in a set. Sometimes the order of elements is irrelevant, only their membership is important. So the unordered arrangements are combinations. For example, a committee of members, namely, A = {Santosh, Harsha, Sujata, Shubdha} is just a set and the order in which the names are listed becomes immaterial. We would like to form subcommittees consisting of three members. These sub committees are {Santosh, Harsha, Sujata}, {Harsha, Sujata, Shubdha}, {Sujata, Shubdha, Santosh}, {Shubdha, Santosh, Harsha}. Each set represents a combination of four elements taken three at a time.

Definition: The number of combinations (selection) of n different things taken r at a time is denoted by C(n, r) or $\binom{n}{r}$. The number of combinations is also called the binomial coefficients.

Example 9.14: Find the number of r combinations of the set {a, b, c}, where $r = 0, 1, 2$ or 3.

(i) For $r = 0$, C(3, 0) = 1. The selection contains exactly one subset which contains zero elements.

(ii) For $r = 1$, C(3, 1) = 3. The selection contains one element each: {a}, {b}, {c}.

(iii) For $r = 2$, C(3, 2) = 3. The selection contains three subsets where each set contains two elements. i.e., {a, b}, {b, c} {c, a}.

(iv) For $r = 3$, C(3, 3) = 1. The arrangement contains exactly one set and it contains 3 elements. Note: {a, b, c},{c, a, b},{b, a, c},{b, c, a},{c, b, a},{a, c, b} are all different arrangements and are regarded as one combination.

The number of r combinations of a set consisting of n elements is given by

$$C(n,r)=\frac{n!}{(n-r)!r!}\quad 0\le r\le n.$$

C(n, r) is the number of combinations of n different things taken r at a time. Each combination contains r things, and combination P(r, r) = r!, which is r permutations. So the total number of r permutations is r! C(n, r). And by definition, these are as follows

$$p(n,r)=\frac{n!}{(n-r)!}\quad r\text{ Permutations.}$$

$$r!\,C(n,r)=p(n,r)=\frac{n!}{(n-r)!}$$

$$\therefore C(n,r)=\frac{n!}{(n-r)!r!}\quad 0\le r\le n.$$

The following observations are made:

(i) $C(n,0)=\dfrac{n!}{(n-0)!0!}=1$. The number of 0 combinations of a set with n elements is one.

(ii) $C(n,n)=\dfrac{n!}{(n-n)!n!}=1$. The number of n combinations of a set with n elements is also one.

(iii) For large values of n, the computation can be made simple as follows.

$$\therefore\quad C(n,r)=\frac{n(n-1)(n-2)\ldots(n-r+1)}{r!}.$$

The following examples make use of the product rule, sum rule, or both.

Example 9.15: There are 12 points in a plane of which four are collinear. Find the number of (a) straight lines (b) triangles, which can be formed from these points.

Solution:

(a) We need two points for a line. If none of the three given points are collinear, then we get C(12, 2) lines. It is given that four points are collinear. So from four points we get only one straight line and we lose C (4, 2) lines.

∴ The number of straight lines = C(12, 2) – C(4, 2) +1

$$= \frac{12.11}{2!} - \frac{4.3}{2!} + 1 = 66 - 6 + 1 = 61.$$

(b) To form a triangle, we need three non-collinear points. In the given 12 points, if none of the points are collinear we get C(12, 3) triangles. Given that four points are collinear, we lose C(4, 3) triangles from the given points.

∴The number of triangles that can be formed = C(12, 3) – C(4, 3).

$$= \frac{12.11.10}{3!} - \frac{4.3.2}{3!} = 220 - 4 = 216.$$

Example 9.16: At an election, a voter may vote for any number of candidates not greater than the number to be chosen. There are ten candidates and five members to be chosen. Find the number of ways in which a voter may vote.

Solution:

Since a voter has to vote for at least one candidate and at the most five candidates. The number of ways in which the voter may vote is as follows.

$$= C(10, 1) + C(10, 2) + C(10, 3) + C(10, 4) + C(10, 5)$$
$$= 10 + 45 + 120 + 210 + 252 = 637$$

Example 9.17: If C(n + 2, 8):p(n – 2, 4) = 57:16, then find n.

Solution:

$$C(n+2, 8) : p(n-2, 4) = 57:16$$

$$\frac{(n+2)!}{8!(n-6)!} \times \frac{(n-6)!}{(n-2)!} = \frac{57}{16}$$

$$\frac{(n+2)(n+1)n(n-1)(n-2)}{8!(n-2)!} = \frac{57}{16}$$

$$(n+2)(n-1)(n+1)n = 143640$$

$$(n^2+n-2)(n^2+n) = 143640$$

$$(n^2+n)^2 - 2(n^2+n) = 143640 \quad \text{[Using the identity } (a+b-1)^2 = (a+b)^2 - 2(a+b)]$$

$$(n^2+n-1)^2 = 143640$$

$$(n^2+n-1) = 379$$

$$n^2+n-380 = 0$$

$(n-19)\ (n+20) = 0.\ n = -20$ or $n = 19$.

n cannot be negative. ∴n = 19.

Example 9.18: A man has seven relatives, 4 of them are women and 3 are men. His wife also has seven relatives, 3 of them are women and 4 are men. In how many ways can they invite the relatives to a dinner party so that 3 of the man's relatives and 3 of the wife's relatives are invited?

Solution:
The different possibilities of selection are as shown in Table 9.1.

Table 9.1 Different possibilities

Man's relatives			Wife's relatives	
Total	**4 women**	**3 men**	**3 women**	**4 men**
1	3-L	0-G	0-L	3-G
2	2-L	1-G	1-L	2-G
3	1-L	2-G	2-L	1-G
4	0-L	3-G	3-L	0-G

The different ways of selection are

(i) $C(4, 3) \times C(3, 0) \times C(3, 0) \times C(4, 3) = 16$
(ii) $C(4, 2) \times C(3, 1) \times C(3, 1) \times C(4, 2) = 324$
(iii) $C(4, 1) \times C(3, 2) \times C(3, 2) \times C(4, 1) = 144$
(iv) $C(4, 0) \times C(3, 3) \times C(3, 3) \times C(4, 0) = 1$

Total number of selections $= 16 + 324 + 144 + 1 = 485$.

9.3.1 Binomial Theorem

In elementary mathematics we have studied the expression $(x+a)^2, (x+a)^3, --$ with actual expansion and by collecting the coefficients of the various powers of x.

Consider,

$$(x+a)^1 = x^1 + a^1 = \binom{1}{0} x + \binom{1}{1} a$$

$$(x+a)^2 = x^2 + 2ax + a^2 = \binom{2}{0} x^2 + \binom{2}{1} x\,a + \binom{2}{2} a^2.$$

$$(x+a)^3 = x^3 + 3x^2a + 3a^2x + a^3 = \binom{3}{0} x^3 + \binom{3}{1} x^2a + \binom{3}{2} xa^2 + \binom{3}{3} a^3$$

$$(x+a)^4 = x^4 + 4x^3a + 6x^2a^2 + 4\,x\,a^3 + a^4 = \binom{4}{0} x^4 + \binom{4}{1} x^3a + \binom{4}{2} x^2a^2 + \binom{4}{3} x\,a^3 + \binom{4}{4} a^4$$

In the above presentation the coefficients of various powers have been written in terms of binomial coefficients $\binom{n}{r}$ or C (n, r). This led to the general expansion of $(x + a)^n$, where n is a positive integer. This was put forth by Sir Isaac Newton.

Binomial Theorem: For a positive integral index, if x and a are real numbers and n is any positive integer then,

$$(x+a)^n = \binom{n}{0}x^n + \binom{n}{1}x^{n-1}a + \binom{n}{2}x^{n-2}a^2 + \binom{n}{3}x^{n-3}a^3 + \text{---} + \binom{n}{n}x^{n-n}a^n$$

$$= \sum_{r=0}^{n} \binom{n}{r} x^{n-r} a^r \qquad \text{(I)}$$

The coefficients $\binom{n}{0}, \binom{n}{1}, \text{----} \binom{n}{n}$ are called binomial coefficients.

9.3.2 Properties of Binomial Expansion

(i) There are $(n + 1)$ terms in the expansion of (I).

(ii) In any terms in the expansion, the sum of the powers of x and a is always n.

(iii) The binomial coefficients of terms equidistant from the beginning and the end are equal, and is given by $C(n, r) = C(n, n - r)$ $(0 \le r \le n)$.

(iv) The term $C(n, r)\, x^{n-r} a^r$, is the $(r + 1)^{th}$ term from the beginning of the expression. It is denoted by T_{r+1} and is called the general term of the expansion.

(v) If n is even, the expansion (I) has just one middle term namely, $\left(\frac{n}{2}+1\right)$ is the middle term.

If n is odd then the expansion (I) has two middle terms namely, $\left(\frac{n+1}{2}\right)$ and $\left(\frac{n+3}{2}\right)$.

9.3.3 Some Particular Cases of Binomial Expansion

(i) Replace a by $(-a)$ in equation (I), and we obtain

$$(x-a)^n = \binom{n}{0}x^n - \binom{n}{1}x^{n-1}a + \binom{n}{2}x^{n-2}a^2 - \binom{n}{3}x^{n-3}a^3 + \text{---} + (-1)^n\binom{n}{n}x^{n-n}a^n_n \qquad \text{(II)}$$

(ii) Putting $x = 1$, in (I) and (II) we get

$$(1+a)^n = \binom{n}{0} + \binom{n}{1}a + \binom{n}{2}a^2 + \binom{n}{3}a^3 + \text{---} + \binom{n}{n}a^n \qquad \text{(III)}$$

$$(1-a)^n = \binom{n}{0} - \binom{n}{1}a + \binom{n}{2}a^2 - \binom{n}{3}a^3 + \text{---} + (-1)^n\binom{n}{n}a^n \qquad \text{(IV)}$$

(iii) The sum of the binomial coefficients is equal to 2^n.
Putting $a = 1$ in equation (III), we get

$$(1+1)^n = \binom{n}{0} + \binom{n}{1} + \binom{n}{2} + \binom{n}{3} + \text{---} + \binom{n}{n}$$

$$\binom{n}{0} + \binom{n}{1} + \binom{n}{2} + \binom{n}{3} + \text{---} + \binom{n}{n} = 2^n$$

(iv) Sum of even binomial coefficients of order n is equal to the sum of the odd binomial coefficients and each is equal to 2^{n-1},
Putting $a = 1$ in (IV)

$$\binom{n}{0} + \binom{n}{2} + \text{---} + = \binom{n}{1} + \binom{n}{3} + \text{---} + = \frac{1}{2}2^n = 2^{n-1}$$

Sum of binomial coefficients is equal to 2^n.

(v) Multinomial theorem: In the expansion of $(x + y + z)^n$, where n is a positive integer and x, y, z are real numbers, the process is to take $y + z = a$, expand the *binomial* expression $(x + a)^n$ and then expand $(y + z)^r$ in each form of the expansion of $(x + a)^n$.

In general we have

$$(a_1 + a_2 + --- + a_m)^n = \sum \frac{n!}{n_1!n_2!---n_m!} a_1^{n_1} a_2^{n_2} -- a_m^{n_m}$$

The summation is taken over all non-negative integers $n_1, n_2, --- n_m$ such that

$$n_1 + n_2 + --- + n_m = n.$$

Example 9.19: Find the coefficient of $x^4 y^5$ in the expansion of $(2x - 3y)^9$.

Solution:

$$(2x-3y)^9 = \sum_{r=0}^{9} \binom{9}{r} (2x)^{9-r} (-3y)^r$$

$$= \sum_{r=0}^{9} \binom{9}{r} (2)^{9-r} x^{9-r} (-3)^r y^r_r$$

Co-efficient of $x^4 y^5 \Rightarrow x^{9-r} y^r = x^4 y^5 \Rightarrow r = 5$.

$\therefore$ Coeff of $x^4 y^5 = \binom{9}{5} 2^4 (-3)^5 = -1{,}469{,}664$.

Example 9.20: Find the coefficient of x^4 in the expansion of $\left(2x^2 - \frac{3}{x}\right)^8$

Solution:

The $(r+1)^{th}$ term is given by

$$T_{r+1} = \binom{n}{r} x^{n-r} a^r$$

$$= \binom{8}{r} (2x^2)^{8-r} \left(\frac{-3}{x}\right)^r$$

$$= \binom{8}{r} 2^{8-r} (-3)^r x^{16-2r} x^{-r}$$

$$= \binom{8}{r} 2^{8-r} (-3)^r x^{16-3r}$$

Coefficient of $x^4 \Rightarrow x^{16-3r} = x^4$, so $r = 4$.

$$T_5 = \binom{8}{4} (-3)^4 2^4 x^4$$

Coefficient of $x^4 = T_5 = \binom{8}{4} (3)^4 2^4 = 90720$.

Example 9.21: Find the coefficient of x^0 in the expansion of $\left(2x^2 - \frac{3}{x^3}\right)^{25}$ or term independent of x.

Solution:

$$T_{r+1} = \binom{n}{r} x^{n-r} a^r$$

$$= \binom{25}{r} (2x^2)^{25-r} \left(\frac{-3}{x^3}\right)^r$$

$$=\binom{25}{r}\ 2^{25-r}(-3)^r\, x^{50-2r}\ x^{-3r}$$

$$=\binom{25}{r}\ 2^{25-r}(-3)^r\, x^{50-5r}$$

Coefficient of $x^0 \Rightarrow 50-5r=0,\ r=10$

Term independent of $x = T_{11} = \binom{25}{10}\ 2^{15}(-3)^{10}$ = eleventh term.

$T_{11} = \binom{25}{10}\ 2^{15}(3)^{10}$

Example 9.22: Find the middle terms in the expansion of $\left(2x^2 - \frac{1}{\sqrt{x}}\right)^{11}$

Solution:

There are 12 terms in the expansion; the sixth and seventh terms are the middle terms. We have to find the sixth and seventh terms.

Sixth term

$$T_{r+1} = \binom{n}{r}\ x^{n-r}\ a^r$$

$$=\binom{11}{5}\left(2x^2\right)^6\left(\frac{-1}{\sqrt{x}}\right)^5$$

$$=\binom{11}{5}(2)^6(x)^{12}(-1)^5(x)^{\frac{-5}{2}}$$

$$=-\binom{11}{5}(2)^6(x)^{12-\frac{5}{2}}$$

$$T_6 = -\binom{11}{5}(2)^6(x)^{\frac{19}{2}}.$$

Similarly seventh term $= T_7 = \binom{11}{6}(2)^5(x)^7$.

Example 9.23: Find the coefficient of $x^3y^3z^2$ in $(2x-3y+5z)^8$

Solution:

Coefficient of $x^3y^3z^2 = \binom{8}{n_1,n_2,n_3}(2x)^{n_1}(-3y)^{n_2}(5z)^{n_3}$

$$=\binom{8}{n_1,n_2,n_3}(2)^{n_1}x_{n_1}^{n_1}(-3)^{n_2}y^{n_2}(5)^{n_3}z^{n_3}$$

$n_1=3,\ n_2=3,\ n_3=2.$

$$=\binom{8}{n_1,n_2,n_3}(2)^3x^3(-3)^3y^3(5)^2z^2$$

$$=\binom{8}{n_1,n_2,n_3}(2)^3(-3)^3(5)^2x^3y^3z^2$$

$$=\left(\frac{8!}{3!3!2!}\right)(2)^3(-3)^3(5)^2x^3y^3z^2$$

$$=\left(\frac{8!}{3!3!2!}(8)(-27)(25)\right)x^3y^3z^2$$

$$=-3024000.$$

9.4 PASCAL'S TRIANGLE

The various binomial coefficients $\binom{n}{r}$ where $(0 \le r \le n)$, can be arranged in the form of a triangle called Pascal's triangle, as given in Fig. 9.10.

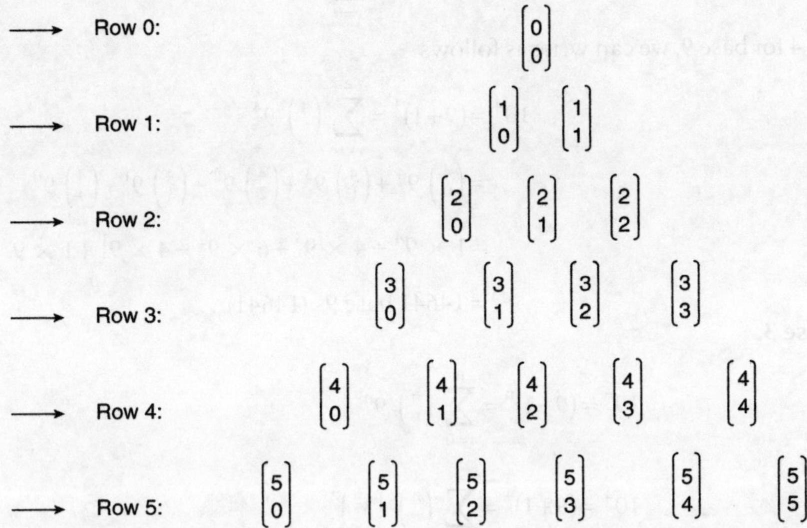

and is summarised as given in Fig. 9.10

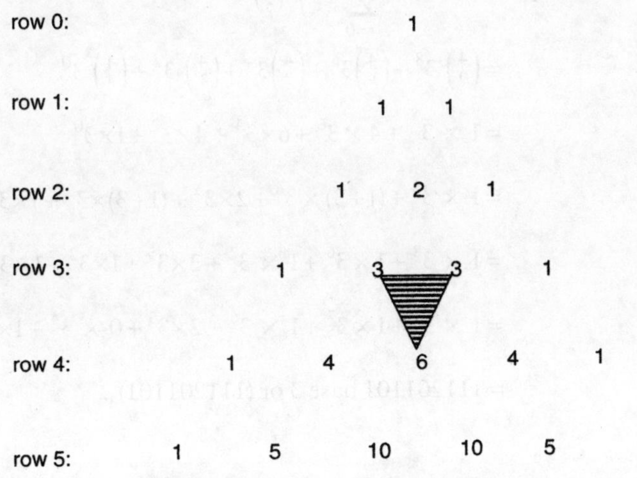

Figure 9.10 Pascal triangle

9.4.1 Properties of Pascal's triangle

(i) Every row begins and ends with 1. This is because $C(n, 0) = C(n, n) = 1$.
(ii) Pascal's triangle is symmetric about a vertical line through the middle. This because $C(n, r) = C(n, n - r)$.
(iii) Any interior number in each row is the sum of the numbers immediately to its left and to its right in the preceding row. This property is known as Pascal's identity.

(iv) The binomial theorem and hence Pascal's triangle can be used to express any positive integer in bases nine and three, as explained by the following.

$$10^n = (9+1)^n = \sum_{r=0}^{n} \binom{n}{r} 9^{n-r}$$

- Let $n = 4$ for base 9, we can write as follows

$$10^4 = (9+1)^4 = \sum_{r=0}^{4} \binom{4}{r} 9^{4-r}$$

$$= \binom{4}{0} 9^4 + \binom{4}{1} 9^3 + \binom{4}{2} 9^2 + \binom{4}{3} 9^1 + \binom{4}{4} 9^0$$

$$= 1 \times 9^4 + 4 \times 9^3 + 6 \times 9^2 + 4 \times 9^1_1 + 1 \times 9^0$$

$$= 14641 \text{ base } 9 \quad (14641)_9.$$

- **For base 3.**

$$10^n = (9+1)^n = \sum_{r=0}^{n} \binom{n}{r} 9^{n-r}$$

$$10^4 = (9+1)^4 = \sum_{r=0}^{4} \binom{4}{r} \left(3^2\right)^{4-r}$$

$$10^4 = (9+1)^4 = \sum_{r=0}^{4} \binom{4}{r} (3)^{8-2r}$$

$$= \binom{4}{0} 3^8 + \binom{4}{1} 3^6 + \binom{4}{2} 3^4 + \binom{4}{3} 3^2 + \binom{4}{4} 3^0$$

$$= 1 \times 3^8 + 4 \times 3^6 + 6 \times 3^4 + 4 \times 3^2 + 1 \times 3^0$$

$$= 1 \times 3^8 + (1+3) \times 3^6 + 2 \times 3^5 + (1+3) \times 3^2 + 1 \times 3^0$$

$$= 1 \times 3^8 + 1 \times 3^6 + 1 \times 3^7 + 2 \times 3^5 + 1 \times 3^2 + 1 \times 3^3 + 1 \times 3^0$$

$$= 1 \times 3^8 + 1 \times 3^7 + 1 \times 3^6 + 2 \times 3^5 + 0 \times 3^4 + 1 \times 3^3 + 1 \times 3^2 + 0 \times 3^1 + 1 \times 3^0$$

$$= 111201101 \text{ base } 3 \text{ or } (111201101)_3.$$

9.5 COMBINATIONS WITH REPETITIONS

The process of permutations contains repetitions. In the same way we can have combinations (selections) with repetitions. Let us look at some simple illustrations.

(i) $S = \{a, b\}$. We can find the number of 3-element combinations of the set. In this set, $n = 2$, and combinations must contain three elements $r = 3$. Since $r > n$, the elements of each combination are repeated. They are as follows.
$\{a, a, a\}$, $\{a, a, b\}$, $\{a, b, b\}$ and $\{b, b, b\}$. There are four 3 combinations of a set of two elements.

(ii) S = {a, b, c}, the number of 3 combinations of the set are as follows.
{a, a, a},{a, a, b},{a, a, c},{a, b, b},{a, b, c},{a, c, c},{b, b, b},{b, b, c},{b, c, c} {c, c, c}.
The set has 10 members with 3-combinations.
Suppose we wish to select, with repetitions, a combinations of r objects from a set of n distinct objects, where ($r \le n$), then the number of such selections is given by

$$C(n+r-1, r) = \frac{(n+r-1)!}{r!(n-1)!} = C(r+n-1, n-1).$$

There are $C(n+r-1, r) = C(n+r-1, n-1)$ r combinations from a set with n elements, when a repetition of elements is allowed.

Example 9.24: Suppose a stationary shop has four different types of body lotions. In how many different ways can six body lotions be chosen?

Solution:
The number of ways to choose six body lotions is the number of 6 combinations of a set with four elements.

$$C(4+6-1,6) = C(9,6).$$

$$C(9,6) = C(9,3) = \frac{9.8.7}{1.2.3} = 84$$

There are 84 different ways to chose six body lotions.

Example 9.25: How many solutions does the equation $x_1 + x_2 + x_3 = 11$ have? x_1, x_2, x_3 are non-negative integers.

Solution:
The process involves selecting 11 items from a set with three elements so that x_1 are of Type 1, x_2 are of Type 2 and x_3 are of Type 3. This is equivalent to the number of 11 combinations with repetition allowed from a set of three elements.

$$C(3+11-1,11) = C(13,11). = C(13,2) = \frac{13.12}{1.2} = 78.$$

It is observed that when the variables are subjected to some constraints, it becomes necessary to find the number of solutions where the variables are integers with ($x_1 \ge 1$), ($x_2 \ge 2$), ($x_3 \ge 3$). The selection of 11 items with x_1 items of Type 1, x_2 items of Type 2 and x_3 items of Type 3 involves at least one item of Type 1, two items of Type 2, and three items of Type 3. So, choose one item of Type 1, two items of Type 2, and three of Type 3. Then, five additional items are selected.

$$C(3+5-1,5) = C(7,5) = C(7,2) = \frac{7.6}{1.2} = 21$$

So there are 21 solutions of the equation subject to the given constraints.

ADDITIONAL SOLVED EXAMPLES

Example W9.1: If ${}^{56}P_{r+6} : {}^{54}P_{r+3} = 30{,}800{:}1$ .Then find r.

Solution:

Since ${}^{n}P_{r} = \frac{n!}{(n-r)!}$

$$\therefore \quad {}^{56}P_{r+6} = \frac{56!}{(56-r-6)!} = \frac{56!}{(50-r)!} \quad \text{(I)}$$

$$ {}^{54}P_{r+3} = \frac{54!}{(54-r-3)!} = \frac{54!}{(51-r)!} \quad \text{(II)}$$

Divide equation (I) by (II) and the given ratio implies

$$\frac{{}^{56}P_{r+6}}{{}^{54}P_{r+3}} = \frac{30,800}{1}$$

$$\frac{56!}{(50-r)!} \times \frac{(51-r)!}{54!} = \frac{30,800}{1}$$

$$\frac{56\times55\times54\,!}{(50-r)!} \times \frac{(51-r)!}{54!} = \frac{30,800}{1}$$

$$\frac{56\times55}{(50-r)!} \times \frac{(51-r)(50-r)!}{1} = \frac{30,800}{1}$$

$$56\times55\,(51-r) = 30,800$$

$$3080\times(51-r) = 30,800$$

$$(51-r) = 10$$

$$\Rightarrow r = 41.$$

Example W9.2: If ${}^{(2n+1)}P_{n-1} : {}^{(2n-1)}P_n = 3:5$, find the value of n.

Solution:

The given problem is written as

$$\frac{{}^{(2n+1)}P_{n-1}}{{}^{(2n-1)}P_n} = \frac{3}{5} \quad \text{(I)}$$

Consider ${}^{(2n+1)}P_{n-1} = \dfrac{(2n+1)!}{(2n+1-n+1)!} = \dfrac{(2n+1)!}{(n+2)!}$

and ${}^{(2n-1)}P_n = \dfrac{(2n-1)!}{(2n-1-n)!} = \dfrac{(2n-1)!}{(n-1)!}$

Then equation (I) becomes

$$= \frac{(2n+1)!}{(n+2)!} \times \frac{(n-1)!}{(2n-1)!} = \frac{3}{5}.$$

$$= \frac{(2n+1)(2n)!}{(n+2)(n+1)n!} \times \frac{(n-1)(n-2)!}{(2n-1)(2n-2)!} = \frac{3}{5}$$

$$= \frac{(2n+1)(2n)}{(n+2)(n+1)n} = \frac{3}{5}$$

$$= \frac{(2n+1)20}{n^2+3n+2} = \frac{3}{5}$$

$$10(2n+1) = 3(n^2+3n+2)$$

$$3n^2 - 11n - 4 = 0$$

$$3n^2 - 12n + n - 4 = 0.$$

$$3n(n-4) + (n-4) = 0$$

$$(3n+1)(n-4) = 0$$

$$3n+1 = 0 \text{ or } n-4 = 0$$

$$n = -1/3 \text{ or } n = 4.$$

n is a non-negative integer $n = -1/3$ is discarded.
Hence, $n = 4$.

Example W9.3: Out of ten pens, 3 are defective but they are unidentified.
In how many ways can

(i) these three pens be selected?
(ii) the selections include at least one defective pen.

Solution:

(i) Three pens can be selected out of ten pens in ${}^{10}C_3 = 120$ ways.
(ii) The selection to include at least one defective. The following three cases arise.
(a) Selection of two pens out of 7 good pens and 1 defective pen out of 3 such pens.

$${}^7C_2 \times {}^3C_1 = \frac{7!}{2!\,5!} \times \frac{3!}{1!\,2!} = 63$$

(b) Selection of 2 defective and 1 good pen

$${}^7C_1 \times {}^3C_2 = \frac{7!}{1!\,6!} \times \frac{3!}{2!\,1!} = 21$$

(c) Selection of 3 defective pens ${}^3C_3 = 1$
Hence the total number of selections which include at least one defective pen $= 63 + 21 + 1 = 85$.

Example W9.4: If ${}^nP_r = 240$, and ${}^nC_r = 120$, then find n and r.

Solution:

Given ${}^nP_r = \dfrac{n!}{(n-r)!} = 240$, ${}^nC_r = \dfrac{n!}{r!\,(n-r)!} = 120$,

$$\frac{{}^nC_r}{{}^nP_r} = \frac{n!}{r!(n-r)!} \times \frac{(n-r)!}{n!} = \frac{120}{240}$$

$$\frac{1}{r!} = \frac{1}{2} \quad \Rightarrow r = 2.$$

Again consider ${}^nP_r = \dfrac{n!}{(n-r)!} = 240$, put $r = 2$

$$\frac{n!}{(n-2)!} = 240,$$

$$\frac{n(n-1)(n-2)!}{(n-2)!} = 240$$

$$n(n-1) = 240 = 16 \times 15.$$

$$\Rightarrow n = 16.$$

Example W9.5: Complementary Combinations

Prove that ${}^nC_r = {}^nC_{n-r}$

Proof: We have, from the combination formula,

$$^nC_r = \frac{n!}{r!\,(n-r)!} \qquad \text{(I)}$$

Change r to $(n-r)$

$$^nC_{n-r} = \frac{n!}{(n-r)!\,(n-(n-r))!}$$

$$= \frac{n!}{(n-r)!\,r!} \qquad \text{(II)}$$

From (I) and (II) we get. ${}^nC_r = {}^nC_{n-r}$

Similarly, we have

(ii) ${}^nC_r + {}^nC_{r-1} = {}^{n+1}C_{n-r}$

(iii) ${}^{n+1}C_{n-r} = {}^{n-1}C_r + 2.{}^{n-1}C_{r-1} + {}^{n-1}C_{r-2}$

SUMMARY

Generation of combinatorial sequences, such as permutations and combinations, has been studied extensively because of the fundamental nature and the importance in practical applications. The combinatorial theory is a major mathematics branch that has extensive applications in many fields such as image analyses, communication networks, languages, graphs, intelligent computing, etc. Many real world problems are mathematically intractable. In these cases, combinatorics techniques are needed to count, enumerate, or represent possible solutions in the process of solving application problems.

EXERCISES

1. In how many different ways can four rings of a lock be combined when each ring has 10 digits 0 to 9? If the lock opens with only one combination of 4 digits, then how many unsuccessful events are there?
2. How many numbers more than 2000 can be formed with the digits 1, 2, 3, 4, 5 without a repetition of digits?
3. If P(n, 4):P(n, 3)= 9:1, then find n.
4. Find the number of arrangements that can be made out of the letters of the following words (a) ASSASSINATION (b) ENGINEERING.
   (i) In the first word, how many 3 A's are together.
   (ii) In the second word find how many of them start with E and end with N.

5. In how many ways can 5 men and 4 women dine at a round table, if no two women are to sit together?
6. Find the number of diagonals in a polygon of n sides.
7. The English language consists of 21 consonants and 5 vowels. How many six-letter words consisting of at least a vowel and two consonants are possible?
8. From a council of 7 lecturers, 5 readers and 3 professors, a committee of five is to be formed with one professor and at least two readers. Show that the number of ways of forming the committee is 85.
9. A team of eleven is to be chosen out of 16 cricketers of whom 4 are bowlers and 2 others are wicket keepers. In how many ways can the team be chosen so that,
   (a) there are exactly 3 bowlers and one wicket keepers
   (b) there are at least 3 bowlers and at least one wicket keeper.
10. Find the number of groups that can be formed from a group of seven marbles if each group must contain at least three marbles.
11. A committee of eight members is to be chosen from 9 teachers and 4 students. In how many ways can this be done if
    (a) the committee contains exactly 3 students
    (b) there is to be a majority of teachers
    (c) there are at least 3 students in the committee
    (d) there are at most 7 teachers in the committee.
12. Show that
    (a) $C(n+1, r) = C(n, r-1) + C(n, r)$
    (b) $C(m+n, 2) - C(m, 2) - C(n, 2) = mn$
    (c) $C(n, r).C(r, k) = C(n, k).C(n-k, r-k)$.
13. How many integral solutions are there to $x_1 + x_2 + x_3 + x_4 + x_5 = 16$, when $x_i \geq 2$.
14. Find the term independent of x in the expansion of $\left(x^2 + \frac{1}{x}\right)^{12}$.
15. Find the coefficients of $x^5 y^{10} z^5 w^5$ in the expansion of $(x + 7y + 3z + w)^{25}$.

10 Introduction to Probability

In Chapter 9, we have discussed the principles of counting. In computer science, we often have random phenomena such as a request for an I/O, a processor encountering an interrupt, etc. Probability theory is concerned with the study of random phenomena, which occur around us in nature too. These phenomena occur in such that their future occurrence is not predictable in a perfect manner, with certainity or regularity. The explanation is often made in the form of a physical model and used for predictive purposes and is further translated into one or more mathematical equations.

10.1 PROBABILITY MODELS

A physical model is an abstraction of a real world problem. Predictions based on the physical model are validated with respect to the actual data collected from physical phenomena. In order to use the model for predictive purposes, it is translated into two or more mathematical equations. These equations are used to determine the value of a specific variable in the model, based on the knowledge of the values assumed by other model variables. For example: The ideal gas equation states that the pressure and volume of a gas may both vary simultaneously, when the temperature of the gas is changed. The verbal model is put into a mathematical equation (1).

$$PV = RT \tag{1}$$

Where P is pressure, V is volume, T is temperature, and R is gas constant.

Based on the perfect gas law model, Vander Waal explained a physical model for real gases by defining some more constants based on realistic models.

$$\left(P + \frac{a}{V^2}\right)(V - b) = RT \tag{2}$$

A correction is applied to the perfect gas equation based on some experiments. Many real gases cannot be described by equation (1), especially at extreme temperatures and pressures. This has led to finding another way to predict the behavior of gases, with some corrections as given in equation (2). This is done with the aid of statistical methods. The statistical methods are designed in such a way that they allow an assessment of the degree of uncertainty present in the results. These methods are classified roughly into three categories (i) descriptive statistics (ii) inferential statistics and (iii) model building. In descriptive statistics, the techniques, both analytic and graphical, allow us to describe or pictorially represent a data set. The inferential statistics is mainly concerned with conclusions about a large group of objects and are based on observation of only a small portion of the objects from a larger group. With this explanation we come across two definitions.

(i) **Population:** The overall group of objects about which conclusions are to be drawn, is called the population.
(ii) **Sample:** A portion of the population that is actually obtained and it is used to draw conclusions about the population.

Model building mainly relates to the development of prediction equations from experimental data. This set of equations is called a statistical model and helps us to predict the behavior of complex systems. The mathematics on which statistical methods rest is called the probability theory. Hence, we begin discussion by considering the basic concepts of probability.

10.2 MEANING OF PROBABILITY

The ability to interpret probabilities is assumed in our culture. One usually hears the words — "the probability of humidity today is about 85%" or "there is 20% change in cold today". It is interesting to note that the general public can interpret these values correctly.

10.2.1 Interpretation of Probabilities

One can interpret the probabilities in the following ways.

(i) Probabilities are numbers between 0 and 1 that indicate the chances of a physical event occurring.
(ii) Probabilities near 1 indicate that the event is extremely likely to occur. It is considered a sure event. They do not mean that the event will occur, but that it most probably will.
(iii) Probabilities near zero indicate that the event is not very likely to occur. It is also called an impossible event. They do not mean that the event will fail to occur, but considered to be rare.
(iv) Probabilities near ½ indicate that the event is just as likely to occur as not.

These properties are the guidelines for interpreting probabilities, once they are available. They will not indicate how to assign probabilities to events. There are three methods which are widely used.

(i) The personal approach
(ii) The relative frequency approach
(iii) The classical approach.

Example 10.1: A cyclone has occurred. A meteorological scientist asks, "What is the probability of the strength of this cyclone causing widespread damage to nearby places?" Many factors come into play, the strength of cyclone, the direction, and the speed of the cyclone over sea. These factors make this cyclone unique. The scientist is called upon to make a value judgment. This is the illustration to assign probability to the event, based on an informed personal opinion. The advantage of the personal approach is that it is always applicable; anyone can have a personal opinion about anything. However, it has a disadvantage that its accuracy depends on the accuracy of information available and the ability of the scientist to assess that information correctly.

Example 10.2: A civil engineer is studying the peak demand at a water distribution reservoir. It is observed that on 80 of the 100 days randomly selected for study from past records, the peak demand occurred between 7AM and 8 AM. It is natural to assume that the probability of this occurring on another day is at approximately $\frac{80}{100}=0.80$. The result obtained is not simply a personal opinion. It is based on repeated experimentation. The technique is known as relative frequency. It can be used whenever the experiment can be repeated many times and the results observed. The probability of occurrence of event A,

$$P[A]=\frac{f}{n}=\frac{\text{Number of times event A occured}}{\text{Number of times experiment performed}}$$

The main disadvantage in this approach is that the experiment cannot be a one-shot situation; it is repeatable. Thus, for a large number of trials, the approximate probability obtained by using the relative frequency approach is quite accurate.

10.2.2 Classical Formula

The probability obtained neither by a personal opinion, nor based on repeated experiments is treated as classical in nature. It is reasonable to apply this when possible outcomes are equally likely.

$$P[A]=\frac{n(A)}{n(S)}=\frac{\text{Number of ways A occur}}{\text{Number of ways the experiment performed}}$$

The classical method does not require experimentation and the event A is not an approximation.

10.3 SAMPLE SPACES AND EVENTS

The first step in analysing most experiments is to make a list of possibilities for the experiment. The list is known as the sample space. A sample space for a random experiment is a set S with the property that each physical outcome of the experiment corresponds to exactly one element of S. Each element of the set S is called a sample point.

Event: Any subset A of a sample space is called an event. The empty set Φ is called the impossible event. The subset S is called the certain event.

$$P[A]=\frac{|A|}{|S|}=\text{Probability that A occurs.}$$

Example 10.3: What is the probability that a child born to a couple heterozygous for eye colour (each gene for both brown and blue eyes), will be brown-eyed.

Solution:

The child receives one gene from each parent, the sample space is

$$S=\{(\text{brown, blue}), (\text{blue, brown}), (\text{blue, blue}), (\text{brown, brown})\}$$

Since each parent is just as likely to contribute a gene for brown eyes as for blue eyes, the sample space S is equally likely. The probability that a child born to a couple heterozygous for eye colour will be brown eyed is denoted as P(A).

$P(A) = P\{(\text{brown, blue}), (\text{blue, brown}), (\text{brown, brown})\}$

$$P(A)=\frac{n(A)}{n(S)}=\frac{3}{4}=0.75$$

The following observations are made:

(i) Suppose $A = \Phi$, the null set, then we have $n(A) = n(\Phi) = 0$.

$$\therefore P(A) = \frac{n(A)}{n(S)} = \frac{0}{n(S)} = 0$$

(ii) Suppose $A = S$, then

$$\therefore P(A) = P(S) = \frac{n(S)}{n(S)} = 1$$

(iii) Suppose A is a proper subset of S, then $0 < n(A) < n(S)$

So that $0 < \frac{0(A)}{n(S)} < 1$ hence $0 < P(A) < 1$.

Based on above conclusions we can also come to conclusion that $(0 \le P(A) \le 1)$.

(iv) If $\bar{A}$ is the complement of A in S, then as given in Fig. 10.1.

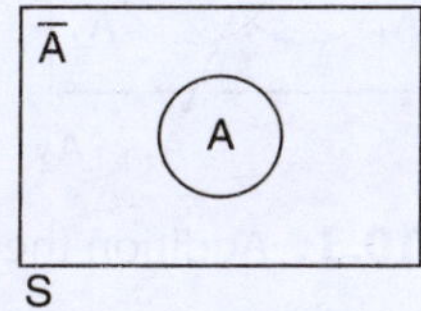

Figure 10.1 Venn diagram for A and $\bar{A}$

$n(\bar{A}) = n(S) - n(A)$ Divide by n(S)

$$\frac{n(\bar{A})}{n(S)} = 1 - \frac{n(A)}{n(S)} = 1 - P(A)$$

Or

$P(\bar{A}) = 1 - P(A)$ and $P(A) + P(\bar{A}) = 1$.

The sum of the probabilities of an event happening and not happening is equal to 1.

10.3.1 Mutually Exclusive Events

The two or more events that do not occur at the same time are called mutually exclusive events. The occurrence of one event prohibiting the occurrence of the other events are said to be mutually exclusive. Two events A_1 and A_2 are mutually exclusive if and only if $A_1 \cap A_2 = \Phi$. Events A_1, A_2 are said to be mutually exclusive if and only if $A_i \cap A_j = \Phi$ for $i \neq j$, as indicated in Fig. 10.2.

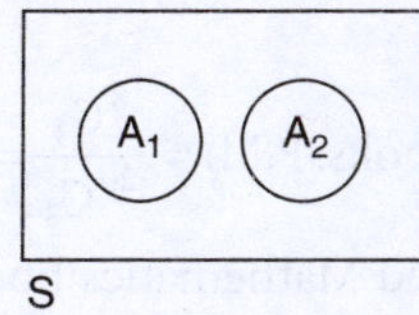

Figure 10.2 Mutually exclusive events A_1 and A_2

10.3.2 Addition Theorem in Probability

We come across some situations wherein we have to find the probability of at least one of two events, which are not mutually exclusive. Let A_1 and A_2 are events in a sample space S, and $A_1 \cup A_2$ is an event in S consisting of outcomes which are in A_1 or A_2 or both, and $A_1 \cap A_2$ is an event in S consisting of outcomes that are common to A_1 and A_2, so that

$$0(A_1 \cup A_2) = 0(A_1) + 0(A_2) - 0(A_1 \cap A_2)$$

Hence,

$$P(A_1 \cup A_2) = P(A_1) + P(A_2) - P(A_1 \cap A_2)$$

The result is known as the *Addition theorem for probability*, as shown in Fig. 10.3.

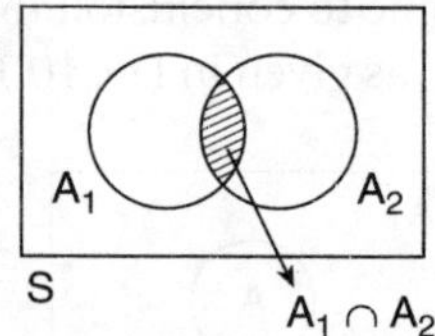

Figure 10.3 Addition theorem

The following are some observations:

(i) Addition theorem is traditionally expressed as
$P(A_1 + A_2) = P(A_1) + P(A_2) - P(A_1 A_2)$

(ii) When A_1 and A_2 are mutually exclusive i.e., when $A_1 \cap A_2 = \phi$ and $P(A_1 \cap A_2) = 0$, then $P(A_1 \cup A_2) = P(A_1) + P(A_2)$.

(iii) The addition rule links the operations of union and intersection.

Example 10.4: A bookshelf contains 20 books of which 12 are on Electronics and eight are on Mathematics. If three books are taken at random, find the probability that all the three are on the same subject.

Solution:

There are 20 books and 3 books are drawn at random. This can be done in ${}^{20}C_3$ different ways. Out of 12 Electronics books, three books can be drawn in ${}^{12}C_3$ different ways. Out of eight Mathematics books, three books can be drawn in ${}^{8}C_3$ different ways.

Probability of drawing three Electronics books: $P(E) = \frac{{}^{12}C_3}{{}^{20}C_3} = \frac{11}{57}$.

Probability of drawing three Mathematics books: $P(M) = \frac{{}^{8}C_3}{{}^{20}C_3} = \frac{14}{285}$

The events of drawing Electronics books and Mathematics books are mutually exclusive. Hence, the probability of drawing three books on the same subject is

$P(E) + P(M) = \frac{11}{57} + \frac{14}{285} = 0.19298 + 0.04912 = 0.2421.$

10.4 CONDITIONAL PROBABILITY

Consider two events A_1 and A_2. It is required to determine the probability that some event A_2 will occur given that some other event A_1 has already occurred. This is denoted as P $[A_2/A_1]$ and indicates the conditional probability of event A_2 occurring given that event A_1 has occurred. Consider this scenario: A factory produces one type of screws (A_2), the other event, namely, the production of defective screws has already occurred (A_1), along with the good screws. Since the finished goods have some percentage of defects of different types.

$$P[A_2/A_1] = \frac{P(A_1 \cap A_2)}{P(A_1)} \text{ as } P(A_1) \neq 0.$$

It is read as the probability of A_2 given A_1.

10.4.1 Independence

Independent objects are the objects acting irrespective of each other. The occurrence or non-occurrence of one does not alter the likelihood of the occurrence or non-occurrence of the other.

Example 10.5: A simple experiment consisting of rolling a single fair die once and then tossing a fair coin once. Let us represent the outcomes in the form of ordered pairs in which the first member of each ordered pair denotes the number appearing on the die and the second the face showing on the coin (H or T). The sample space is given by the outcomes which are equally likely.

S = {(1,H), (1,T), (2,H), (2,T), (3,H), (3,T), (4,H), (4,T), (5,H), (5,T), (6,H), (6,T)}.

Let us consider the following events, (outcomes)

A: The die shows three or four.

B: The coin shows tails.

$A \cap B$ the die shows 3 or 4 and the coin shows tails. The die roll gives no additional information about the landing of coin. Hence A and B are independent of each other.

$P(A) = P[(3,T), (3,H), (4,H), (4,T)] = 4/12 = 1/3.$

$P(B) = P[(1,T), (2,T), (3,T), (4,T), (5,T), (6,T)] = 6/12 = 1/2.$

$P(A \cap B) = P[(3,T), (4,T)] = 2/12 = 1/6.$

We can conclude that $P(A \cap B) = P(A)\,P(B) = (1/3)\,(1/2) = 1/6.$

Events A_1 and A_2 are said to be independent if and only if

$P[A_1 \cap A_2] = P[A_1]\,P[A_2].$

10.4.2 Multiplication Rule

Among the events A_1 and A_2, if at least one of the events has non-zero probabilities, then their characterisation as independence is justified.

$P[A_1 \cap A_2] = P[A_1]\,P[A_2]$. Dividing by $P(A_1)$ we get,

$$\frac{P(A_1 \cap A_2)}{P(A_1)} = P(A_2)$$

$$= P[A_2/A_1]$$

Hence,

$P[A_2/A_1] = P[A_2]$ if $P[A_1] \neq 0$

$P[A_1/A_2] = P[A_1]$ if $P[A_2] \neq 0$

$P[A_1 \cap A_2] = P[A_2/A_1]\,P[A_1]$. This is known as the multiplication rule.

Example 10.6: Consider an orbital satellite launch. The primary computer system is backed up by three secondary systems. They operate independently (3-stage rocket) and the system is 95% reliable. What is the probability that all three systems will be operable at the time of the launch?

Solution:

Let us consider the outcomes,

A_1: The main system is operable

A_2: The first backup is operable

A_3: The second backup is operable

A_4: The third backup is operable.

Given that $P[A_1] = P[A_2] = P[A_3] = P[A_4] = 0.95$.

We need to show that

$$\begin{aligned} P[A_1 \cap A_2 \cap A_3 \cap A_4] &= P[A_1]\, P[A_2]\, P[A_3]\, P[A_4] \\ &= (0.95)\,(0.95)\,(0.95)\,(0.95) \\ &= 0.81451. \end{aligned}$$

Example 10.7: The probability that machine A will be performing satisfactorily in five years is 1/4, while that machine B will be performing is 1/3. Find the probability of the following in five years time.

(i) both machines will be performing satisfactorily

(ii) neither will be performing

(iii) only machine B will be performing

(iv) at least one of the machines will be performing.

Solution:

Let P(A) = probability of machine A performing satisfactorily = 1/4.

Then, q(A) = Probability of machine A not performing satisfactorily

$$= 1 - \frac{1}{4} = \frac{3}{4}.$$

P(B) = probability of machine B performing satisfactorily = 1/3.

Then, q(B) = probability of machine B not performing satisfactorily.

$$= 1 - \frac{1}{3} = \frac{2}{3}$$

(i) Probability of both machines performing satisfactorily = $P(A \cap B)$ (machines work independently)

$$= P(A)\,P(B) = \frac{1}{4} \times \frac{1}{3} = \frac{1}{12}.$$

(ii) That neither will be performing = 1 – P(A) P(B) = q(A) q(B).

$$= \frac{3}{4} \times \frac{2}{3} = \frac{1}{2}$$

(iii) $P(\text{only B performing}) = P(B)\,q(A) = \frac{1}{3} \times \frac{3}{4} = \frac{1}{4}.$

(iv) P(at least one of the machines performing)

$$\begin{aligned} &= 1 - P(\text{none of them performing}). \\ &= 1 - \frac{1}{2} = \frac{1}{2}. \end{aligned}$$

Example 10.8: A problem of statistics is given to three students A, B and C, whose chances of solving it are ½, ¾, and ¼ respectively. What is the probability that the problem will be solved?

Solution:

Let P(A) = The probability of the problem being solved by A = 1/2.

$q(A)$ = The probability of the problem not being solved by $A = 1-\frac{1}{2}=\frac{1}{2}$

$P(B)$ = The probability of the problem being solved by B = 3/4

$q(B)$ = The probability of the problem not being solved by $B = 1-\frac{3}{4}=\frac{1}{4}$

$P(C)$ = The probability of the problem being solved by C = 1/4

$q(C)$ = The probability of the problem not being solved by $C = 1-\frac{1}{4}=\frac{3}{4}$

The probability that A, B and C will not be able to solve the problem is:

$$q(A \cap B \cap C) = q(A)\, q(B)\, q(C) = \frac{1}{2}\times\frac{1}{4}\times\frac{3}{4}=\frac{3}{32}$$

The probability of the problem solved $=1-q(A\cap B\cap C)$

$$=1-\frac{3}{32}=\frac{29}{32}.$$

Example 10.9: A husband and wife appear for an interview for two vacancies in the same post. The probability of the husband's selection is 1/7 and that of the wife's selection is 1/5. What is the probability that

(i) both of them will be selected?
(ii) only one of them will be selected?
(iii) none of them will be selected?

Let P(H) = Husband's selection $=\frac{1}{7}$: and $q(H)=1-\frac{1}{7}=\frac{6}{7}=P(\bar{H})$

$P(W)$ = Wife's selection $=\frac{1}{5}$: and $q(W)=1-\frac{1}{5}=\frac{4}{5}=P(\bar{W})$

(i) P (both selected) $= P(H \cap W) = P(H)\, P(W) = \frac{1}{7}\times\frac{1}{5}=\frac{1}{35}$

(ii) P (only one selected) $=P\left(H\cap\bar{W}\right)+P\left(W\cap\bar{H}\right)$

$$=P(H)\, P(\bar{W})+P(W)\, P(\bar{H})$$

$$=\frac{1}{7}\times\frac{4}{5}+\frac{1}{5}\times\frac{6}{7}=\frac{10}{35}=\frac{2}{7}$$

(iii) P (none will be selected) $=P\left(\bar{H}\cap\bar{W}\right)=P(\bar{H})P(\bar{W})$

$$=\frac{6}{7}\times\frac{4}{3}=\frac{24}{35}$$

10.5 BAYE'S THEOREM

The theorem enables us to determine the conditional probability of E, knowing A, if we know the individual probabilities of E and $A \cap E$ as well as the conditional probabilities of A, knowing E, and A, knowing $\bar{E}$.

Probabilities of E assuming A = P(E/A)

$$= \frac{P(E \cap A)}{P(A)}$$

$$= \frac{P(A \cap E)}{P(A)} = \frac{P(A/E)P(E)}{P(A)} \qquad (1)$$

by using multiplication rule.

Consider that A and E are shown as typical sets of outcomes. The outcomes in A are split into two kinds, those that are also in E, hence are in $A \cap E$, and those that are not in E, hence are in $A \cap \bar{E}$. Further, $A \cap E$ and $A \cap \bar{E}$ have no outcomes in common.

$\because P(A) = P(A \cap E) + P(A \cap \bar{E})$

Using product rule

$$P(A) = P(A/E)P(E) + P(A/\bar{E})\, P(\bar{E}) \qquad (2)$$

Using this in (1),

$$P(E/A) = \frac{P(A/E)P(E)}{P(A/E)P(E) + P(A/\bar{E})P(\bar{E})}$$

The theorem enables us to determine the conditional probability of E, knowing A, if we know the probability of E. The probability $\bar{E}$, the probability of A knowing E, and the probability of A knowing $\bar{E}$. This is summarised in Fig. 10.4.

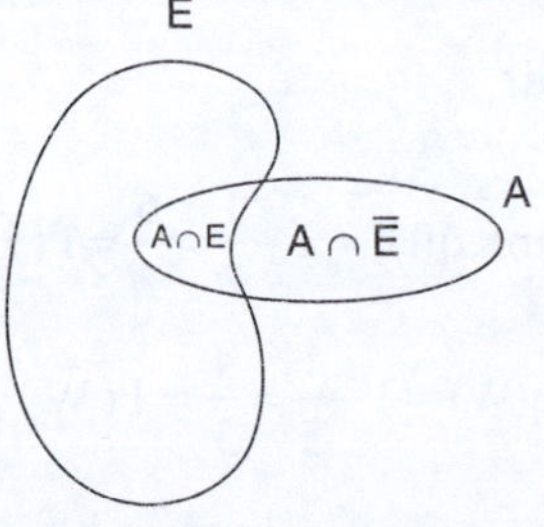

Figure 10.4 Bayes' theorem

Example 10.10: A ship manufacturing company receives spare parts from five companies. The following Table 10.1 gives the percentage of the total requisite parts supplied by each company and the probability for each company that a part is defective. Find the total probability of defective. What is the probability that it came from Companies 4 and 5?

Solution:

Table 10.1 Company details

Company	Items in %	Probability of defective
1	15	0.03
2	2	0.04
3	4	0.01
4	48	0.02
5	27	0.08

Let E_i, be the event that the part come from company i, where $i = 1, 2, 3, 4, 5$.
Let D be the event that the part is defective, P(D) and we have to determine $P(E_4/D)$ and $P(E_5/D)$.
$P(E_1) = 0.15$, $P(E_2) = 0.02$, $P(E_3) = 0.04$, $P(E_4) = 0.48$, $P(E_5) = 0.27$, and $P(D/E_1)$= defective from $E_1 = 0.03$ similarly,
$P(D/E_2) = 0.04,\ P(D/E_3) = 0.01,\ P(D/E_4) = 0.01,\ P(D/E_5) = 0.08.$
Probability of total defective: P(D): = $P(E_1)P(D/E_1) + P(E_2)P(D/E_2) + P(E_3)P(D/E_3) + P(E_4)P(D/E_4) + P(E_5)$ $P(D/E_5)$

$$P(D) = (0.15)(0.03)+(0.02)(0.04)+(0.04)(0.04)+(0.48)(0.02)+(0.27)(0.08).$$
$$= 0.0381.$$
$$\therefore\ P(E_4/D) = \frac{P(E_4)P(D/E_4)}{P(D)} = \frac{(0.48)(0.02)}{0.0381}$$
$$= 025197\ (25.19\%)$$
$$\therefore\ P(E_5/D) = \frac{P(E_5)P(D/E_5)}{P(D)} = \frac{(0.27)(0.08)}{0.0381}$$
$$= 0.56693\quad (0.57 \text{ or } 57\%).$$

The disparity is due to the fact that while company 5 supplies a little more than half the parts company 4 does, it has four times the failure rate on its parts.

Example 10.11: An urn A contains three white and four red balls, and an urn B contains five white and six red balls. One ball is drawn at random from one of the urns and is found to be white. Find the probability that it was drawn from i) urn A, and ii) urn B.

Solution: Let us consider the outcomes,

A_1: ball drawn from urn A
B_1: ball drawn from urn B
W: The ball is white.

We have to find $P(A_1/W)$ and $P(A_2/W)$

Example 10.12: Target guns are manufactured in six cities. The following Table 10.2 gives the percentage of the total production that comes from each city, together with the probability that a gun explodes upon firing.

Table 10.2 Production details

City	% of total production	Probability of Exploding
1	15	0.02
2	07	0.01
3	21	0.06
4	04	0.02
5	09	0.03

Suppose a gun explodes. Calculate for each city the probability that the gun came from that city.

Example 10.13: Three urns contain 6 red, 4 black; 4 red, 6 black; 5 red, 5 black balls respectively. One of the urns is selected at random. If the ball drawn is red, find the probability that it was drawn from the i) first urn ii) second urn iii) third urn.

10.6 RANDOM VARIABLES

In Chapter 9, the concept of an event is described as the outcome of an experiment. Events are more flexible in determining properties of an experiment in terms of outcomes. An event can be varied from descriptive (drawing a white ball in a sample space containing white and red balls) to numerical (throwing a dice and outcome is the face numbering 2). In this chapter, we introduce a new concept that will allow events to be defined in a more consistent manner and will be always be numerical. The new concept is that of a 'random variable'.

Consider a random experiment; let S be the outcome set (sample space), and consider that a unique real number X is associated with respect to some rule within S. Then X is called a random variable (or variant) or stochastic variable.

i.e., $f : S \rightarrow R$. We deal with random variables—variables whose obtained value is determined by a chance.

Example 10.14: Consider that a coin is tossed twice. We shall associate two different random variables X, Y as follows.
The sample space is given by S = {HH, HT, TH, TT}.

Let X = number of heads in the outcome set S, as shown in Table 10.3.

Table 10.3 Outcomes of X

Out come	HH	HT	TH	TT
Random Variable: X	02	01	01	00

Now Range: X = {0, 1, 2}

Suppose Y = Number of 'tails' in the outcome set. The Range: Y = {0, 1, 2}.

Random variables usually fall into two categories.

(i) Discrete random variable (ii) Continuous random variable

10.6.1 Discrete Probability Densities

The behaviour of a random variable is governed by chance; these predictions must be made in the face of a great deal of uncertainty. The behaviour of the random variable is defined in terms of probability. These are referred to as density function and cumulative distribution function. For each value X_i of a discrete random variable X, a real number $p(x_i)$ is assigned such that

i) $p(x_i) \geq 0$ ii) $\sum p(x_i) = 1$.

The function $p(x)$ is called a probability function. The set of values

$[x_i, p(x_i)]$ is called a discrete probability distribution of the discrete random variable X. The above conditions i) and ii) can be termed as necessary and sufficient conditions for a function to be a discrete function.

10.7 BERNOULLI'S TRIALS

Let us consider determining the probability by considering a very practical problem. This system involves only two possible outcomes on any trial. They are categorised as success and failure. The probability of success p is same for each trial. The examples are, flipping a coin, hitting the target, outcome of a **0** or **1** in a computer bit system.

The probability of x successes in n trials is equal to $^{n}C_{x}p^{x}q^{n-x}$, where p is probability of success and q is the probability of failure and $q = 1 - p$.

$$p(x) = {}^{n}C_{x}p^{x}q^{n-x}$$

10.7.1 Binomial Distribution

Let p be the probability of success and q the probability of failure. Consider the finite probability distribution of a random variable x described in the following Table 10.4.

Table 10.4 Distribution of variable X

X	0	1	2	3	--	--	n
P(x)	q^{n}	$^{n}C_{1}\,p\,q^{n-1}$	$^{n}C_{2}\,p^{2}\,q^{n-2}$				p^{n}

$$\sum p(x) = q^{n} + {}^{n}C_{1}q^{n-1}p + {}^{n}C_{2}q^{n-2}p^{2} + ------ + p^{n}$$

$$(p+q)^{n} = 1 \quad \because p+q=1$$

Hence, $p(x) = {}^{n}C_{x}p^{x}q^{n-x}$ is a probability distribution.

(i) Mean of the Binomial distribution:

$$\mu = \sum_{x=0}^{n} x\,p(x) = np,$$ where n is the number of trials and p is the probability of success.

(ii) Variance $V = \sum_{x=0}^{n} x^{2}\,p(x) - \mu^{2} = npq \quad q =$ Probability of failure.

(iii) Standard Deviation: $\sqrt{V} = \sigma = \sqrt{npq}$

Following are some observations:

(i) The experiment consists of a fixed number n, of 'Bernoulli trials', that result in either a "success" (p) or a "failure" (q).
(ii) The trials are identical and independent, and therefore the probability of success p, remains the same from trial to trial.
(iii) The random variable X denotes the number of successes obtained in the n trials.
(iv) The repeated trials are independent.

Example 10.15: If 10% of the bolts produced by a machine are found to be defective, determine the probability that out of ten bolts chosen at random,
(i) 1 is defective (ii) none of them are defective (iii) at the most two bolts are defective.

Solution:
Probability of defective bolts: $p = 10\% = 10/100 = 0.1$
Probability of non-defective bolts : $q = 1 - p = 0.9$
Total number of bolts $= n = 10$

(i) Probability of 1 defective bolt: $p(x=1) = {}^{10}C_{1}(0.1)^{1}(0.9)^{9} = 0.3874.$

1. Probability that none of the bolts are defective:

$$p(x=0) = {}^{10}C_{0}(0.1)^{0}(0.9)^{10} = 0.3487.$$

2. Probability of 2 defective: $p(x=2) = {}^{10}C_2 (0.1)^2 (0.9)^8 = 0.1937.$
3. Probability of at the most 2 defective $= p(x=0) + p(x=1) + p(x=2)$
$= 0.3487 + 0.3874 + 0.1937 = 0.9298.$

Example 10.16: In tunnelling work carried out in a hilly place, six pumps are installed for pumping out rain water. The probability that any one the pumps fail to operate during a storm is 1/7. What is the probability that (i) at least three pumps will be working, (ii) all the pumps will be working, during a particular storm?

Solution:

Probability of a pump failing to work $= \frac{1}{7}$

Probability of a pump working $= 1 - \frac{1}{7} = \frac{6}{7}$

$\therefore p = \frac{1}{7}, q = \frac{6}{7}, n = 6.$

(i) P(at least 3 pumps working) =

$= P(3 \text{ or } 4 \text{ or } 5 \text{ or } 6 \text{ pumps working})$

$= P(3) + P(4) + P(5) + P(6).$

$= {}^6C_3 p^3 q^3 + {}^6C_4 p^4 q^2 + {}^6C_5 p^5 q^1 + {}^6C_6 p^6 q^0_0$

$= 0.0367208 + 0.1652413 + 0.3965715 + 0.3965618$

(ii) P(all the 6 pumps working) $= P(6) = {}^6C_6 p^6 q^0 = 1 \times (0.85714)^6$

$= 0.3965623.$

Example 10.17: The probability distribution function p(x) of a variant X (= x) is given in Table 10.5.

Table 10.5 Distribution of random variable X

x	0	1	2	3	4	5	6
P(x)	k	$3k$	$5k$	$7k$	$9k$	$11k$	$13k$

1. Find the value of k, so that it may represent a valid probability distribution.
2. Find $p(x < 4)$, $p(x \geq 5)$ and $p(2 < x \leq 6)$.

Solution:

1. In order that the given p(x) may be a probability distribution function, we have $p(x_i) \geq 0$ and

$\sum p(x_i) = 1$. The first condition is satisfied by the given definition and

$\sum p(x_i) = k + 3k + 5k + 7k + 9k + 11k + 13k = 1.$

$\therefore k = \frac{1}{49}.$

2. $p(x<4) = p(x=0) + p(x=1) + p(x=2) + p(x=3).$

$= k + 3k + 5k + 7k = 16k = \frac{16}{49}$

$P(x \geq 5) = p(x=5) + p(x=6) = 11k + 13k = 24k = \frac{24}{49}.$

$P(2 < x \leq 6) = p(3) + p(4) + p(5) + p(6)$

$= 7k + 9k + 11k + 13k = 40k = \frac{40}{49}$

Example 10.18: A fair coin is tossed three times. Let x denote the number of times it turns up heads. Find the distribution of x. Also, find its mean, variance and standard deviation.

Solution:

When a fair coin is tossed three times, the following are the 8 possible outcomes.

S = {HHH, HHT, HTH, THH, HTT, THT, TTH, TTT}.

Now, X denotes the number of heads showing up

$P(x=0)$ = Probability of no heads showing up $= \frac{1}{8}$

$P(x=1)$ = Probability of one head showing up $= \frac{3}{8}$

$P(x=2)$ = Probability of two heads showing up $= \frac{3}{8}$

$P(x=3)$ = Probability of three heads showing up $= \frac{1}{8}$

So we have $p(x_i) \geq 0$ and $\sum p(x_i) = 1$. The distribution is given in Table 10.6.

Table 10.6 Probability values

x	0	1	2	3
P(x)	$\frac{1}{8}$	$\frac{3}{8}$	$\frac{3}{8}$	$\frac{1}{8}$

$$\text{Mean}: \mu = \sum_{i=0}^{3} x_i p(x_i) = (x_0 p(x_0)) + (x_1 p(x_1)) + (x_2 p(x_2)) + (x_3 p(x_3)).$$

$$= \frac{3}{8} + \frac{6}{8} + \frac{3}{8} = \frac{12}{8} = 1.5$$

$$\text{Variance } \sigma^2 = \sum_{i=0}^{3} (x_i - \mu)^2 p(x_i) = (0-1.5)^2 \times \frac{1}{8} + (1-1.5)^2 \times \frac{3}{8} + (2-1.5)^2 \times \frac{3}{8} + (3-1.5)^2 \times \frac{1}{8}$$

$$= \frac{1}{8}[2.25 + 0.75 + 0.75 + 2.25] = \frac{6}{8} = 0.75$$

$$\text{S.D} = \sigma = \sqrt{V} = \sqrt{0.75} \approx 0.8660254.$$

Example 10.19: The number of telephone lines busy at an instant of time is a binomial variant with $p = 0.3$. If at an instant 20 lines are chosen at random, what is the probability that (i) six lines are busy (ii) at the most, three lines are busy (iii) all lines are busy.

Example 10.20: The probability of a patient recovering from a chronic disease is 0.4. If 15 people are known to have contracted this disease, what is the probability that (i) at least ten survive (ii) from three to eight survive, (iii) exactly five survive. Find also the mean and variance of the binomial random variable.

10.8 AREAS OF APPLICATION

The binomial distribution finds applications in many scientific fields. An engineer in an industry is interested in the proportion of defective products in the process of manufacturing. In the process, the quality control measures and sampling schemes are based on the binomial distribution. The method is applicable for any industrial situation and the results of the process are independent, with the probability of a success being constant from trial to trial. The binomial distribution is also used extensively for medical and military applications. In such cases, a success or failure result is important. In the medical domain, 'contagious' or 'non-contagious' is important. In the military, 'target' or 'miss' is often used to interpret the results in case of firing of a guided missile. The probability distribution of any binomial random variable depends only on the values assumed by the parameters, namely *n*, p, q. Also, the mean and variance of a binomial random variable depend on the parameters stated.

10.8.1 Poisson Distribution

It is the limiting form of the binomial distribution: When *n* is large ($n \to \infty$) and the probability of success is very small ($p \to 0$), the product *np* tends to a fixed finite constant. Let the constant be m. The Binomial distribution function $p(x) = {}^{n}C_{x} p^{x} q^{n-x}$ reduces to a Poisson distribution function as shown by the following.

$p(x) = \dfrac{m^{x}\, e^{-m}}{x\,!}$ where $m = np$. The function p(*x*) is non-negative. To see that its sum equals to 1. Let us consider the Maclaurin series.

$$e^{x} = 1 + x + \frac{x^{2}}{2!} + \frac{x^{3}}{3!} + - - - - + \frac{x^{n}}{x\,!} + - - - \text{to } \infty.$$

$$\sum p(x) = \sum \frac{e^{-m}\, m^{x}}{x\,!} = e^{-m} \sum \frac{m^{x}}{x\,!}$$

$$= e^{-m}\left[1 + m + \frac{m^{2}}{2\,!} + \frac{m^{3}}{3!} + - - - -\right]$$

$$= e^{-m} e^{m} = 1$$

$$\therefore \sum_{i=0}^{\infty} p(x_{i}) = 1 \text{ and } p(x) \geq 0$$

Poisson random variables usually arise in connection with what is called a Poisson process. It involves observing discrete events in a continuous interval of time, length or space. The word interval in describing the general Poisson process is not an interval in the usual mathematical sense. The given time interval may be of any length such as a minute, a day, a month or even a year. The random variable X in the discussion may be the number of days a school is closed during summer, the number of postponed games due to rain during a tennis play season. The specified region could be a line segment, an area, a volume, or perhaps a piece of material. In such instances X might represent the number of bacteria in a given culture, the number of air-planes landing in a given time, etc.

1. Mean of Poisson distribution: $\mu = \sum_{i=0}^{\infty} x_{i}\; p(x_{i}) = m$
2. Variance: $V = \sum_{i=0}^{\infty} x_{i}^{2}\; p(x_{i}) - \mu_{2}^{2} = m$

   Mean and variance are equal in a Poisson distribution.
3. Standard deviation: $S.D = \sqrt{V} = \sigma = \sqrt{m}.$

Example 10.21: A manufacturer knows that the condensers he makes contain on an average of 1% defectives. He packs them in boxes of 100. What is the probability that a box picked at random will contain four or more faulty condensers?

Solution:

$p = 1\% = 0.01, n = 100, m = np = 100 \times 0.01 = 1.$

$$p(x) = \frac{e^{-x} m^x}{x!} = \frac{e^{-1}(1)^x}{x!} = \frac{e^{-1}}{x!}$$

$$\begin{aligned} P(\text{4 or more faulty condensers}) &= p(4) + p(5) + \text{------} + p(100) \\ &= 1 - [p(0) + p(1) + p(2) + p(3)] \\ &= 1 - \left[\frac{e^{-1}}{0!} + \frac{e^{-1}}{1!} + \frac{e^{-1}}{2!} + \frac{e^{-1}}{3!}\right] \\ &= 1 - e^{-1}\left[1 + 1 + \frac{1}{2} + \frac{1}{6}\right] \\ &= 1 - \frac{8}{3e} = 1 - 0.981 = 0.019. \end{aligned}$$

Example 10.22: The probability that an individual suffers an adverse reaction from a certain injection is 0.001. Determine the probability that out of 2000 individuals (i) exactly three (ii) more than two individuals (iii) more than one individual (iv) none, will suffer a reaction.

Solution:

$p = 0.001, n = 2000, m = np = 2000 \times 0.001 = 2.$

$$p(x) = \frac{m^x e^{-m}}{x!} = \frac{2^x e^{-2}}{x!} = \frac{2^x}{e^2 x!}$$

(i) Exactly 3, $(p(x = 3)) = \frac{2^3 e^{-2}}{3!} = (0.135)\frac{4}{3} = 0.18$

(ii) More than 3 $(p(x > 2)) = p(x = 3) + p(x = 4) + \text{---} + p(x = 2000).$

$$\begin{aligned} &= 1 - p(x \le 2). \\ &= 1 - [p(x = 0) + p(x = 1) + p(x = 2)]. \\ &= 1 - 5 \times (0.135) \\ &= 0.325 \end{aligned}$$

(iii) More than 1 $= (p(x > 1)) = 1 - [p(x \le 1)]$

$$\begin{aligned} &= 1 - [p(x = 0) + p(x = 1)] \\ &= 0.595. \end{aligned}$$

(iv) None $= p(x = 0) = \frac{m^0 e^{-m}}{0!} = e^{-2} = 0.135$

10.8.2 Continuous Probability Distributions

In discrete probability distributions the random variable varies only over discrete values, and a function (f) is called the density and is used to compute probabilities associated with the random variable X. Let us consider the probability distributions where the random variable varies continuously over an interval. Distributions of this type are known as continuous probability distributions. The function is given by

$$F(X) = P[X = x] : x \text{ is real.}$$

Definition: Let X be a continuous random variable. A function f is called a density function and has the following properties.

1. $f(x) \geq 0$, where x is a real number
2. $\int_{-\infty}^{\infty} f(x)\,dx = 1$
3. $p[a \leq x \leq b] = \int_{a}^{b} f(x)\,dx$; a and b real.

The necessary and sufficient conditions for a function to be a density for a continuous random variable are as follows,

1. $f(x) \geq 0$, 2. $\int_{-\infty}^{\infty} f(x)\,dx = 1.$

Example 10.23: Verify that $f(x) = e^{-x}$ defined by,

$$f(x) = \begin{cases} e^{-x} \; ; \text{for } x \geq 0 \\ 0 \; ; \text{for } x < 0 \end{cases} \quad \text{is a probability density function.}$$

Find the probability that the random variable $X = x$, having density in the range 0.1 to 0.5.

Solution:

From the given expression, it is evidently clear that $f(x) \geq 0$.

$$\text{Further } \int_{-\infty}^{\infty} f(x)\,dx = \int_{-\infty}^{\infty} e^{-x}\,dx$$

$$= \int_{-\infty}^{0} e^{-x}\,dx + \int_{0}^{\infty} e^{-x}\,dx = 0 + \int_{0}^{\infty} e^{-x}\,dx = 1.$$

Hence $f(x) = p(x)$ is a probability density function.

$$\text{Then } P(0.1 \leq x \leq 0.5) = \int_{0.1}^{0.5} f(x)\,dx = \int_{0.1}^{0.5} e^{-x}\,dx = -(e^{-0.5} - e^{-0.1}) = 0.2983.$$

Example 10.24: A random variable x has the density function

$$f(x) = \begin{cases} kx^2 \; : 0 \leq x \leq 4 \\ 0 \quad : \text{else where} \end{cases}$$

Evaluate k, and find i) $p(x \leq 1)$ ii) $p(1 \leq x \leq 2)$ iii) $p(x \leq 2)$ iv) $p(x > 1)$ v) $p(x > 2)$.
Also determine the mean, variance and the standard deviation of the distribution.

Solution:

Now $f(x) \geq 0$, and $k \geq 0$, the second condition is $\int_{-\infty}^{\infty} f(x)\,dx = 1.$

$$\therefore \int_{0}^{4} f(x)\,dx = \therefore \int_{0}^{4} k\,x^2\,dx = k\left(\frac{x^3}{3}\right)_0^4 = 1 \;\therefore k = \frac{3}{64}.$$

(i) $P(x \le 1) = \therefore \int_{-\infty}^{1} f(x)\,dx = \therefore \int_{-\infty}^{1} \frac{3}{64}x^2\,dx = \frac{1}{64}.$

(ii) $P(1 \le x \le 2) = \int_{1}^{2} f(x)\,dx = \int_{1}^{2} \frac{3}{64}x^2\,dx = \frac{7}{64}$

(iii) $P(x \le 2) = \int_{-\infty}^{2} f(x)\,dx = \int_{-\infty}^{2} \frac{3}{64}x^2\,dx = \frac{1}{8}$

(iv) $P(x > 1) = 1 - p(x \le 1) = 1 - \frac{1}{64} = \frac{63}{64}$

(v) $P(x > 2) = 1 - p(x \le 2) = 1 - \frac{1}{8} = \frac{7}{8}$

$$\text{Mean} = \mu = \int_{-\infty}^{\infty} x\ f(x)dx = \frac{3}{64}\int_{0}^{4} x^3 dx = \frac{3}{64}\left[\frac{x^4}{4}\right]_0^4 = 3$$

$$\text{Variance V} = \int_{-\infty}^{\infty} x^2\ f(x)dx - \mu^2 = \frac{3}{64}\int_{-\infty}^{\infty} x^3 dx - 9 = \frac{3}{64}\left[\frac{x^4}{4}\right]_0^4 = \frac{3}{5} = 0.6$$

Standard deviation: $\sigma = \sqrt{0.6} = 0.77459.$

Example 10.25: The probability density function of the time to failure of an electronic component in a copier (in hour) is

$f(x) = \frac{e^{-x/1000}}{1000}\quad x > 0.$ Determine the probability that,

(i) a component lasts more than 3000 hours before failure.
(ii) a component fails in the interval from 1000 to 2000 hours.
(iii) a component fails before 1000 hours.
(iv) Determine the number of hours after which 10% of all components will have failed.

Solution:

(i) Component lasts more than 3000 hours.

$$P(X > 3000) = \int_{3000}^{\infty} \frac{e^{-x/1000}}{1000}\,dx > 0 = -e^{-x/1000}\Big|_{3000}^{\infty} = e^{-3} = 0.05$$

(ii) Component fails in the interval from 1000 to 2000 hours.

$$P(1000 < X > 3000) = \int_{1000}^{2000} \frac{e^{-x/1000}}{1000}\,dx > 0 = -e^{-x/1000}\Big|_{3000}^{2000} = e^{-1} - e^{-2} = 0.233$$

(iii) Component fails before 1000 hours:

$$P(X < 1000) = \int_{0}^{1000} \frac{e^{-x/1000}}{1000}\,dx > 0 = -e^{-x/1000}\Big|_{0}^{1000} = 1 - e^{-1} = 0.632$$

(iv) The number of hours after which 10% of all components will have failed.
$P(X < x) = 0.10.$

$$P(X<x)=\int_0^x \frac{e^{-x/1000}}{1000}\,dx>0=-e^{-x/1000}\Big|_0^x=1-e^{-x/1000}$$

$$1-e^{-x/1000}=0.10$$

$$e^{-x/1000}=0.90$$

$(0.9)e^{x/1000}=1$ and $e^{x/1000}=10/9$ taking log both sides.

$$\log(e^{x/1000})=\log(10/9)$$

$$\frac{x}{1000}=0.105360 \text{ or } x=105.36.$$

10.8.3 Exponential Distribution

The continuous probability distribution having the probability density function f(x) is given by,

$$f(x)=\begin{cases}\alpha\, e^{-\alpha x}, & \text{for } x\geq 0\\ 0\ ; & \text{for } x<0\end{cases}$$

The constant $\alpha>0$ is known as an exponential distribution. Evidently

$$f(x)>0, \text{ and } \int_{-\infty}^{\infty} f(x)\,dx = \int_{-\infty}^{\infty}\alpha e^{-\alpha x}\,dx = \alpha\left[\frac{e^{-\alpha x}}{\alpha}\right]_0^{\infty}=1$$

The function f(x) satisfies the conditions required for continuous probability function/probability density function.

1. Mean: $=\mu=\int_{-\infty}^{\infty} x\ f(x)dx=\int_{-\infty}^{\infty} x\ \alpha^{-\alpha x}\ dx=\frac{1}{\alpha}$
2. Variance : $V=\sigma^2=\int_{-\infty}^{\infty} x^2\ f(x)dx-\mu^2=\frac{1}{\alpha^2}$
3. Standard deviation: $\sigma=\sqrt{V}=\frac{1}{\alpha}$

Example 10.26: In a certain town, the duration of showers is exponentially distributed with a mean time of 5 minutes. What is the probability that a shower will last for i) 10 minutes ii) less than 10 minutes iii) Between 10 and 12 minutes.

Solution:

The p.d.f of the exponential distribution is given by $f(x)=\alpha\, e^{-\alpha x}$, $x>0$, $\alpha=1/5$.

$$f(x) = \frac{1}{5}e^{-x/5}.$$

(i) Probability of a shower for 10 mins or more: $p(x\geq 10)=\int_{-\infty}^{\infty} f(x)\,dx =$

$$\int_{10}^{\infty}\frac{1}{5}e^{-x/5}dx=-\left[e^{-x/5}\right]_{10}^{\infty}=e^{-2}=0.1353$$

(ii) Probability of a shower less than 10 minutes: $p(x<10)$:

$$= \int_0^{10} \frac{1}{5} e^{-x/5} dx = -\left[e^{-x/5}\right]_0^{10} = -[e^{-2} - 1] = 0.8647.$$

(iii) Probability of a shower between 10 and 12 minutes: $p(10 < x < 12)$:

$$p(10 < x < 12) = \int_{10}^{12} \frac{1}{5} e^{-x/5} dx$$

$$= -\left[e^{-x/5}\right]_{10}^{12}$$

$$= \frac{1}{e^2} - \frac{1}{e^{2.4}}$$

$$= 0.0446171$$

Example 10.27: The time between calls to a plumbing supply business is exponentially distributed with a mean time of 15 minutes between calls.

(a) What is the probability that there are no calls within a 30 min interval?
(b) What is the probability that at least one call arrives within a 10 minute interval?
(c) What is the probability that the first call arrives within 5 and 10 minutes after opening?
(d) Determine the length of an interval of time such that the probability of at least one call in the interval is 0.90?

Solution:

Let X denote the time until the first call. Then X is exponential, with $\alpha = 1/15$.

(a) The probability that there are no calls within a 30 min interval

$$p(X \geq 30) = \int_{30}^{\infty} \frac{1}{15} e^{-x/15} dx = -e^{-x/15}\Big|_{30}^{\infty} = e^{-2} = 0.1353$$

(b) The probability of at least one call in a 10 min interval $= 1 - P(x > 10)$.

$$P(X > 10) = \int_{10}^{\infty} \frac{1}{15} e^{-x/15} dx = -e^{-x/15}\Big|_{10}^{\infty} = e^{-2/3} = 0.5134$$

∴ Probability of at least one call in a 10 min interval $= 1 - P(x > 10)$.
$= 1 - 0.5134 = 0.4856$

(c) The probability that the first call arrives within 5 and 10 minutes after opening, is given by;

$$P(5 < X < 10) = \int_5^{10} \frac{1}{15} e^{-x/15} dx = -e^{-x/15}\Big|_5^{10} = e^{-1/3} - e^{-2/3} = 0.2031.$$

(e) Given $P(X < x) = 0.90$, and to determine the length of an interval x.

$$P(X < x) = \int_0^x \frac{1}{15} e^{-x/15} dx = -e^{-x/15}\Big|_0^x = -(e^{-x/15} - 1) = 0.90$$

$e^{-x/15} = 0.1$, or $(0.1)\, e^{x/15} = 1$

$e^{x/15} = 10$. Taking log both sides

$\log e^{x/15} = \log 10$,

$x = 15 \times \log 10 = 15 \times 2.3025 = 34.5389$ minutes.

10.8.4 Normal Distribution

The continuous probability distribution having the probability density function f(x) is given by,

$f(x) = \dfrac{1}{\sigma\sqrt{2\pi}} e^{-(x-\mu)^2/2\sigma^2}$ where $-\infty < \mu < \infty : -\infty < \mu < \infty$, and $\sigma > 0$ is known as a normal distribution, as given in Fig. 10.5.

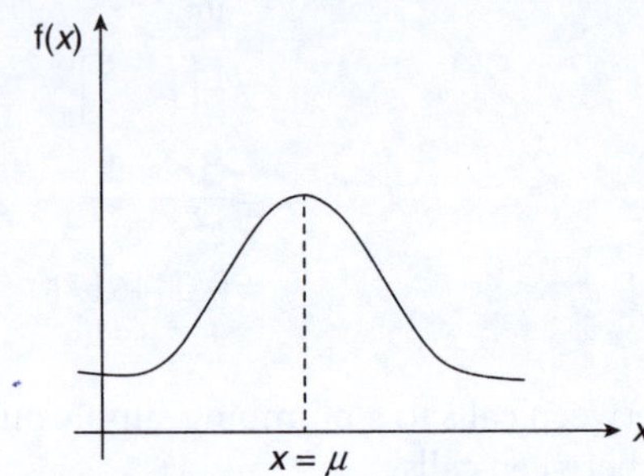

Figure 10.5 Normal distribution

The line $x = \mu$ divides the total area under the curve which is equal to 1, into two equal parts. The area to the right as well as to the left of the line $x = \mu$ is 0.5.

10.8.5 Standard Normal Distribution

The normal random variables are uniquely characterised by the two parameters (μ and σ). To calculate the probabilities associated with a specific normal curve, it is required to integrate the normal density over a particular interval. One cannot integrate normal density to a closed form. To find areas under the normal curve requires the use of numerical integration techniques. A simple transformation is employed to overcome this problem. The transformation process is called the standardisation procedure. Any question about any normal random variable can be transformed to an equivalent question concerning a normal random variable with mean 0 and standard deviation 1. The particular normal random variable is denoted by z, and is called the standard normal variable.

We have $p[a \le x \le b] = \int_a^b f(x)\,dx$; a and b real. In case of a normal distribution, we have

$$P(a \le x \le b) = \frac{1}{\sigma\sqrt{2\pi}} \int_a^b e^{-(x-\mu)^2/2\sigma^2}\,dx. \qquad (1)$$

Let $z = \dfrac{x-\mu}{\sigma}$ or $x = \mu + \sigma z$ and $dx = dz\,\sigma$

Let $z_1 = \dfrac{x_1 - \mu}{\sigma}$ $z_2 = \dfrac{x_2 - \mu}{\sigma}$ be the values of z, corresponding to $x = a$, and $x = b$. Hence from equation (1),

$$P(a \le x \le b) = \frac{1}{\sqrt{2\pi}} \int_{z_1}^{z_2} e^{-z^2/2}\,dz. \qquad (2)$$

and $F(z) = \frac{1}{\sqrt{2\pi}} e^{-z^2/2}$. Geometrically, (2) represents the area bounded by the standard normal curve F(z), between $z = z_1$ and $z = z_2$. In particular, if $z_1 = 0$,

$$\Phi(z) = \frac{1}{\sqrt{2\pi}} \int_0^z e^{-z^2/2} dz,$$ as given in Fig. 10.6.

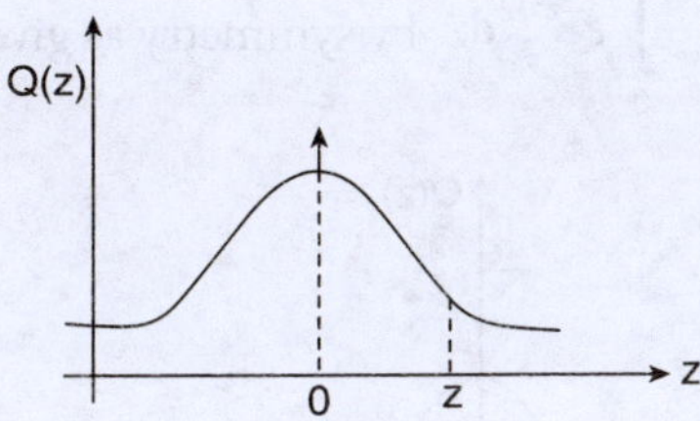

Figure 10.6 A typical normal distribution

Hence,

(i) $\int_{-\infty}^{\infty} \Phi(z)\, dz = 1$

(ii) $\int_{-\infty}^{0} \Phi(z)\, dz = \int_{0}^{\infty} \Phi(z)\, dz = \frac{1}{2}$

Equivalently, one can write as follows,

1. $P(-\infty \le z \le \infty) = 1$
2. $P(-\infty \le z \le 0) = 1/2$
3. $P(0 \le z \le \infty)$ or $P(z \ge 0) = 1/2$.
4. $P(-\infty \le z \le z_1) = P(-\infty < z \le 0) + P(0 \le z < z_1)$

   $P(z \le z_1) = 0.5 + \Phi(z_1)$
5. $P(z > z_2) = P(z \ge 0) - P(0 \le z \le z_2) = 0.5 - \Phi(z_2)$

Example 10.28: To find the area under the standard normal curve between $z = 0$, $z = 1.65$, as given in Fig. 10.7.

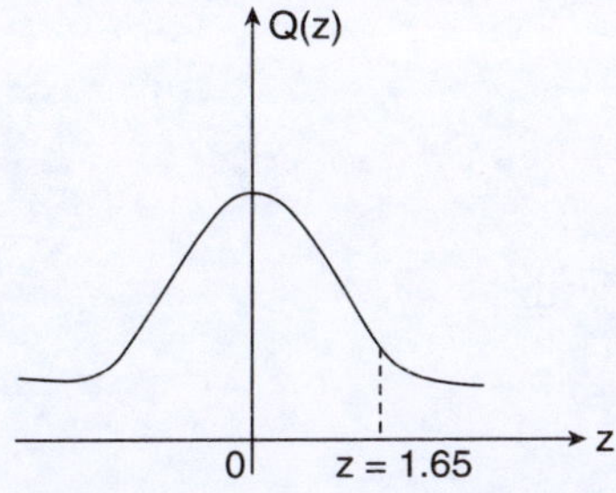

Figure 10.7 Typical normal distribution

Theoretically the area $= \Phi(z) = \frac{1}{\sqrt{2\pi}} \int_{0}^{1.65} e^{-z^2/2} dz = \Phi(1.65) = 0.4505$

Equivalently: $P(0 \le z \le 1.65) = \Phi(1.65) = 0.4505$

Example 10.29: To find the area bounded by $z = -0.85$, $z = 0$.

$$\Phi(z) = \frac{1}{\sqrt{2\pi}} \int_{-0.85}^{0} e^{-z^2/2} dz = \frac{1}{\sqrt{2\pi}} \int_{0}^{0.85} e^{-z^2/2} dz$$ by symmetry as given in Fig. 10.8.

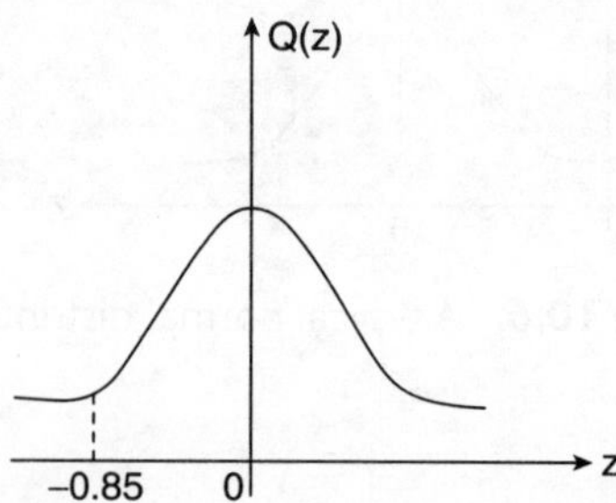

Figure 10.8 Typical normal distribution

The required area $= \Phi(0.85) = 0.3023$. Equivalently, it is written as follows.

$P(-0.85 \le z \le 0) = \Phi(0.85) = 0.3023$.

Example 10.30: To find the area of the standard normal curve between $z = -0.40$ and $z = 1.85$, as given in Fig. 10.9.

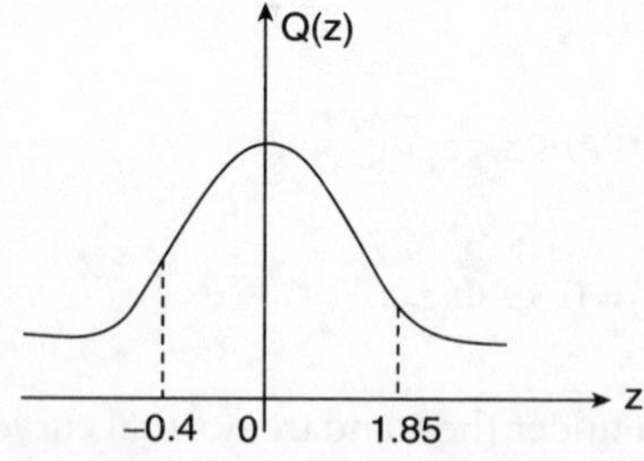

Figure 10.9 Typical normal distribution

$$\text{Area} = \Phi(z) = \frac{1}{\sqrt{2\pi}} \int_{-0.4}^{1.85} e^{-z^2/2} dz.$$

$$= \frac{1}{\sqrt{2\pi}} \int_{-0.4}^{0} e^{-z^2/2} dz + \frac{1}{\sqrt{2\pi}} \int_{0}^{1.85} e^{-z^2/2} dz.$$

$$= \Phi(0.4) + \Phi(1.85) = 0.6232$$

Equivalently: $P(-0.4 \le z \le 1.85) = 0.6232$.

Example 10.31: To find the area of the standard normal curve between $z = 0.55$ and $z = 1.80$, as given in Fig. 10.10.

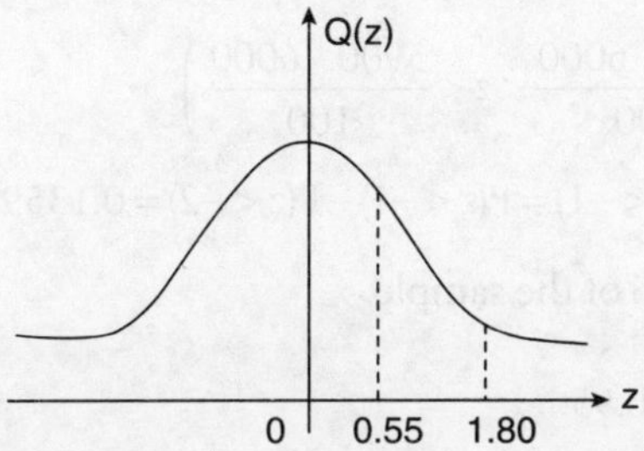

Figure 10.10 Typical normal distribution

The required area = (area between $z = 0$ and $z = 1.80$) – (area between $z = 0$ and $z = 0.55$)

$$= \Phi(1.80) - \Phi(0.55) = 0.4641 - 0.2088 = 0.2553$$

Equivalently, it is written as, $P(0.55 \le z \le 1.80) = 0.2553$.

Example 10.32: To find the area of the standard normal curve to the right of $z = -1.80$, as given in Fig. 10.11.

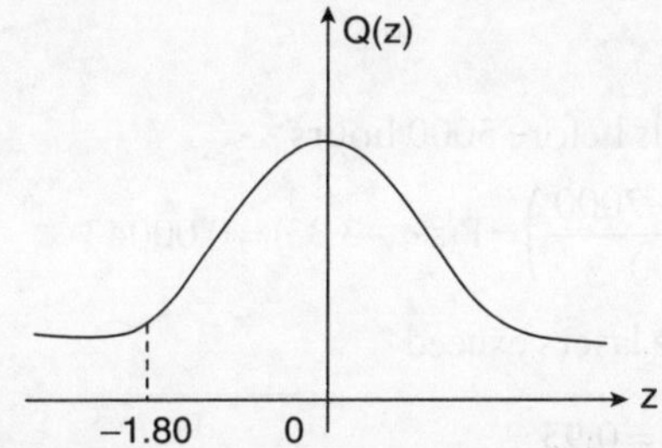

Figure 10.11 Typical normal distribution

Required area = (Area between $z = -1.80$ to 0) + (Area to the right of $z = 0$)

$$= (\text{Area between } z = 0 \text{ to } 1.80) + 0.5$$

$$= \Phi(1.80) + 0.5 = 0.4641 + 0.5 = 0.9641.$$

Equivalently: $P(z \ge -1.80) = 0.9641$

Example 10.33: The compressed strength of samples of cement can be modelled by a normal distribution with a mean of 6000 kilograms per square centimeter and standard deviation 100.

(i) What is the probability that a sample's strength is less than 6250kg/cm^2?
(ii) What is the probability that a sample's strength is between 5800 and 5900 kg/cm^2?
(iii) What is the strength exceeded by 95% of the sample?

Solution:

Standard normal variable : $z = \dfrac{x - \mu}{\sigma}$ $\mu = 6000$ and $\sigma = 100$

(i) If $x < 6250$, then $P(X < 6250) =$

$$= P\left(z < \frac{6250-6000}{100}\right) = P(z < 2.5) = 0.5 + \Phi(2.5) = 0.99379$$

(ii) $P(5800 < z < 5900) = P\left(\frac{5800-6000}{100} < z < \frac{5900-6000}{100}\right)$

$$= P(-2 < z < -1) = P(z < -1) - P(z < -2) = 0.13591$$

(iii) The strength exceeded by 95% of the sample.

$$P(z > x) = P\left(z > \frac{x-6000}{100}\right) = 0.95$$

Therefore, $\frac{x-6000}{100} = -1.65$, and $x = 5835$.

Example 10.34: The life of a semiconductor laser at a constant power is normally distributed with a mean of 7000 hours and a standard deviation of 6000 hours.

(a) What is the probability that the laser fails before 5000 hours?
(b) What is the life in hours that 95% of the lasers exceed?
(c) If three lasers are used in a product and they are assumed to fail independently, what is the probability that all three are still operating after 7000 hours?

Solution:

(a) Probability that the laser fails before 5000 hours

$$= P(X < 5000) = P\left(z < \frac{5000-7000}{600}\right) = P(z < -3.33) = 0.00043$$

(b) The life span that 95% of the lasers exceed

$$= P(X > x) = P\left(z > \frac{x-7000}{600}\right) = 0.95$$

$\frac{x-7000}{600} = -1.64$, consequently $x = 6016$.

(c) $P(X > 7000) = P\left(z > \frac{7000-7000}{600}\right) = P(z > 0) = 0.5$

P (three lasers operating after 7000 hours) $= (1/2)^3 = 1/8$.

ADDITIONAL SOLVED EXAMPLES

Example W10.1: A box contains 10 screws, three of which are defective. Two screws are drawn at random. Find the probability that none of the two screws is defective.

Solution:

The following two cases will be considered.

A: First drawn screw is non-defective.

B: Second drawn screw is non-defective.

Seven out of 10 screws are non-defective. Random sampling is done and each screw has the same probability (1/10) of being picked.

$$\therefore\ P(A) = \frac{7}{10}$$

If we continue to pick the second with replacement, the situation before second drawing is the same as that at the beginning and $P(B) = \frac{7}{10}$

The two events are independent.

$$\therefore P(A \cap B) = P(A)P(B) = \frac{7}{10} \times \frac{7}{10} = \frac{49}{100}$$

If the sampling is done without replacement, then $P(A) = \frac{7}{10}$

If A occurs then there are 9 screws left in the box, 3 of which are defective.

Then $P(B/A) = P(B) = \frac{6}{9} = \frac{2}{3}$

$$\therefore P(A \cap B) = P(A)P(B) = \frac{7}{10} \times \frac{2}{3} = \frac{7}{15} = 46.66\%$$

Note: The event without replacement is smaller than the event with replacement.

Example W10.2: On an average, 2 cars enter a certain parking lot per minute. What is the probability that during any given minute 4 or more cars will enter the lot?

Solution:

We are dealing with binomial distribution with very large n and very small p.
(Poisson distribution).
$\mu = np = 2.$ i.e., 2 cars enter on the average.
Probability of 4 cars or more during a given minute $= P\ (x \geq 4)$

$$= 1 - P(x < 3)$$

$$= 1 - [P(x=0) + P(x=1) + P(x=2) + P(x=3)]$$

$$= 1 - e^{-2}\left[\frac{2^0}{1!} + \frac{2^1}{2!} + \frac{2^2}{3!} + \frac{2^3}{3!}\right]$$

$$= 1 - 0.857$$

$$= 0.143$$

$$= 14.3\%.$$

Example W10.3: Out of 800 families with four children each, how many families would be expected to have (i) 2 boys and 2 girls (ii) at least one boy (iii) no girl and (iv) at most two girls.

Solution:

Let us assume equal probability for boys and girls.
Let p = probability of having a boy, q = probability of having a girl.
$p = ½, q = ½, n = 4, N = 800.$

(i) The number of family who expect 2 boys and 2 girls.

$$= 800 \times {}^4C_2 \left(\frac{1}{2}\right)^2 \left(\frac{1}{2}\right)^2 = 800 \times 6 \times \frac{1}{16} = 300$$

(ii) The number of families expecting at least one boy

p (boy)	**q (girl)**	
1	3	${}^4C_1\left(\frac{1}{2}\right)^3\left(\frac{1}{2}\right)$
2	2	${}^4C_2\left(\frac{1}{2}\right)^2\left(\frac{1}{2}\right)^2$
3	1	${}^4C_3\left(\frac{1}{2}\right)\left(\frac{1}{2}\right)$
4	0	${}^4C_4\left(\frac{1}{2}\right)^4\left(\frac{1}{2}\right)^0$

$$=800\left[{}^4C_1\left(\frac{1}{2}\right)^3\left(\frac{1}{2}\right)+{}^4C_2\left(\frac{1}{2}\right)^2\left(\frac{1}{2}\right)^2+{}^4C_3\left(\frac{1}{2}\right)\left(\frac{1}{2}\right)^3+{}^4C_4\left(\frac{1}{2}\right)^4\left(\frac{1}{2}\right)^0\right]$$

$$=800\times\frac{1}{16}[4+6+4+1]=750.$$

(iii) Number of family having no girl, i.e., having 4 boys.

$$={}^4C_4\left(\frac{1}{2}\right)^4\left(\frac{1}{2}\right)^0=50$$

(iv) Number of family having at least two girls (having at least 2 boys).

p (boy)	**q (girl)**	
2	2	${}^4C_2\left(\frac{1}{2}\right)^2\left(\frac{1}{2}\right)^2$
3	1	${}^4C_3\left(\frac{1}{2}\right)\left(\frac{1}{2}\right)^3$
4	0	${}^4C_4\left(\frac{1}{2}\right)^4\left(\frac{1}{2}\right)^0$

$$=800\left[{}^4C_2\left(\frac{1}{2}\right)^2\left(\frac{1}{2}\right)^2+{}^4C_3\left(\frac{1}{2}\right)\left(\frac{1}{2}\right)^3+{}^4C_4\left(\frac{1}{2}\right)^4\left(\frac{1}{2}\right)^0\right]=550$$

Example W10.4: Suppose that the time to failure of fans in a personal computer can be modelled by an exponential distribution with mean 0.0003.

(i) What proportion of the total fans will last at least 10,000 hours?
(ii) What proportion of the fans will last at most 7000 hours?

Solution:

Let X denote the time to failure of fans in a personal computer. Then, X is an exponential random variable. Mean = 0.0003.

(i) $P(X > 10{,}000) = \int_{10{,}000}^{\infty} (0.0003)e^{-(0.0003)x}\,dx = e^{-3} = 0.0498$

(ii) $p(X < 7000) = \int_{0}^{7000} (0.0003)e^{-(0.0003)x}\,dx = 1 - e^{-2.1} = 0.8775$

Example W10.5: Assume X is normally distributed with a mean of 10 and a standard deviation of 2. Determine the value of x for the followings.

(i) $P(X > x) = 0.5.$ (ii) $P(X > x) = 0.95.$

(iii) $P(x < X < 10) = 0.2$ (iv) $P(-x < X - 10 < x) = 0.95.$

Solution:

(i) $P(X > x) = P\left(Z > \frac{x-10}{2}\right) = 0.5$. Therefore, $\frac{x-10}{2} = 0$ and $x = 10$

(ii) $P(X > x) = P\left(Z > \frac{x-10}{2}\right) = 1 - P\left(Z < \frac{x-10}{2}\right) = 0.95.$

Therefore $P\left(Z < \frac{x-10}{2}\right) = 0.05.$ and $\frac{x-10}{2} = -1.64$ and $x = 6.72$

(iii) $P(x < X < 10) = P\left(\frac{x-10}{2} < Z < 0\right) = P(Z < 0) - P\left(Z < \frac{x-10}{2}\right)$

$= 0.5 - P\left(Z < \frac{x-10}{2}\right) = 0.2$

Therefore $P\left(Z < \frac{x-10}{2}\right) = 0.3$ and $\frac{x-10}{2} = -0.52$

Consequently $x = 8.96$.

(iv) $P(10 - x < X < 10 + x) = P(-x/2 < Z < x/2) = 0.99$

Therefore $x/2 = 2.58$ and $x = 5.16$.

Example W10.6: Assume X is normally distributed with a mean of 5 and a standard deviation of 4. Determine the following.

(i) $P(X < 11)$ (ii) $P(X > 0)$ (iii) $P(3 < X < 7)$ (iv) $P(-2 < X < 9)$.

Solution:

(i) $P(X < 11) = P\left(Z < \frac{11-5}{4}\right)$

$= P(Z < 1.5) = 0.93319.$

(ii) $P(X > 0) = P\left(Z > \frac{0-5}{4}\right)$

$= P(Z > -1.25)$

$= 1 - P(Z < -1.25) = 0.89435$

(iii) $P(3 < X < 7) = P\left(\frac{3-5}{4} < Z < \frac{7-5}{4}\right)$

$$= P(-0.5 < Z < 0.5)$$

$$= P(Z < 0.5) - P(Z < -0.5) = 0.38292$$

(iv) $P(-2 < X < 9) = P\left(\frac{-2-5}{4} < Z < \frac{9-5}{4}\right)$

$$= P(-1.75 < Z < 1)$$

$$= P(Z < 1) - P(Z < -1.75) = 0.80128$$

SUMMARY

Students usually find the concept of probability somewhat difficult to understand. It requires readers to think before they attempt to solve a problem. In this chapter, we have covered the topic on probability fairly, from the basics to the concept of conditional probability. Practical examples have been solved to make the reader comfortable with the topic. A sub topic on continuous probability distribution has been covered and different types of distributions have also been dealt with.

EXERCISES

Probability Theory

1. The probability that an integrated circuit chip will have defective etching is 0.12; the probability that it will have a crack defect is 0.29, and the probability that it will have both defects is 0.07. What is the probability that a newly manufactured chip will have i) an etching or crack defect ii) neither of the defects.
2. A problem is given to four students A, B, C, D, whose chances of solving it are $\frac{1}{2}, \frac{1}{3}, \frac{1}{4}, \frac{1}{5}$ respectively. Find the probability that the problem is solved.
3. A computer company manufactures two parts, mainly processor chips and RAM. In the manufacturing process 9% of the processor chips and 5% of the the RAM chips are likely to be defective. If a unit is assembled with these two, what is the probability that it is not defective?
4. The probability that a communication system will have high fidelity is 0.81 and the probability that it will have fidelity and high selectivity is 0.18. Find the probability that a system will have high selectivity, given that it has high fidelity.
5. Let A and B be independent events. Prove that (i) $\bar{A}$ and B and (ii) A and $\bar{B}$ are independent.
6. Prove that $P(\bar{A}|E) = 1 - P(A|E)$.
7. A student takes his examination in four subjects α, β, γ, δ. He estimates that his chances of passing in α are 4/5, in β are 3/4, in γ are 5/6 and in δ are 2/3. To qualify, he must pass in α and at least two other subjects. What is the probability that he qualifies?
8. If A, B, C are mutually independent events. Show that the events $A \cup B$ and C are also independent. (i.e., prove that $P[(A \cup B) \cap C] = P(A \cup B)\, P(C)$.)
9. Let P_1, P_2,----P_n be the probabilities of n independent events. Find the probability of the occurrence of at least one of them.
10. Three urns contain six red, four black; four red, six black; and five red, five black balls respectively. One of the urns is selected at random and a ball is drawn from it. If the ball drawn is red, find the probability that it was drawn from the i) first urn ii) second urn iii) third urn.

Random Variables

1. The random variable X has a binomial distribution with $n = 10$, and $p = 0.5$. Determine the following probabilities.
   (a) $P(X \le 2)$ (b) $P(X = 4)$ (c) $P(X > 8)$ (d) $P(5 \le X \le 7)$.

2. The phone lines to an airline reservation are occupied 40% of the time. Assume that the events that the lines are occupied on successive calls, are independent. Assume that 10 calls are made to the airline.
   (a) What is the probability that for exactly three calls the lines are occupied?
   (b) What is the probability that for at least one call the lines are not occupied?
   (c) What is the expected number of calls in which the lines are all occupied?

3. Supposing X has a Poisson distribution with a mean life of 4, determine the following probabilities.
   (a) $P(X = 0)$ (b) $P(X \le 2)$ (c) $P(X = 4)$ (d) $P(X = 8)$.

4. The number of telephone calls reaching a phone exchange is often modeled as a Poisson random variable. Assume that on an average there are 10 calls per hour.
   (a) What is the probability that there are exactly five calls in one hour?
   (b) What is the probability that there are three or fewer calls in one hour?
   (c) What the probability that there are exactly 15 calls in two hours?
   (d) What is the probability that there are exactly five calls in 30 minutes.

5. Suppose that $f(x) = e^{-x}$, for $0 < x$. Determine the following probabilities.
   (a) $P(1 < X)$ (b) $P(1 < X < 2.5)$ (c) $P(X = 3)$ (d) $P(X < 4)$ (e) $P(3 \le X)$ (f) Determine x such that $P(x < X) = 0.10$.

6. Suppose the counts recorded by a Geiger counter follow a Poisson process, with an average of two counters per minute.
   (a) What is the probability that there are no counts in a 30 second time interval?
   (b) What is the probability that the first count occurs in less than 10 seconds?
   (c) What the probability that the first occurs between 1 and 2 minutes after start-up?

7. The life of automobile voltage regulators has an exponential distribution with a mean life of six years. An automobile is purchased which is six years old, with a working voltage regulator. You plan to own it for six years.
   (a) What is the probability that the voltage regulator fails during your ownership?
   (b) If your regulator fails after you own the automobile for three years, and it is replaced, what is the mean time until the next failure?

8. The time between arrivals of taxis at a busy intersection is exponentially distributed with a mean of 10 minutes.
   (a) What is the probability that you wait longer than one hour for a taxi?
   (b) Suppose you have already been waiting for a taxi for one hour, what the probability that one arrives within the next 10 minutes.
   (c) Determine x such that the probability that you wait less than x minutes is 0.10.

9. The line width for semiconductor manufacturing is assumed to be normally distributed with a mean of 0.5 micrometre and standard deviation of 0.05 micrometer.
   (a) What is the probability that a line width is greater than 0.62 micrometer?
   (b) What is the probability that a line width is between 0.47 and 0.63 micrometer?
   (c) The line width of 90% of samples is below what value?
10. The speed of file transfer from a server on campus to a personal computer at a student's home on a weekday evening is normally distributed, with a mean of 60 kilobits per seconds.
   (a) What is the probability that the file will be transferred at a speed of 70 kilobits per second or more?
   (b) What is the probability that the file will be transferred at a speed less than 58 kilobits per second?
   (c) If the file is 1 megabyte, what is the average time it will take to transfer the file ? (assume 8-bits per byte)

11

Finite State Machine

A finite state machine is the simplest of different types of automata. This type of automaton does not have temporary storage and accepts the regular set generated by the right linear grammar. It is basically a recognition device. It is also called a finite state machine in the literature and abbreviated as FSA or FSM or even FA. We find certain variations in this machine, namely Deterministic Finite State Automata (DFA) and Non Deterministic Finite State Automata (NDFA). We shall deal with these machines in this chapter.

11.1 DETERMINISTIC FINITE AUTOMATA (DFA)

We define DFA formally as a 5-tuple, as these machines (M) are deterministic in their operations.

$$M = (Q, \Sigma, \delta, q_0, F), \text{ where}$$

Q: The finite set of states called internal states

Σ: The input alphabet, which is a finite set of symbols

δ: Transition function, $\delta: Q \times \Sigma \to Q$

q_0: Initial state and $q_0 \in Q$.

F: Set of final states and $F \subseteq Q$.

The machine starts from the initial state q_0. As it accepts symbols from the input string, it changes the internal states and at the end of the input it reaches the final state, if the string input is accepted. Otherwise the machine remains at a non-final state at the end of the input. The formal model of FSM is given in Fig. 11.1

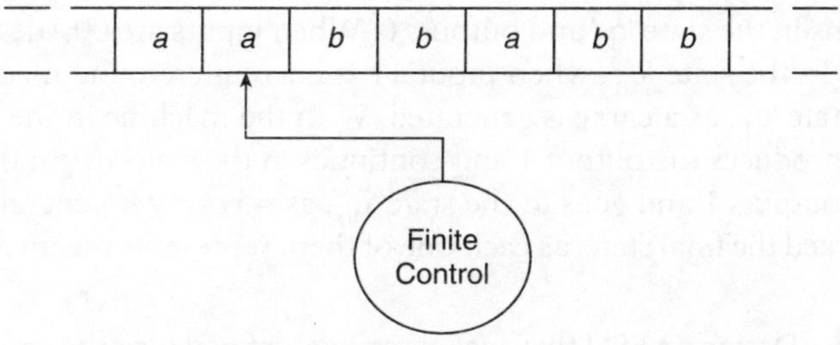

Figure 11.1 Basic model of finite state machine

The input is stored on the infinite tape which is divided into cells, and each cell holds one input symbol. The finite control scans the cells from left-to-right, one symbol at a time, for an input. The current state of the machine and the input read from the tape together determine the next state for the machine. A typical move in FSM is given as follows.

$$\delta\,(q_1, a) = q_2 \tag{1}$$

Here, the machine is in the state 'q_1' and reads an input symbol a on the tape and makes a transition to a new state 'q_2'. The states are shown as circles with a state label inside. The edges are shown across the states; with an arrow head and labels on them. The label on the edge indicates the input upon which the machine makes the transition from the given state to the next. A typical move of FSM is shown in Fig. 11.2 corresponding to (1)

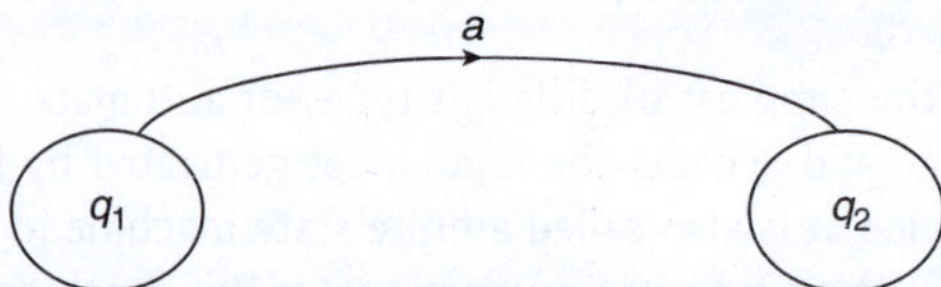

Figure 11.2 A typical move in a finite state machine

The finite control is capable of sensing the end of the input. Say, a blank cell at the end of the input means the end of the input string. At the end of the input, if the machine is in the final state, then the input is said to be accepted. Otherwise, it is said to be rejected.

Example 11.1: Design an FSM to perform serial addition of binary digits.

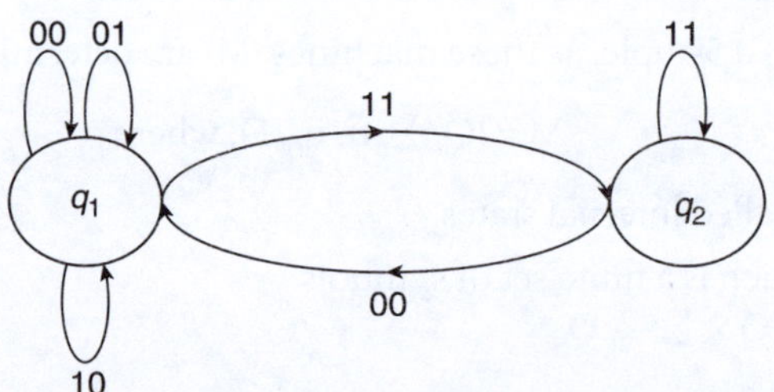

Figure 11.3 FSM for serial addition

The serial adder accepts a pair of bits, namely 00, 01, 10 and 11. The output of the serial adder is 0 or 1. The machine is in the state 'q_0' corresponding to a 'no carry' situation during addition. The state 'q_1' corresponds to a 'carry generated' situation during addition. When 00 is input to the machine, it (the machine) remains in the state 'q_0' and outputs 0. When inputs are 00, 01, and 10, the machine remains in the state 'q_0'. In the state 'q_0', when input 11 is encountered, the machine outputs 0 and makes a transition to a state 'q_1', as a carry is generated. With the machine in the state q_1, when an input 11 is encountered, it produces the output 1 and continues in the state 'q_1'. In the state 'q_1', when input 00 is encountered, it outputs 1 and goes to the state 'q_0', as no carry is generated. In this FSM, we have not deliberately marked the final state, as each one of them represents the final state.

Example 11.2: Design an FSM that accepts strings defined over the alphabet {0, 1} that end with 011. The simplest string that the FSM should accept is 011 and the corresponding FSM is given in Fig. 11.4.

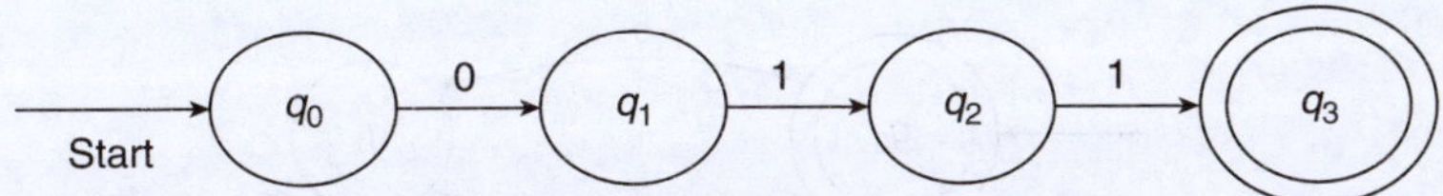

Figure 11.4 FSM to accept string 011

Here 'q_0' is the initial state and 'q_3' is the final state. In the state 'q_0', any number of 1's occurring in the string that is input to FSM should not make any state transition. When 0 is encountered with an input string, the machine changes state from 'q_0' to 'q_1'. In the state 'q_1', any number of 0's encountered in the input should not alter the state of the machine and it continues in state 'q_1'. When the input 1 is encountered, the state of the machine changes to 'q_2' so as to facilitate the acceptance of the substring 011 in the input string. The complete FSM is given in Fig. 11.5

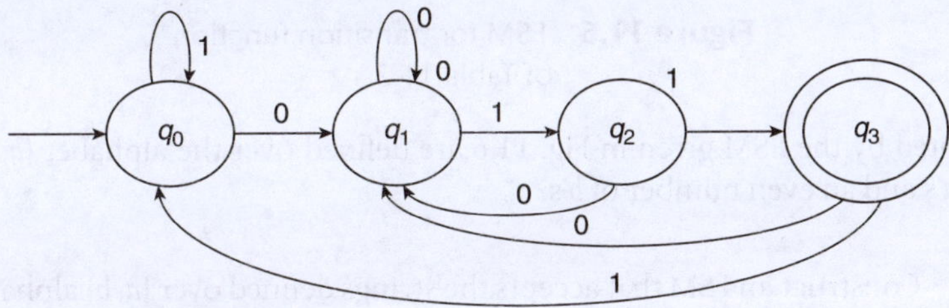

Figure 11.5 FSM to accept strings ending with 011 defined over the Alphabet {0, 1}

For the FSM in Fig. 11.5, the transition table depicting the state changes is given in Table 11.1.

Table 11.1 State transition table for Fig. 11.5

State / Input	0	1
q_0	q_1	q_0
q_1	q_1	q_2
q_2	q_1	q_3
q_3	q_1	q_0

Example 11.3: Consider the transition Table 11.2 and write the FSM. The state 'q_0' is both the initial and the final state.

Table 11.2 Transition function for states

Sate / Input	*a*	*b*
q_0	q_2	q_1
q_1	q_3	q_0
q_2	q_0	q_3
q_3	q_1	q_2

The FSM corresponding to the transition of states given in Table 11.2 is given in Fig. 11.6

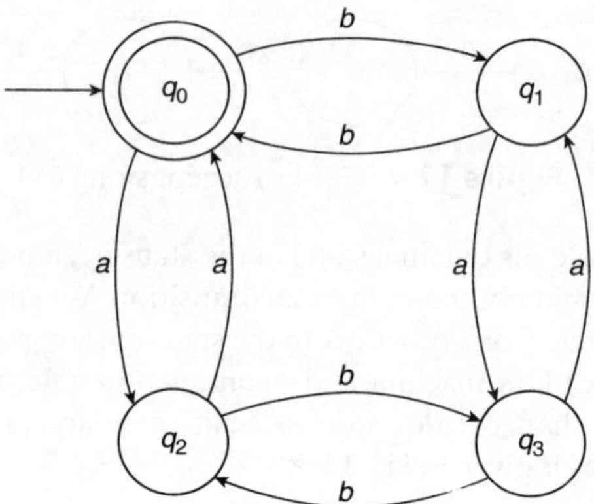

Figure 11.6 FSM for transition function of Table 11.2

The strings accepted by the FSM given in Fig. 11.6 are defined over the alphabet $\{a, b\}$ and contain an even number of *a*'s and an even number of *b*'s.

Example 11.4: Construct an FSM that accepts the strings defined over $\{a, b\}$ alphabet and contain all *a*'s and an ending *b*. i.e. $L = \{a^n b \mid n \geq 0\}$

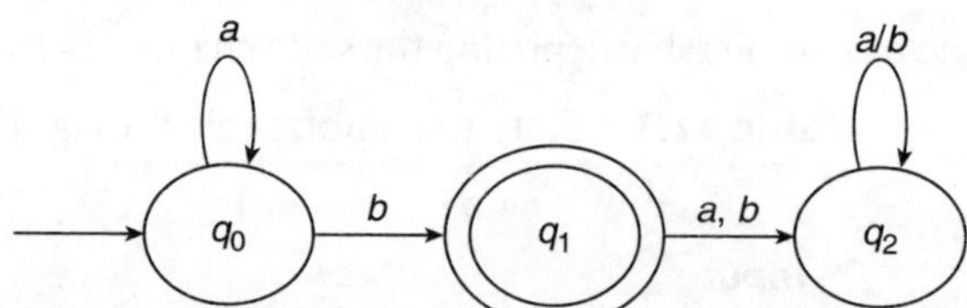

Figure 11.7 FSM for Example 11.4

In this FSM diagram, the edges have more than one label. This indicates that the FSM makes a transition from a given state to the next state upon both the labels. For example, in the solution given in Fig. 11.7 (solution for Example 11.4), the FSM makes a transition from the state 'q_1' to 'q_2' upon the input *a* or *b*. A similar argument holds good for the state 'q_2'.

Example 11.5: Write an FSM for testing whether the given binary number is divisible by 5.
Any binary number when divided by 5 will produce a remainder in the range 0 – 4. We shall consider a state corresponding to each of the remainders. When the remainder is 0, the number is said to be divisible by 5. Hence q_0 is the final state and also the initial state. The state transition function δ is defined as:

$$\delta\,(q_i, d) = q_j, \text{ where } j = (2i + d) \bmod 5.$$

The state transitions are given in Table 11.3. The FSM corresponding to this is given in Fig. 11.8.
All the FSMs explained in this section represent Deterministic Finite Automata (DFA) as the transitions from one state to another are definite and well defined upon an input symbol.

Table 11.3 State transition for Example 11.5

Sate / Input	0	1
q_0	q_0	q_1
q_1	q_2	q_3
q_2	q_4	q_0
q_3	q_1	q_2
q_3	q_1	q_2
q_4	q_3	q_4

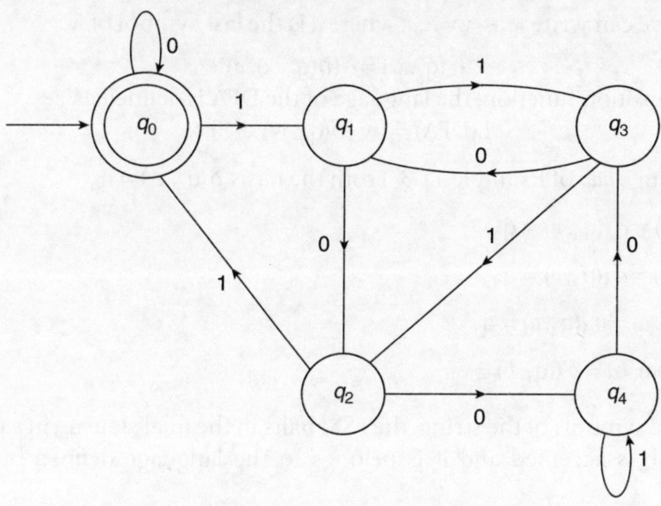

Figure 11.8 FSM for Example 11.5

Theorem 11.1: When M = (q, Σ, δ, q_0, F) is an FSM and G is the corresponding transition graph, for every q_i, and q_j belonging to Q and $w \in \sum^+$, $\delta^*(q_i, w) = q_j$ iff there is a walk w from q_i to q_j.

Proof: The proof is through induction on the length (*w*) of the input string.

Basis: For $n = 1$ and $n = 2$ the theorem holds good.

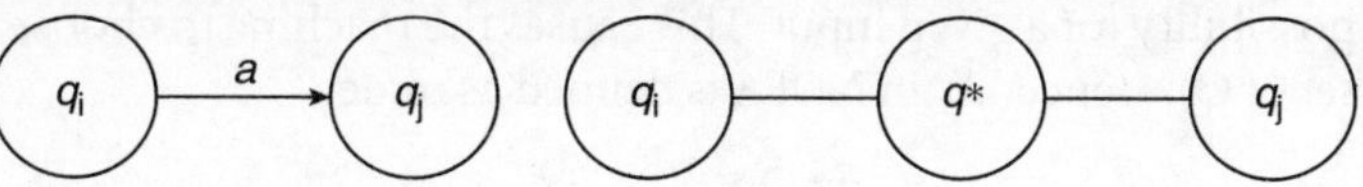

Induction: Assume that the claim is true for all strings of length *v* with $|v| \le n$ and prove for *w* of length $(n + 1)$. We shall write *w* of length $n + 1$ as a concatenation of sub strings *v* and *a*.

$$w = va.$$

Suppose $\delta^*(q_i, v) = q_k$ and $|v| = n$. There is a walk in G corresponding to *v* and leveled q_i to q_k

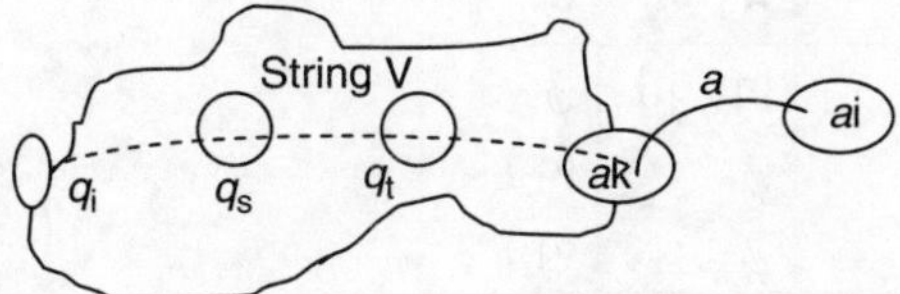

But $\delta^*(q_i, w) = q_j$ and the machine M must have a transition $\delta(q_k, a) = q_j$ upon the input '*a*'. There must be an edge from 'q_k' to 'q_j' with label '*a*'. Thus, there is a walk from 'q_i' to 'q_j' and hence the theorem is proved.

11.1.1 Language of the DFA

We need to define the language of the DFA. In order to do this, we must know how the DFA processes the given input string. First, we define the extended transition function $\hat{\delta}$ using induction on the length of the string.

Basis: If $|w| = 0$, i.e., empty string $w = \in$

$$(q_0, \in) = q_0$$

Induction: If $|w| \neq 0$, we can write w as $w = x.a$, where a is the last symbol of w.

$$\hat{\delta}(q_0, w) = \delta(\hat{\delta}(q_0, x), a)$$

Using the extended transition function, the language of the DFA is defined as

$$L(DFA) = \{w \mid \hat{\delta}(q_0, w) \in F\}.$$

Consider the input string *abab* of Example 11.3. From the basis $\hat{\delta}(q_0, \in) = q_0$

$\hat{\delta}(q_0, a) = \delta(\hat{\delta}(q_0, \in), a) = \delta((q_0, a) = q_2$

$\hat{\delta}(q_0, ab) = \delta(\hat{\delta}(q_0, a), b) = \delta((q_2, b) = q_3$

$\hat{\delta}(q_0, aba) = \delta(\hat{\delta}(q_0, ab), a) = \delta((q_3, a) = q_1$

$\hat{\delta}(q_0, abab) = \delta(\hat{\delta}(q_0, aba), b) = \delta((q_1, b) = q_0$

After processing all the symbols of the string, the FSM halts in the final state q_0, in case of Example 11.3. Hence, the string abab is accepted and it is belongs to the language defined by the FSM given in Example 11.3.

11.2 NON-DETERMINISTIC FINITE AUTOMATA (NDFA)

Computer systems are deterministic in nature. However, the concept of non-determinism is a useful tool in the construction of DFA, as the direct construction of a DFA is more complex. In DFA, the transition function δ is a mapping from $Q \times \Sigma \to Q$. This means that for a given state of the machine, upon reading an input symbol from the tape, the machine has a unique or definite move. But in NDFA, there exists more than one possibility for a given input. This causes the machine to choose one of 2^Q possibilities (2^Q is the power set of Q). Hence 'δ' in NDFA is defined as under

$$\delta : Q \times \Sigma \to 2^Q$$

Example 11.6: Consider the machine $M = (Q, \Sigma, \delta, q_0, F)$, where $Q = \{q_0, q_1, q_2\}, \Sigma = \{a, b\}, F = \{q_2\}$ and δ is given in Table 11.4.

Table 11.4 Transition function for Example 11.6

	A	B
q_0	$\{q_0, q_1\}$	ϕ
q_1	ϕ	$\{q_2\}$
q_2	ϕ	$\{q_2\}$

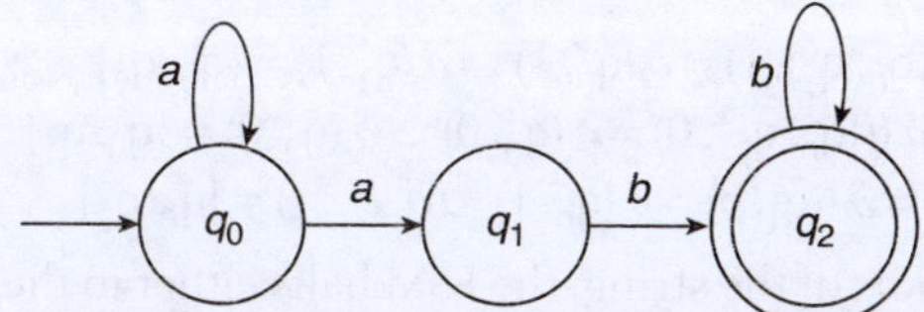

Figure 11.9 NDFA for Example 11.6

The NDFA for the transition function given in Table 11.4 is given in Fig. 11.9. For the state 'q_0', the machine has two possibilities on the input *a*, either to stay in the state 'q_0' or move to the state 'q_1'. This makes the machine nondeterministic in nature. The language accepted by the machine is:

$$L = \{a^n b^m : n, m \geq 1\}$$

Example 11.7: Consider $M = (Q, \Sigma, \delta, q_0, F)$, where $Q = \{q_0, q_1, q_2\}$, $\Sigma = \{0, 1\}$, $F = \{q_2\}$ and 'δ' is given in Table 11.5. The NDFA for the transition function given in Table 11.5 is illustrated in Fig. 11.10.

Table 11.5 Transition function for example 11.7

	0	1
q_0	$\{q_0, q_1\}$	$\{q_0\}$
q_1	ϕ	$\{q_2\}$
q_2	ϕ	ϕ

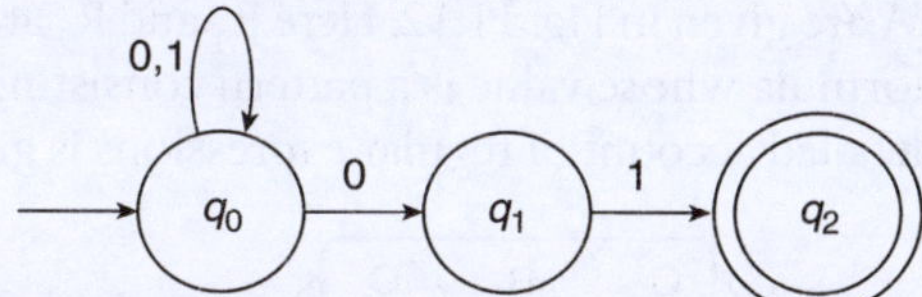

Figure 11.10 NDFA for Example 11.7

11.2.1 Language of the NFA

The language of the NFA using the extended transition function of NFA is defined as follows. First, we will define the extended transition function based on induction on the length of the string.

Basis: If $|w| = 0$, i.e. the empty string $w = \in$

$\hat{\delta}(q_0, w) = \{q_0\}$ i.e., without reading any symbol the NFA will not changes its state.

Induction: If $|w| \neq 0$. We can write w as $w = x. A$.

Then,

$\hat{\delta}(q_0, w) = \delta(\hat{\delta}(q_0, x), a)$

Suppose $\hat{\delta}(q_0, x) = \{p_1, p_2, \ldots p_n\}$ then

$\delta(\{p_1, p_2, \ldots p_n\}, a)$

$= \cup_i^n = \delta(\text{pi}, a)$

$= \{r_1, r_2, \ldots r_n\}$

So, language of the NFA is defined as

$L(N) = \{w \mid \hat{\delta}(q_0, w) \cap F \neq 0\}$

Consider the input string 0101 of Example 11.7. From the basis $\hat{\delta}(q_0, \in) = q_0$

$\hat{\delta}(q_0, 0) = \delta(\hat{\delta}(q_0, \in), 0) = \delta((q_0, a) = \{q_0, q_1\}$

$\hat{\delta}(q_0, 01) = \delta(\hat{\delta}(q_0, 0), 1) = \delta(\{q_0, q_1\}, 1) = \delta(q_0, 1) \cup \delta(q_1, 1) = \{q_0, q_2\}$
$\hat{\delta}(q_0, 010) = \delta(\hat{\delta}(q_0, 01), 0) = \delta((\{q_0, q_2\}, 0) = \delta(q_0, 0) \cup \delta(q_2, 0) = \{q_0, q_1\}$
$\hat{\delta}(q_0, 0101) = \delta(\hat{\delta}(q_0, 010), 1) = \delta((q_1, b) = \delta(q_0, 1) \cup \delta(q_1, 1) = \{q_0, q_2\}$

After processing all the symbols of the string, the FSM halts either in the state q_2 or q_0 in case of Example 11.7. Here, one of the states, namely, q_2 is the final state. Hence, the string 0101 is accepted and it is belongs to the language defined by the particular FSM.

11.2.2 NDFA and Regular Expressions

We know that regular expressions, regular sets, regular grammar and regular language are analogous concepts. Given a regular language, we can write or define an FSA or DFSA for recognition. We can write an NDFA for the regular sets namely, ϕ, $\{\in\}$ and $\{a\}$. Figure 11.11 gives the corresponding finite state machines.

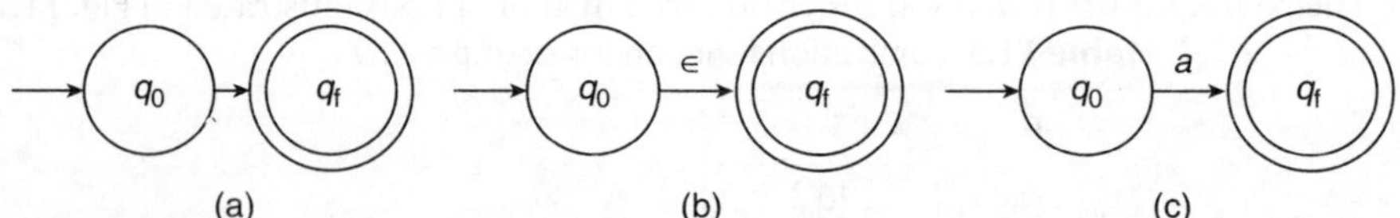

Figure 11.11 (a) NDFA accepts ϕ (b) NDFA accepts {∈} (c) NDFA accepts {a}

Extending the same notion for other regular expressions, we can write for the regular expressions $R_1 + R_2$, $R_1 \cdot R_2$, and R_1^*, the NDFA are given in Fig. 11.12. Here R_1 and R_2 are regular expressions. [A regular expression is an algebraic formula whose value is a pattern consisting of a set of strings, called the language of the expression. A detailed account of regular expressions is given in Chapter 4.]

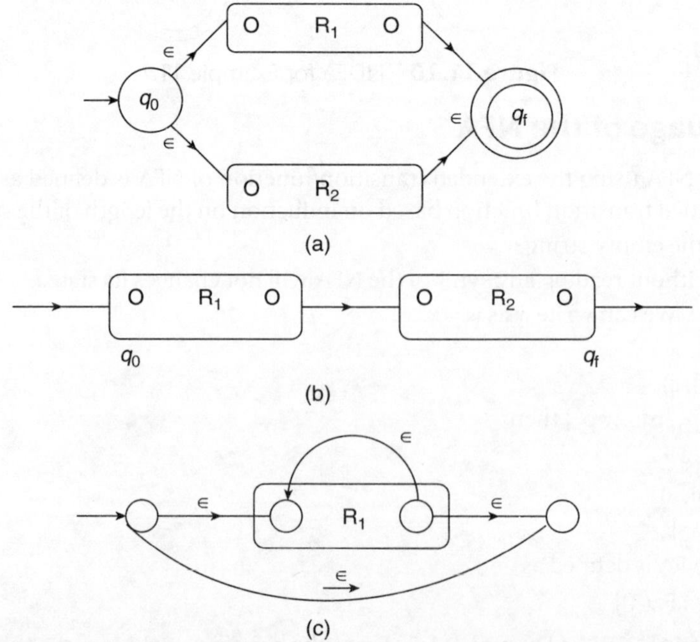

Figure 11.12 (a) NDFA for $R_1 + R_2$ (b) NDFA for $R_1.R_2$ (c) NDFA for $(R_1)^*$

The regular expression R_1 with initial state 'q_0' and final state 'q_f' is as shown in the following. In the case of the regular expression $R_1 + R_2$, the input is accepted by either R_1 or R_2.

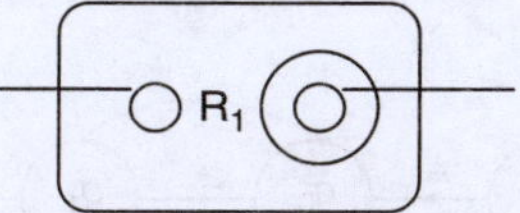

The finite state machine is obtained by adding new initial and final states with ∈ moves. In the regular expression $R_1 \cdot R_2$, the initial state of R_1 is the initial state of R_1R_2 and the final state of R_2 is the final state of $R_1 \cdot R_2$. The final state of R_1 is combined with the initial state of R_2. The regular expression $(R_1)^*$ is called Kleene's closure of R_1 and accepts the null string '∈' in addition to the strings accepted by R_1. This is facilitated by ∈ transition from the new initial state to the final state. We also have ∈ transition from the final state of R_1 to its initial state.

Example 11.8: Write NDFA for accepting the following regular expression

$$(a + bb)^* (ba^* + \in)$$

The regular expression given in Example 11.7 is viewed as comprising two parts $(a + bb)^*$ and $(ba^* + \in)$. The RE $(a + bb)^*$ comprises REs a and bb. The (a + bb) is a RE. The machines written for accepting these REs are given in Fig. 11.13

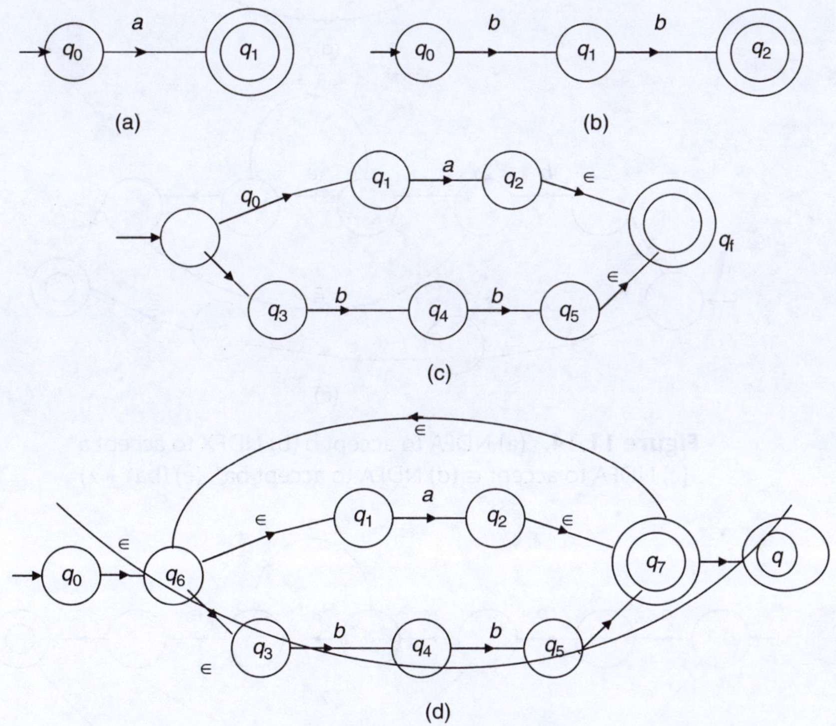

Figure 11.13 (a) NDFA to accept $\{a\}$, (b) NDFA to accept $\{bb\}$ (c) NDFA to accept $(a + bb)$ (d) NDFA to accept $(a + bb)^*$.

The second part of the RE $(ba^* + \in)$ is represented with an equivalent NDFA as given in Fig. 11.14.

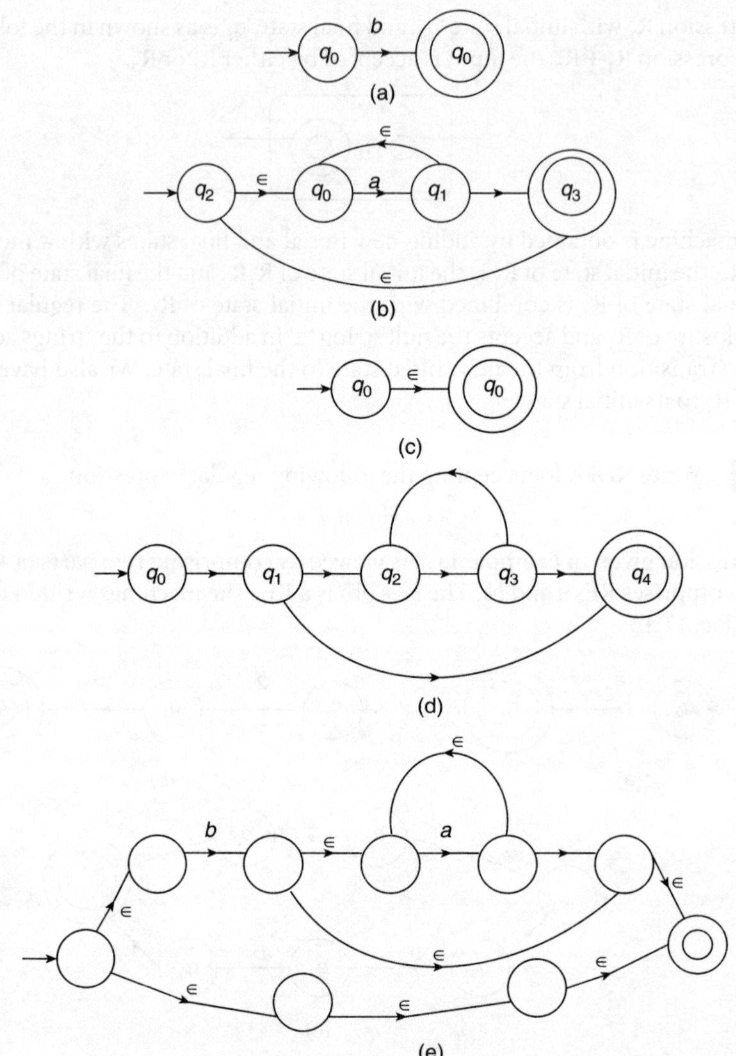

Figure 11.14 (a) NDFA to accept b (b) NDFX to accept a* (c) NDFA to accept ∈ (d) NDFA to accept ba*. (e) (ba* + λ)

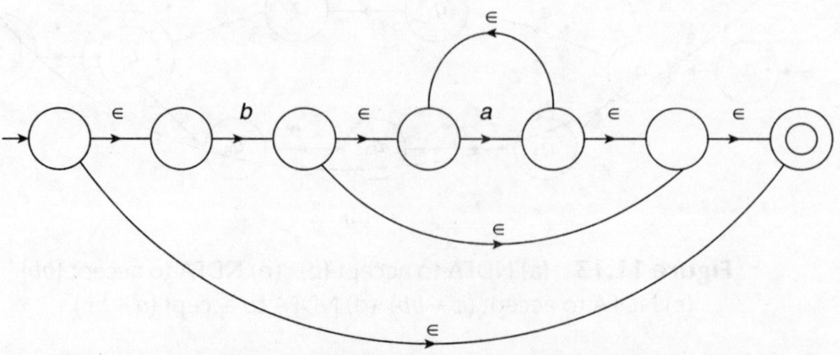

Figure 11.15 NDFA to accept (ba^{*} + ∈)

The Fig. 11.14 (e) can be rewritten as in Fig. 11.15. The complete NDFA to accept the RE $(a + bb)^* (ba^* + \lambda)$ is given in Fig. 11.16

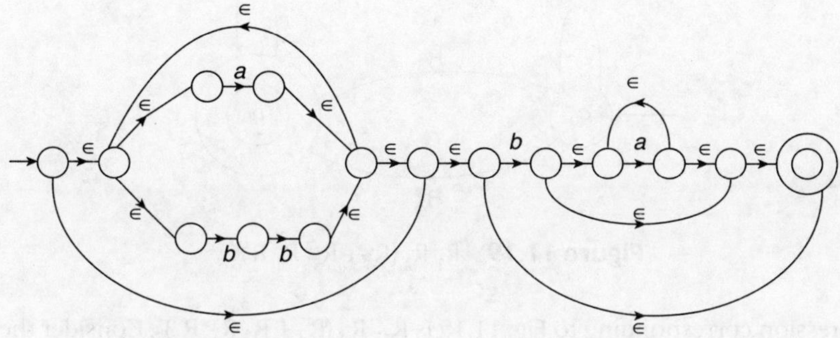

Figure 11.16 NDFA to accept $(a + bb)^* (ba^* + \lambda)$

The NDFA given in Fig. 11.16 is further simplified to the form given in Fig. 11.17. The ∈ transitions are combined here. The ∈ transitions are those which lead to the new state without any input. These NDFAs are called NDFAs with ∈ moves. The transition function is defined as follows. The '∈' is considered a valid symbol in the alphabet and $\Sigma = \Sigma \cup \{\in\}$.

$$\delta: Q \times \Sigma \cup \{\in\} \rightarrow 2^Q$$

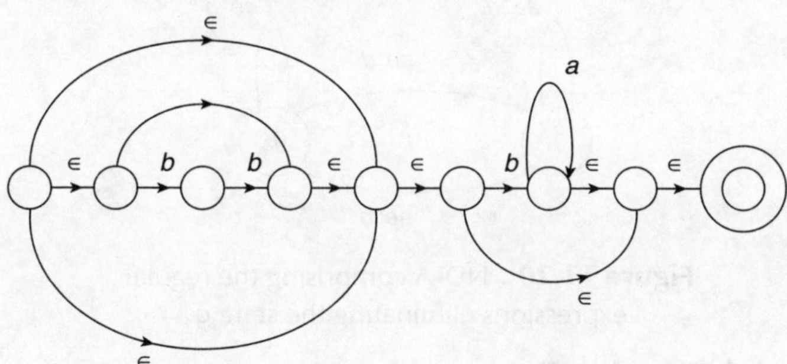

Figure 11.17 Equivalent NDFA of $(a + bb)^* (ba^* + \lambda)$.

Example 11.9: Consider the regular expression $a^* + a^* (a + b) c^*$ and represent it with an NDFA. The equivalent NDFA corresponding to $a^* + a^* (a + b)c^*$ is given in Fig. 11.18

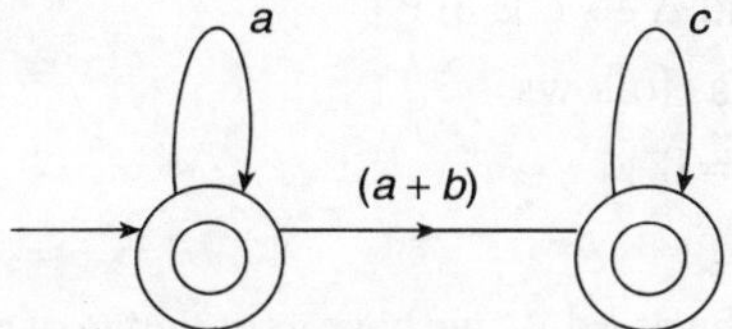

Figure 11.18 NDFA for $a^* (a + b) c^*$

Consider the NDFA given in Fig. 11.19, where R_1, R_2, R_3 and R_4 are regular expressions (RE).

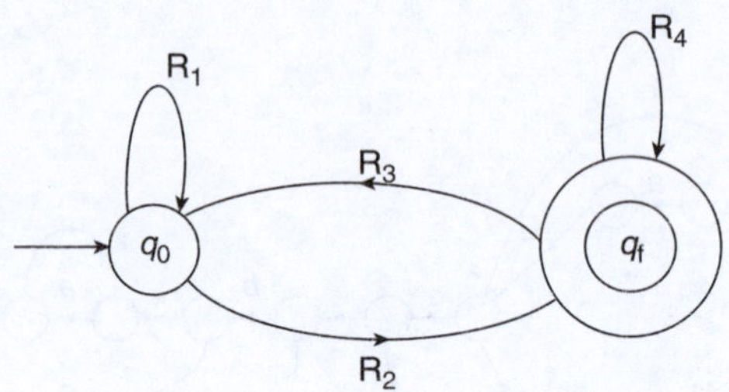

Figure 11.19 $R_1^*R_2(R_4 + R_3\ R_1^*R_2)^*$

The regular expression corresponding to Fig. 11.19 is $R_1^*\ R_2\ (R_4 + R_3 R_1{}^*R_2)^*$. Consider the NDFA given in Fig. 11.20 (a). The equivalent NDFA can be written as in Fig. 11.20 (b) by using the regular expressions.

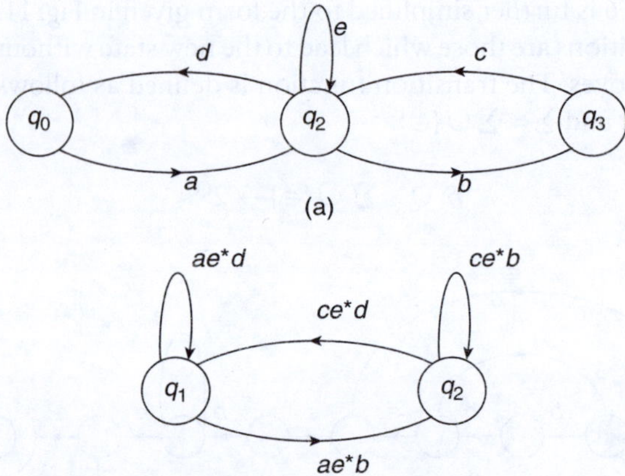

Figure 11.20 NDFA comprising the regular expressions eliminating the state q_2

11.2.3 NDFA and Regular Expressions

Consider the following theorems, which show the analogy between the regular languages or regular expressions and the non-deterministic finite automata (NDFA).

Theorem 11.2: There exist finite state automata (FSA) for a regular language.

Proof: Let 'L' be a regular language generated by a right linear grammar given by G = (N, T, P, S). Let 'M' be a finite state automaton accepting this 'L' and defined as $M = (Q, \Sigma, \delta, q_0, F)$. Where $Q = N \cup \{q\}$ and $q \notin N$, $q_0 = S$. $F = \{q\} \cup \{A \mid A \in N \text{ and } A \to \epsilon \text{ is in } P\}$

The δ transition function is defined as follows

If $A \to a\,B$ is a rule in P then $\delta(A, q) = B$

If $A \to a$ is a rule in P then $\delta(A, a) = q$

Corresponding to the derivation of the word W, we have a sequence of moves made by M to accept the W. Conversely, corresponding to a sequence of move to accept W by M, these exists a derivation for W in G. Hence, the theorem L(G) = T(M) and the machine is non-deterministic in nature.

Theorem 11.3: If M is a finite state automata, then there exists a right linear grammar G such that L(G) = L(M).
Proof: Let us consider the machine 'M' to be deterministic automaton and given by $M = (Q, \Sigma, q_0, \delta, F)$ and $G = (N, T, P, S)$ be the corresponding right linear grammar.
Here, $N = Q$, $T = \Sigma$, $S = q_0$ and the productions are $P = \{A \to aB \mid \delta(A, a) = B\} \cup \{A \to a \mid \delta(A, a) \in q$ where $q \in F\}$. We add the rule $S \to \in$ if $q_0 \in F$ and hence the theorem.

Example 11.10: Let G = (N, T, P, S) be a right linear grammar, where N = {S, A}, T = {*a*, *b*}

$P = \{S \to aA,$
$A \to bS$
$A \to b\}$

$L(G) = (ab)^+ = \{ab, abab, ababab, \ldots\}$ is the regular set. Let $M = (Q, \Sigma, \delta, q_0, F)$ be the NFSM, where $Q = \{S, A, q\}$, $\Sigma = \{a, b\}$, $q_0 = S$ and $F = \{a\}$. The 'δ' is defined as under

$\delta(q_0, a) = A$ $\qquad \delta(A, a) = \varphi$ $\qquad \delta(q, a) = \varphi$
$\delta(q_0, b) = \varphi$ $\qquad \delta(A, b) = \{q_0, q\}$ $\qquad \delta(q, b) = \varphi$

Hence, M also accepts L(G). The corresponding NFSM is given in Fig. 11.21

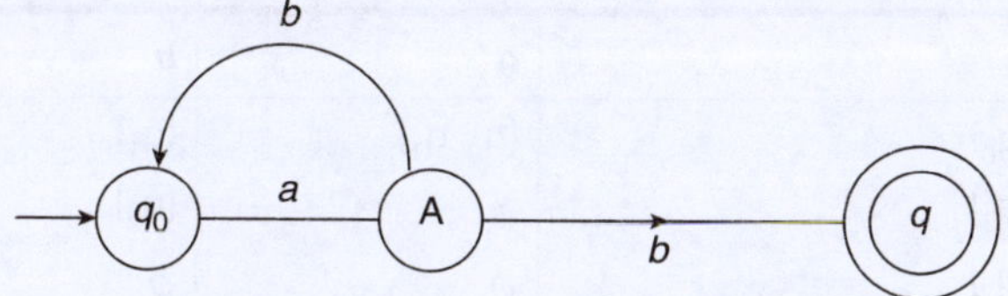

Figure 11.21 NFSM accepting $(ab)^+$

Example 11.11: Write NFSM accepting the strings of $(a + b)^*$, such that every *a* has a *b* immediately to its right. The NFSM is given in Fig. 11.22.

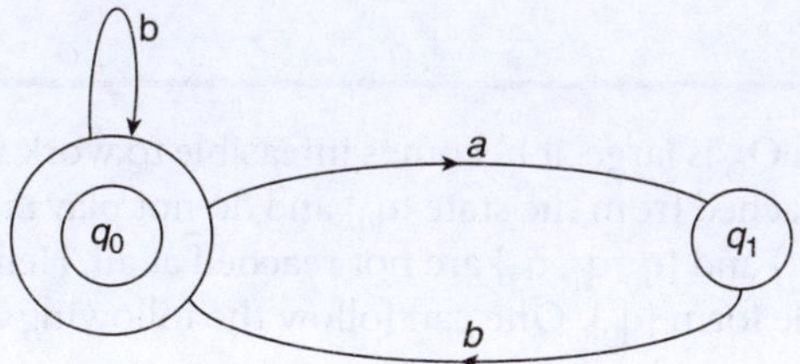

Figure 11.22 NFSM accept the strings of Example 11.10

11.3 NFA TO DFA CONVERSION

It is easy to construct the NFA for a given problem. The NFA has transitions defined over null string (∈). This means that the machine makes transitions without an input. It corresponds to 'no symbol stored on the tape'. We have seen ∈ transitions during the conversion of regular expressions into NDFA. The transition function for ∈ NDFA is given as follows. For every NFA, one can write an equivalent DFA.

$$\delta: \Sigma \cup \{\in\} \times Q \to 2^Q$$

11.3.1 Equivalent of NFA and DFA

In case of NFA, if there are 'Q' number of states, then 2^Q possibilities exist for transitions. Let Q_N and Q_D represent the number of states in the NFA and DFA respectively. The transition function δ_D of the DFA is expressed in terms of δ_N of the NFA as δ_D (S ∪ *a*) = ∪ δ_N(P, *a*), where S is the set of states in Q_D and *a* ∈ It is obtained by taking the union of all transitions from every state of S upon the input symbol *a*. Consider the NFA given in Table 11.6.

Table 11.6 Transition table

	a	***b***
q_0	$\{q_0, q_1\}$	$\{q_1\}$
q_1	ϕ	$\{q_2\}$
q_2	ϕ	ϕ

No. of states is $\{q_0, q_1, q_2\}$. The power set of this is $\{Q, \{q_0\}, \{q_1\}, \{q_2\}, \{q_0\ q_1\}, \{q_0, q_2\}, \{q_1, q_2\}, \{q_0, q_1, q_2\}\}$, The transition function δ_D of the DFA is given in Table 11.7.

Table 11.7 Transition table

	a	***b***
$\{q_0\}$	$\{q_0, q_1\}$	$\{q_1\}$
$\{q_1\}$	ϕ	$\{q_2\}$
$\{q_2\}$	ϕ	ϕ
$\{q_0, q_1\}$	$\{q_0, q_1\}$	$\{q_1, q_2\}$
$\{q_0, q_2\}$	$\{q_0, q_1\}$	$\{q_1\}$
$\{q_1, q_2\}$	ϕ	$\{q_2\}$
$\{q_0, q_1, q_2\}$	$\{q_0, q_1\}$	$\{q_1, q_2\}$
ϕ	ϕ	ϕ

When the number of states in Q_D is large, it becomes infeasible to work with those many states. Sometimes certain states are not reached from the state $\{q_0\}$ and do not play any role during computation. In this example, the states $\{q_0, q_2\}$ and $\{q_0, q_1, q_2\}$ are not reached at all. Hence, the set of states is always a subset of 2^Q that is a reachable form $\{q_0\}$. One can follow the following Algorithm 11.1 to obtain such a set.

Algorithm 11.1: Procedure to minimise the number of states
Start

1. Make the set Q_D so as to contain q_0 as the initial state i.e. $Q_D = \{q_0\}$.
2. For each subset S in Q_D, for each symbol $a \in \Sigma$, compute the new states and add to Q_D. If the state is already in the set, ignore it.
3. Repeat Step 2 until all the states in 2^Q are exhausted.

Stop

For the above example, we get Q_D using the aforesaid Algorithm 11.1 and the elements of Q_D are given as under.

$$Q_D = \{\{q_0\}\}\ \delta_0(\{q_0\}, a) \rightarrow \{q_0, q_1\}$$
$$\delta_0(\{q_0\}, b) \rightarrow \{q_1\}$$

$Q_D = \{\{q_0\}, \{q_0, q_1\}, \{q_1\},\}\ \delta_0(q_1, a) = \{q_2\}$
$Q_D = \{\{q_0\}, \{q_0, q_1\}, \{q_1\}, \{q_2\}\}$
$Q_D = \{\{q_0\}, \{q_0, q_1\}, \{q_1\}, \{q_2\}, \{q_1, q_2\}\}$
No other states are reachable from q_0 and hence Q_D has 5 elements. The remaining three states are redundant states.

Example 11.12: Consider the transition function (δ) given in Table 11.9 and construct a DFA. NDFA $= \{Q, \Sigma, \delta, q_0, F\}, F = \{q_1\}$

Table 11.8 Transition table

δ	*a*	*b*
q_0	$\{q_0, q_1\}$	$\{q_1\}$
q_1	ϕ	$\{q_0, q_1\}$

Let DFA $= (Q^1, \Sigma, \delta^1, q_0{}^1, F^1)$, where $Q^1 = \{\{q_0\}\ \{q_1\}, \{q_0, q_1\}, \{Q\}\}$, i.e., $Q^1 = 2^Q$ and $q_0{}^1 = \{q_0\}$. $F^1 = \{\{q_1\}, \{q_0, q_1\}\}$, δ^1 is given in Table 11.9

Table 11.9 Transition table

$\delta 1$	*a*	*b*
$\{q_0\}$	$\{q_0, q_1\}$	$\{q_1\}$
$\{q_1\}$	$\{\phi\}$	$\{q_0, q_1\}$
$\{q_0, q_1\}$	$\{q_0, q_1\}$	$\{q_0, q_1\}$
Q	Q	Q

The first two rows are easy to obtain by merely looking at the transition Table 11.8. The third row is obtained using the procedure

$\delta^1(\{q_0, q_1\}, a) = \{q_0, q_1\}$
$\delta^1(\{q_0, q_1\}, a) = \delta(q_0, a) \cup \delta(q_1, a)$
$= \{q_0, q_1\} \cup \varphi$
$= \{q_0, q_1\}$

Similarly, $\delta^1(\{q_0, q_1\}, b) = \{q_0, q_1\}$
$= \delta(q_0, b) \cup \delta(q_1, b)$
$= \{q_1\} \cup \{q_0, q_1\}$
$= \{q_0, q_1\}.0$

Algorithm 11.2 helps to convert a given NFA to DFA.

Algorithm 11.2: Conversion from NFA to DFA

Start

1. Create a graph G_D with vertex $\{q_0\}$ which is the initial vertex.
2. Repeat the following steps until no more edges are missing.

2.1 Take any vertex q_j of G_D that has no outgoing edge for some $a \in \Sigma$ and compute the transition $\delta(q_j, a)$. Say $\delta(q_k, a)$, $\delta(q_i, a)$ for (q_i, q_j, q_k).

2.2 Take the union of all the transitions to give the new set (q_l, q_m, q_n).

2.3 Create a vertex if it does not exist.

2.4 Add an edge from $\{q_i, q_j, q_k\}$ to $\{q_l, q_m, q_n\}$ and label it a.

3. Every state of G_D whose label contains. If $\in$ F is n as the final state.
4. If λ is acceptable then the vertex $\{q_0\}$ is also made a final vertex or final state.

Stop

Example 11.13: Consider the following NFA and convert it into a deterministic machine. The transition function is given in Table 11.10.

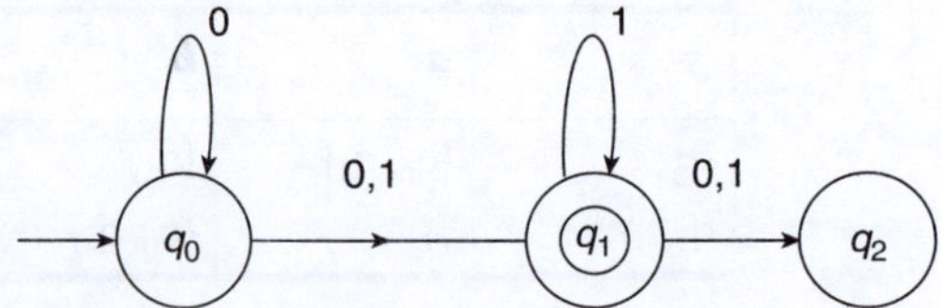

Figure 11.23 NFA for Example 11.13

Table 11.10 Transition function

δ	0	1
q_0	$\{q_0, q_1\}$	$\{q_1\}$
q_1	$\{q_2\}$	$\{q_2\}$
q_2	ϕ	$\{q_2\}$

Table 11.11 NFA to DFA conversion table

δ	0	1
$\{q_0\}$	$\{q_0, q_1\}$	$\{q_1\}$
$\{q_1\}$	$\{q_2\}$	$\{q_2\}$
$\{q_2\}$	ϕ	$\{q_2\}$
$\{q_0, q_2\}$	$\{q_0, q_1, q_2\}$	$\{q_1, q_2\}$
$\{q_1, q_2\}$	$\{q_2\}$	$\{q_2\}$
$\{q_0, q_2\}$	$\{q_0, q_1\}$	$\{q_1, q_2\}$
$\{q_0, q_1, q_2\}$	$\{q_0, q_1, q_2\}$	$\{q_1, q_2\}$
ϕ	ϕ	ϕ

The equivalent DFA is given in Fig. 11.24. In Table 11.11, the state $\{q_0, q_2\}$ is not shown in the final DFA because the state has no transition from other states.

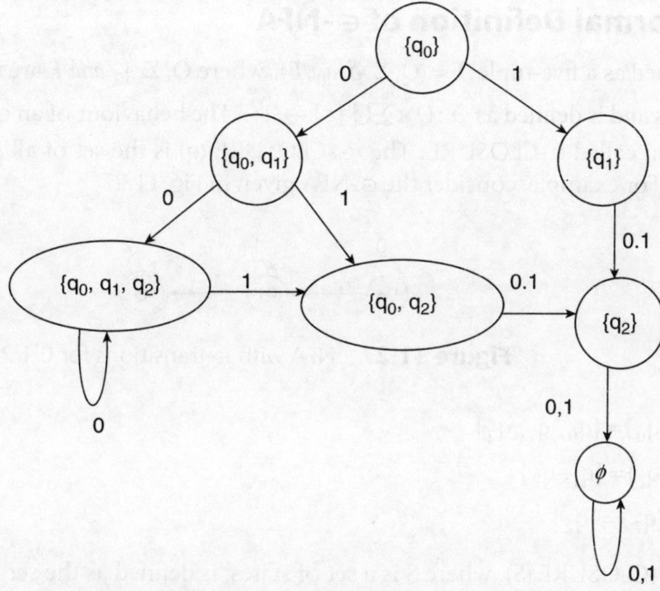

Figure 11.24 Equivalent DFA for Example 11.13

Hence, it is a redundant state. Further, it is common to have ∈ transitions while converting regular expressions to equivalent NFAs. These null string transitions (∈) are to be eliminated while obtaining equivalent DFAs. Consider the following NFA with ∈ transitions.

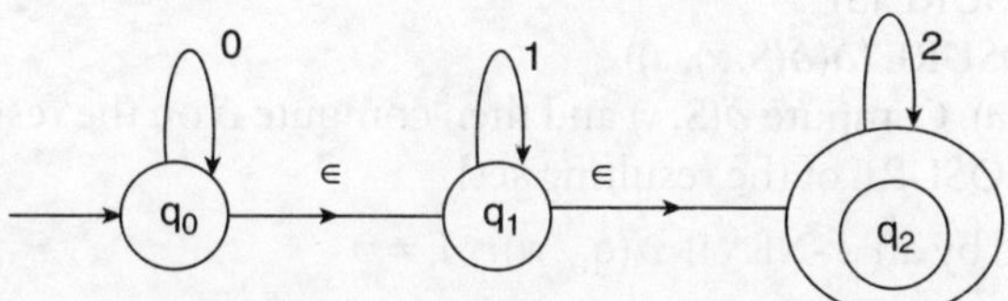

Figure 11.25 NFA with ∈ transitions

11.4 NFA WITH ∈ –TRANSITIONS (∈ -NFA)

The ∈-NFA allows the automata to have transitions on an empty string (∈) called ∈-transitions. This means that, without any input, the FSM changes its states. For example ∈-transitions represented by $\delta(q_0, \in) = q_1$ means that the FSM changes its state from q_0 to q_1 without any input. Thus, it introduces a hidden non-determinism in an FSM. However, ∈-transitions are useful in specifying optional inputs and render the design of FSM simple.

For example, in a typical programming language, while specifying a numeric constant, the sign is optional. As an example, an ∈-NFA to recognise and accept an integer constant is given in Fig. 11.26:

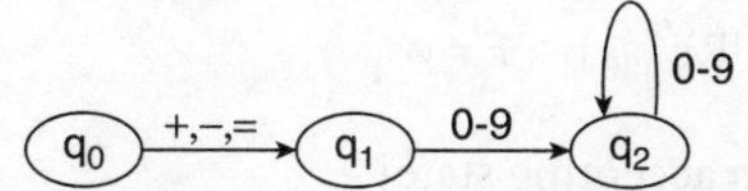

Figure 11.26 NFA with ∈ transitions for integer constant with optional sign

11.4.1 Formal Definition of ∈-NFA

∈-NFA is defined as a five-tuple $E=(Q,\Sigma,\delta,q_0,F)$, where Q,Σ,q_0 *and* F are the same as in an NFA. But δ has ∈-moves and is defined as $\delta: Q\times\Sigma\cup\{\in\}\rightarrow 2^Q$. The behaviour of an ∈-NFA with strings is based on the function called ∈-CLOSURE. The ∈-CLOSURE (q) is the set of all states reachable from q on ∈-transitions. For example, consider the ∈-NFA given in Fig. 11.27

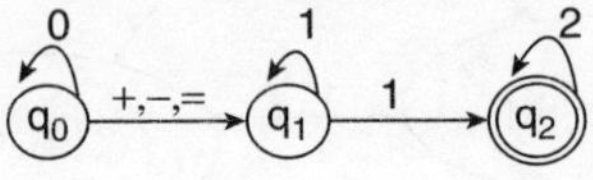

Figure 11.27 NFA with ∈-transitions for $0^*1^*2^*$

∈-CLOSURE $(q_0)=\{q_0, q_1, q_2\}$

∈-CLOSURE $(q_1)=\{q_1, q_2\}$

∈-CLOSURE $(q_2)=\{q_2\}$

Definition: ∈-CLOSURE (S), where S is a set of states, is defined as the set of all states reachable from any state *p* in S via ∈-transitions. Computation of an ∈-CLOSURE (S) is defined recursively as follows:

1. Basis: S is in the ∈-CLOSURE
2. Induction step: For each state *q* in ∈-CLOSURE, for each transition $\delta(q,\in)=p$, add *p* until no more can be added.

   Definition of $\hat{\delta}$ i.e., transitions or strings of an ∈-NFA:

   (i) $\hat{\delta}(S,\in)=\in$-CLOSURE (S)

   (ii) $\hat{\delta}(S,xa)=\in$-CLOSURE $(\hat{\delta}(\hat{\delta}(S,x),a))$

   [i.e., to obtain $\hat{\delta}$ (S, *xa*), Compute $\hat{\delta}(S,x)$ and then compute $\hat{\delta}$ on the resulting set on the input and then obtain an ∈-CLOSURE of the resulting set]

   A string w is accepted by an ∈-NFA if $\hat{\delta}(q_0,w)\cap F\neq\emptyset$

11.5 CONVERSION FROM ∈-NFA TO NFA

It turns out that given an ∈-NFA E, we can construct an NFA N without ∈-moves. The method of construction is described as follows.

Let $E=(Q,\Sigma,\delta_E,q_0,F)$ be an ∈-NFA. We can construct an NFA

$$N=(Q,\Sigma,\delta_N,q_0,F^1),$$

where δ_N and F^1 are defined as:

$$\delta_N(q,a)=\hat{\delta}_E(q,a)$$

$F^1=F$: if ∈-CLOSURE $(q_0)\cap F=\emptyset$

$= F\cup\{q_0\}$ if ∈-CLOSURE $(q_0)\cap F\neq\emptyset$

[*i.e.,∈−CLOSE* (q_0) contains an accepting state]

Example 11.14: Convert the $\in$-NFA given below to an NFA, without $\in$–move.

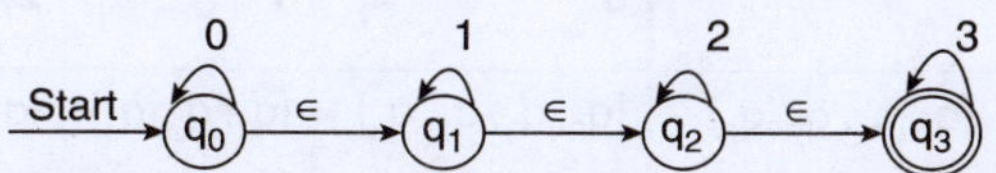

Figure 11.28 NFA with $\in$-transitions for $0^*1^*2^*3^*$

(i) $\in$- CLOSURE $(q_0) = \{q_0, q_1, q_2, q_3\}$

(ii) $\in$- CLOSURE $(q_1) = \{q_1, q_2, q_3\}$

(iii) $\in$- CLOSURE $(q_2) = \{q_2, q_3\}$

(iv) $\in$- CLOSURE $(q_3) = \{q_3\}$

$$\delta(q_0, 0) = \{q_0, q_1, q_2, q_3\}$$

$$\delta(q_0, 1 = \{q_1, q_2, q_3\}$$

$$\delta(q_0, 2) = \{q_2, q_3\}$$

$$\delta(q_0, 3) = \{q_3\}$$

Similarly, $\in$-CLOSURE operations are performed on other states namely, q_1, q_2, and q_3. The complete transitions are given in Table 11.12.

Table 11.12 Transition table for Fig. 11.28

	0	1	2	3
$\rightarrow {}^*q_0$	$\{q_0, q_1, q_2, q_3\}$	$\{q_1, q_2, q_3\}$	$\{q_2, q_3\}$	$\{q_3\}$
${}^*q_1$	ϕ	$\{q_1, q_2, q_3\}$	$\{q_2, q_3\}$	$\{q_3\}$
q_2	ϕ	ϕ	$\{q_2, q_3\}$	$\{q_3\}$
q_3	ϕ	ϕ	ϕ	$\{q_3\}$

11.6 CONVERSION FROM $\in$-NFA TO DFA

We can convert a given $\in$-NFA to an NFA without $\in$-transitions. Further, it is possible to convert an NFA to a DFA. Given an $\in$-NFA, we can always obtain an equivalent DFA. It is also possible to convert a given $\in$-NFA to a DFA directly.

Given an $\in$-NFA E = $(Q_E, \Sigma, \delta_E, q_D, F_E)$ we construct an equivalent DFA D=$(Q_D, \Sigma, \delta_D, q_D, F_D)$ as follows:

(i) $Q_D \subseteq 2Q^{QE}$

(ii) $\delta D(S,a) = \hat{\delta}_E(S,a)$ for each S in Q_D

(iii) start state $q_D = \in$-CLOSURE (q_D)

(iv) $F_D = \{S \mid S \cap F_E \neq \varphi\}$

The DFA (D) constructed from the above example using this method is given in Table 11.13.

Table 11.13 Transition table for Fig. 11.28

	0	**1**	**2**	**3**
$^*\{q_0, q_1, q_2, q_3\}$	$\{q_0, q_1, q_2, q_3\}$	$\{q_1, q_2, q_3\}$	$\{q_2, q_3\}$	$\{q_3\}$
$^*\{q_1, q_2, q_3\}$	ϕ	$\{q_1, q_2, q_3\}$	$\{q_2, q_3\}$	$\{q_3\}$
$^*\{q_2, q_3\}$	ϕ	ϕ	$\{q_2, q_3\}$	$\{q_3\}$
$^*\{q_3\}$	ϕ	ϕ	ϕ	$\{q_3\}$
ϕ	ϕ	ϕ	ϕ	ϕ

Example 11.15: Convert the $\in$-NFA to accept real decimal numbers.
A real decimal constant in a typical language is defined as a decimal string with an optional sign and containing a '.'

The following conditions are specified.

(i) Sign (+/−) is optional
(ii) Decimal string preceding '.' String following '.' Should contain at least one digit.

$\in$-NFA for the same is given as follows:

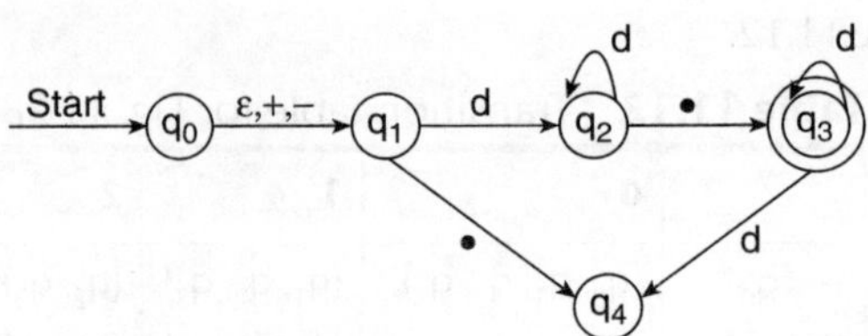

Figure 11.29 NFA with $\in$-transitions for floating point constant with optional sign

Here d denotes any decimal digit 0-9.
In the equivalent DFA, the start state is $\{q_0, q_1\}$ which is $\in$-CLOSURE (q_0)

$$\delta_D(\{q_0, q_1\}, +) = \{q_1\}$$

and

$$\delta_D(\{q_0, q_1\}, -) = \{q_1\}$$

$$\delta_D(\{q_0, q_1\}, d) = \{q_2\}$$

$$\delta(\{q_0, q_1\}, .) = \{q_4\}$$

Similarly, we construct other states. We get

$$Q_D = \{\{q_0, q_1\}, \{q_1\}, \{q_2\}, \{q_3\}, \{q_4\}\}$$

and $F_D = \{q_3\}$. The transition diagram is shown in Fig. 11.29 (a).

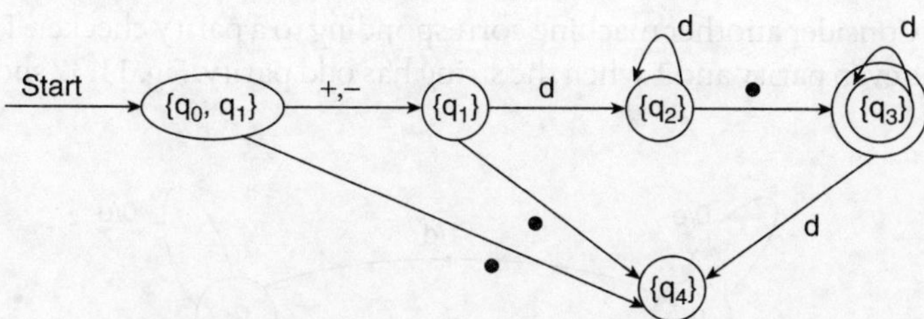

Figure 11.29(a) NFA with ∈-transitions for floating point constant with optional sign

11.7 OUTPUT ASSOCIATED FINITE AUTOMATA

The finite automata is not only used as acceptor but is also used to produce output at the end of the input. The labels on the edges indicate the output. Consider the Example 11.11, which is a serial adder as shown in Fig. 11.30.

Example 11.16:

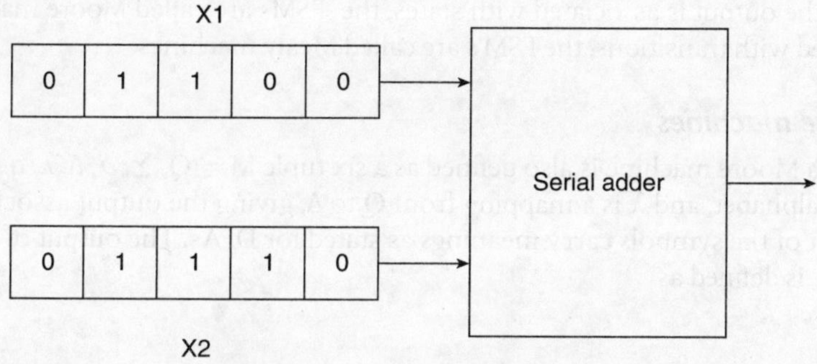

Figure 11.30 Serial adder

In this case $z = 1\,1\,0\,1\,0$. This can be realized with a FSM as shown in Fig. 11.31.

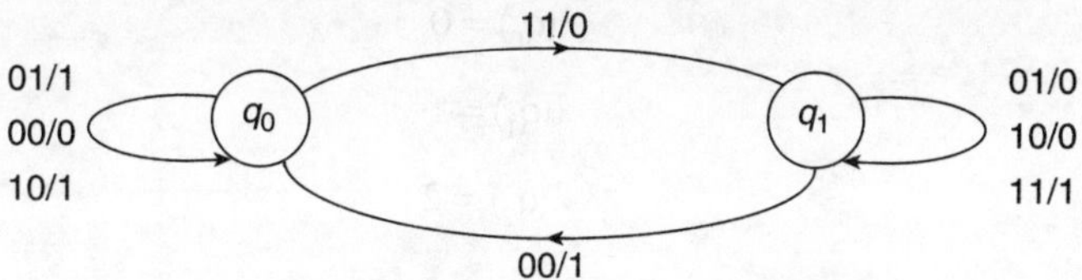

Figure 11.31 FSM for serial adder

The state 'q_0' corresponds to the state 'no carry' generated after adding the bits. The state 'q_1' refers to the state 'carry generated' after the addition of bits. When bits 01, 00 and 10 are added in the state 'q_0', no carry is generated. When bits 01 and 10 are added with 1, the output is 0 and a carry is generated. When bits 11 are added in state q_1, , a carry is generated (carry bit is set) and the output is 1. In the state q_1, when 00 is encountered, the output is 1 and no carry is generated and hence the machine makes a transition to the state q_0.

Example 11.17: Consider another machine corresponding to a parity checker. The machine outputs 0 when the string has even parity and 1 when the string has odd parity. Fig. 11.32 shows the corresponding FSM.

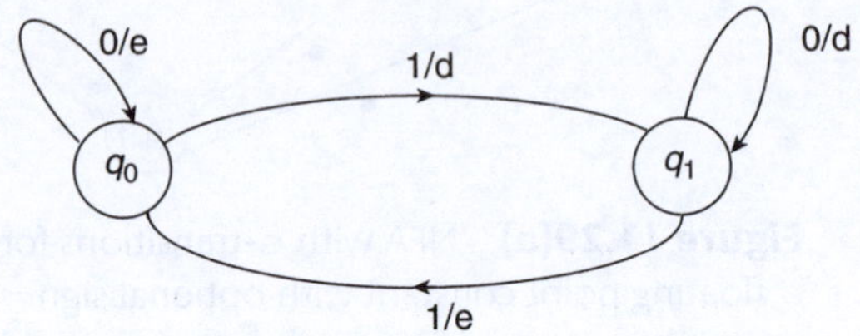

Figure 11.32 Parity checker

At the end of the input, if the machine is in q_0, then the string has even parity. If the machine is in q_1 then it has odd parity. The label 'e' corresponds to even parity and 'd' corresponds to odd parity.

11.7.1 Moore and Mealy Machines

There are two different types of output associated with FSMs, and are called Moore machines and Mealy machines. When the output is associated with states, the FSMs are called Moore machines. When the output is associated with transitions, the FSM's are called Mealy machines.

11.7.1.1 Moore machines

Similar to a DFA, a Moore machine is also defined as a six tuple $M = (Q, \Sigma, \delta, \Delta, \lambda, q_0)$. The symbol Δ is called the output alphabet, and λ is a mapping from Q to Δ, giving the output associated with each of the states. The rest of the symbols carry meanings as stated for DFAs. The output of a Moore machine for a given input a_i is defined as

(i) $\delta(q_i, a_i) = q_j$
(ii) $\lambda(q_j) \in \Delta$

Consider the Moore machine given in Fig. 11.33 and the input string 1010. The set of states $Q = \{q_0, q_1, q_2\}$, input alphabet $\Sigma = \{0,1\}$ and output alphabet $\Delta = \{0,1,2\}$. The mapping λ is defined as follows:

$$\lambda(q_0) = 0,$$

$$\lambda(q_1) = 1$$

$$\lambda(q_2) = 2$$

Figure 11.33 Example of a Moore machine

From the extended transition function of DFA
$\hat{\delta}(q_0, 1) = \delta(\hat{\delta}(q_0, \in), 1) = q_1$

$\hat{\delta}(q_0, 10) = \delta(\hat{\delta}(q_0, 1), 0) = q_2$

$\hat{\delta}(q_0, 101) = \delta(\hat{\delta}(q_0, 10), 1) = q_2$

$\hat{\delta}(q_0, 1010) = \delta(\hat{\delta}(q_0, 101), 0) = q_1$

So, on the input 1010 the sequence of states is $q_0\, q_1\, q_2\, q_2\, q_1$ and the output is $(q_0)\, \lambda(q_1)\, \lambda(q_2)\, \lambda(q_2)\, \lambda(q_1)$, which is 01221.

11.7.1.2 Mealy machines

Similar to a Moore machine, a Mealy machine is also defined as a six-tuple $M = (Q, \Sigma, \Delta, \delta, q_0, \lambda)$, where $Q, \Sigma, \Delta, \delta, q_0$ stand for the same elements as in the case of a Moore machine. λ is the mapping from $Q \times \Sigma$ to Δ. $\lambda(q, a)$ gives the output associated with the transition from state q on input *a*.

The output of M in response to input $a_1, a_2 \ldots a_n$ is $\lambda(q_0, a_1)\, \lambda(q_1, a_2) \ldots \lambda(q_n - 1, a_n)$ where $q_0, q_1 \ldots q_{n-1}$ is the sequence of states such that $\delta(q_{i-1}, a_i) = q_i$ for $1 \le i \le n$.

Consider the example given in Fig. 11.34.

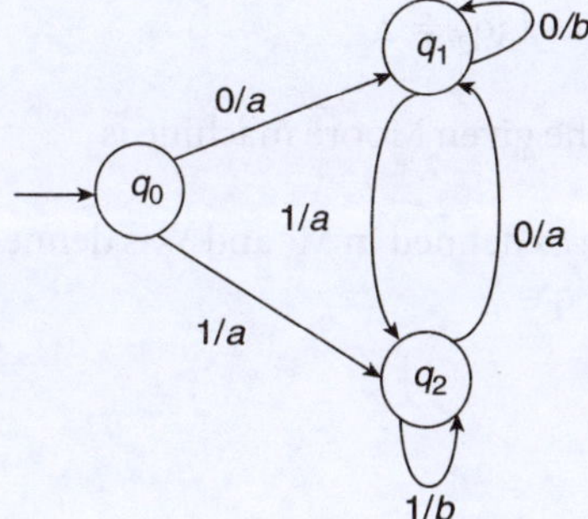

Figure 11.34 Example of a Mealy machine

The label 0/*a* on the transition from state q_0 to q_1 indicates that $\delta(q_0, 0) = q_1$ and $\lambda(q_0, 0) = a$. For the given input 01100, the output of the machine is *aabab*, which is described as follows

$\hat{\delta}(q_0, \in) = \lambda(q_0, \in) = \in$ i.e., for any $\in$ transition, the associated output is $\in$

$\hat{\delta}(q_0, 0) = \delta(\hat{\delta}(q_0, \in), 0) = \lambda(q_0, 0) = a$

$\hat{\delta}(q_0, 01) = \delta(\hat{\delta}(q_0, 0), 1) = \lambda(q_1, 1) = a$

$\hat{\delta}(q_0, 011) = \delta(\hat{\delta}(q_0, 01), 1) = \lambda(q_1, 1) = b$

$\hat{\delta}(q_0, 0110) = \delta(\hat{\delta}(q_0, 011), 0) = \lambda(q_2, 0) = a$

$\hat{\delta}(q_0, 01100) = \delta(\hat{\delta}(q_0, 0110), 0) = \lambda(q_1, 0) = b$

11.7.1.3 Equivalence of moore and mealy machines

Theorem 11.4: illustrates the equivalence between the Moore and Mealy machines. If $M_1 = (Q, \Sigma, \Delta, \delta, q_0, \lambda)$ is a Moore machine, then there is an equivalent Mealy machine M_2 for all inputs *w* of M_1

Proof: Let $M_2 = (Q, \Sigma, \Delta, \delta, q_0, \lambda')$ where $Q, \Sigma, \Delta, \delta$, and q are as the same as defined for a Moore machine, and λ' is defined as follows

$\lambda(q, a) = \lambda(\delta(q, a))$ for all states $q \in Q$ and input symbol $a \in \Sigma$, then M_1 and M_2 enter the same sequence of states on the same input.

Example 11.18: Convert the following Moore machine to a Mealy machine.

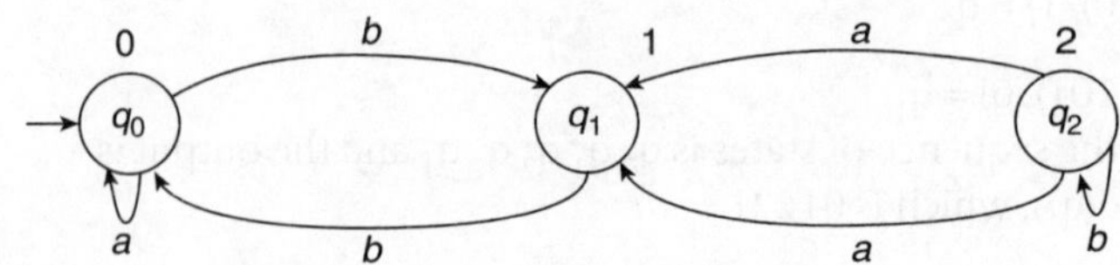

Figure 11.35 Moore machine

Solution:
The given Moore machine is $M_1 = (Q, \Sigma, \Delta, \delta, q_0, \lambda)$
where $Q = \{q_0, q_1, q_2\}$
$\Sigma = \{a, b\}$
δ is defined through a transition diagram
$\Delta = \{0, 1, 2\}$
λ is defined as $\lambda(q_0) = 0$, $\lambda(q_1) = 1$ and $\lambda(q_2) = 2$
q_0 is the initial state.
The equivalent Mealy machine for the given Moore machine is
$M_2 = (Q, \Sigma, \Delta, \delta, q_0, \lambda^{l})$
where Q, Σ, δ, Δ and q_0 remain same as defined in M_1 and Y^{l} is defined as in M_1 and λ^{l} is defined as
$\lambda^{l}(q, a) = \lambda(\delta(q, a))$ for all $q \in Q$ and $q \in \Sigma$
$\lambda^{l}(q_0, a) = \lambda(\delta(q_0, a)) = \lambda(q_0) = 0$
$\lambda^{l}(q_0, b) = \lambda(\delta(q_0, b)) = \lambda(q_1) = 1$
$\lambda^{l}(q_1, a) = \lambda(\delta(q_1, a)) = \lambda(q_2) = 2$
$\lambda^{l}(q_1, b) = \lambda(\delta(q_1, b)) = \lambda(q_0) = 0$
$\lambda^{l}(q_2, a) = \lambda(\delta(q_2, a)) = \lambda(q_1) = 1$
$\lambda^{l}(q_2, b) = \lambda(\delta(q_2, b)) = \lambda(q_2) = 2$
The equivalent transition diagram is given in Fig. 11.36.

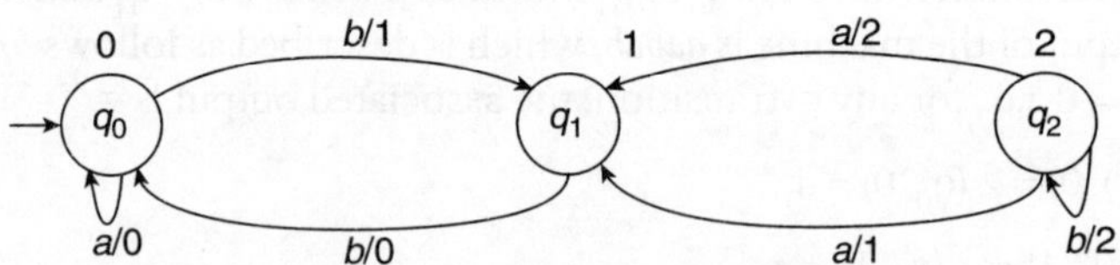

Figure 11.36 Equivalent Mealy machine for the Moore machine shown in Fig. 11.35

Theorem 11.5: If $M_1 = (Q, \Sigma, \Delta, \delta, q_0, \lambda)$ is a Mealy machine, then there is an equivalent Moore machine M_2 for all inputs *w* of M_1.
Proof: Let $M_2 = (Q \times \Delta, \Sigma, \Delta, \delta, [q_0, b], \lambda^{l})$
where, *b* is an arbitrarily selected symbol from Δ. That is, the states of M_2 are pairs [q, *b*] consisting of a state of M_1 and the output symbol Δ and is defined as

$$\delta^{1}([q, b], a) = [\delta(q, a), \lambda(q, a)] \text{ and } \lambda^{l}([q, b]) = b.$$

The second element of state [q, *b*] of M_2 is the output made by M_1 on some transition into state q. Only the first element of the states of M_2 determines the moves made by M_2. By induction on *n*, it can be seen that if M_1 enters states $q_0, q_1 \ldots q_n$ on inputs $a_1, a_2, \ldots a_n$ and gives outputs $b_1, b_2, \ldots b_n$, then M_2 enters states $[q_0, b_0], [q_1, b_1] \ldots [q_n, b_n]$ and gives outputs $b_0, b_1, \ldots b_n$.

Example 11.19: Convert the Mealy machine given in Fig. 11.37 to a Moore machine.

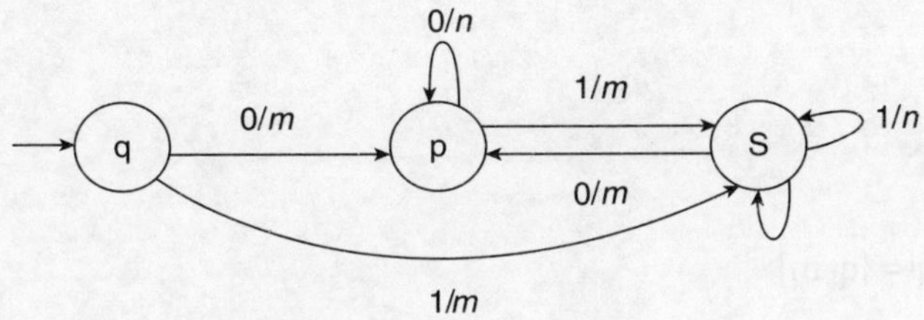

Figure 11.37 Mealy machine

Solution:

Let the Mealy machine be defined by $M_1 = (Q, \delta, \Delta, \delta, q, \lambda)$, where

$Q = \{q, p, s\}$

$\Sigma = \{0, 1\}$

$\Delta = \{m, n\}$

δ and λ are the transition functions.

q is the initial state.

The equivalent Moore machine for the given Mealy is defined as

$M_2 = (Q^|, \delta, \Delta\, \delta, q^|, \lambda^|)$ where

$Q^1 = \{[q, m], [q, n], [p, m], [p, n], [s, m], [s, n]\}$

$\Sigma = \{0, 1\}$

$\Delta = \{m, n\}$

The transition functions are defined as

$\delta([q, m], 0) = [\delta(q, 0), \lambda(q, 0)]$
$= [p, m]$

$\delta([q, m], 1) = [\delta(q, 1), \lambda(q, 1)]$
$= [s, m]$

$\delta([q, n], 0) = [\delta(q, 0), \lambda(q, 0)]$
$= [p, m]$

$\delta([q, n], 1) = [\delta(q, 1), \lambda(q, 1)]$
$= [s, m]$

$\delta([p, m], 0) = \delta(p, 0), \lambda(p, 0)]$
$= [p, n]$

$\delta([p, m], 1) = [\delta(p, 1), \lambda(p, 1)]$
$= [s, m]$

$\delta([p, n], 0) = [\delta(p, 0), \lambda(p, 0)]$
$= [p, n]$

$\delta([p, n], 1) = [\delta(p, 1), \lambda(p, 1)]$
$= [s, m]$

$\delta([s, m], 0) = [\delta(s, 0), \lambda(S, 0)]$
$= [p, m]$

$\delta([s, m], 1) = [\delta(s, 1), \lambda(s, 1)]$
$= [s, n]$

$\delta([s, n], 0) = [\delta(s, 0), \lambda(s, 0)]$
$= [p, m]$
$\delta([s, n], 1) = [\delta(s, 1), \lambda(s, 1)]$
$= [s, n]$

Outputs associated with the states are defined as

$\lambda([q, m]) = m$
$\lambda([q, n]) = n$
$\lambda([p, m]) = m$
$\lambda([p, n]) = n$
$\lambda([s, m]) = m$
$\lambda([s, n]) = n$

and the initial state is $q^{|} = [q, m]$

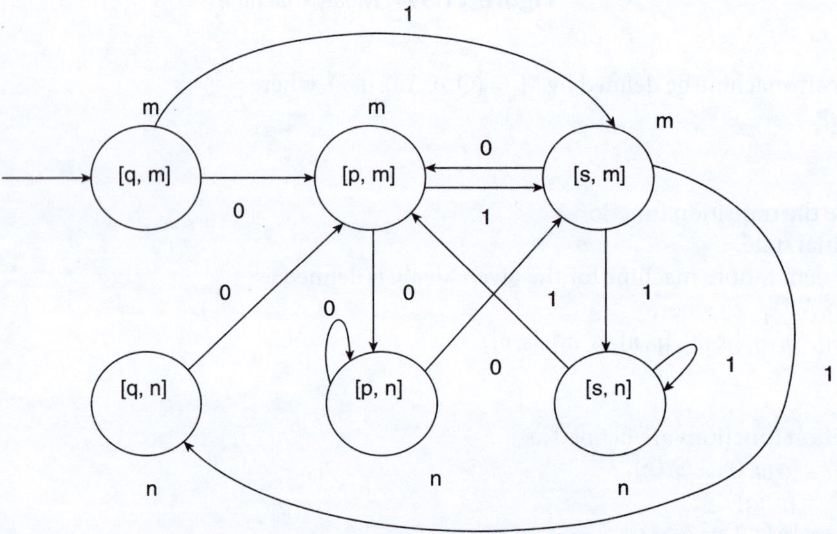

Figure 11.38 Equivalent Moore machine for the Mealy machine given in Fig. 11.40

11.8 MINIMISATION OF AUTOMATA

In this section, we discuss how to test whether the two descriptors of a regular language are equivalent or whether both the descriptions define the same language. Given any DFA, it is possible to find its equivalent with the minimum number of states. This DFA is unique. Given two DFAs, one can determine how to state that the two DFA's are equivalent. This is possible by discriminating between distinguishable and indistinguishable states.

11.8.1 Testing Equivalence of States

In a given DFA, we can determine whether the two states, say p and q, can be replaced by a single state. If so, then the states are equivalent. The states p and q are said to be equivalent if and only if, for all input strings $w \in \Sigma^*$, both $\delta(p, w)$ and $\delta(q, w)$ lead to an accepting or non-accepting state, i.e.,

If $\delta(p, w) \in F$ and $\delta(q, w) \in F$ or if $\delta(p, w) \notin F$ and If $\delta(q, w)$ if $\delta(p, w) \notin F$, then the states p and q are equivalent and are also called **indistinguishable**

If there is at least one string w such that $\delta(p, w) \in F$ is the accepting states and $\delta(q, w) \in F$ is non-accepting states then the states p and q are not equivalent and are also called distinguishable states. The **distinguishable** and **indistinguishable pairs of** states can be obtained by using the **table filling algorithm** (also called the mark algorithm). The algorithm is defined as follows.

Algorithm 11.1: Table filling algorithm

Start

Step 1: For each pair (p, q), if $p \in F$ and $q \notin F$ or vice versa then the pair (p, q) is distinguishable and the corresponding cell (p, q) is to be marked using X.

Step 2: If a pair of states (p, q) transit to a new par (r, s) for some input $a \in \Sigma$, where, $\delta(p, q) = r$ and $(q, a) = s$, and the pair (r, s) is already marked as distinguishable then the pair (p, q) is also distinguishable and hence mark corresponding cell (p, q) using x.

Step 3: Repeat Step 2 for each pair (p, q) till no corresponding cell (p, q) remains unmarked.

Stop

Example 11.20: Consider the DFA whose transitions are given in Table 11.14

Table 11.14 Transitions for Example 11.22

	Σ	
δ	a	b
A	B	F
B	G	C
*C	A	C
D	C	G
E	H	F
F	C	G
G	G	E
H	G	C

The transition diagram without minimisation of states is given in Fig. 11.39.

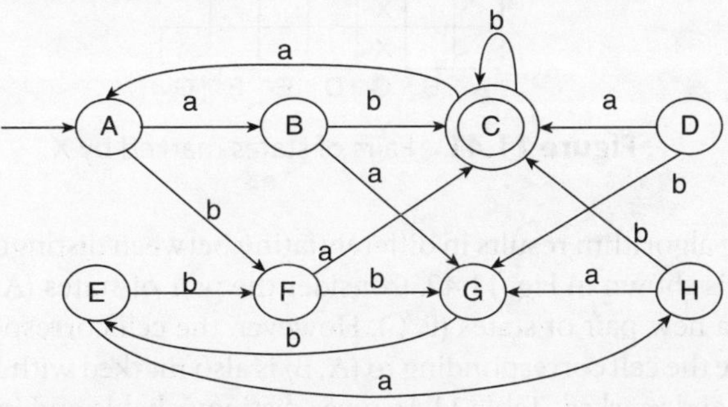

Figure 11.39 DFA before minimisation of states

In order to minimise the states, we shall first find out whether the table contains any distinguishable or indistinguishable pairs of states, as given in Table 11.15. This is done for different states of the given DFA, namely A, B, C, D, E, F, G, H. The possible pairs of states are written as follows:

(A, B) (A, C) (A, D) (A, E) (A, F) (A, G) (A, H)
(B, C) (B, D) (B, E) (B, F) (B, G) (B, H)
(C, D) (C, E) (C, F) (C, G) (C, H)
(D, E) (D, F) (D, G) (D, H)
(E, F) (E, G) (E, H)
(F, G) (F, H)
(G, H)

These pairs are represented in the form of a table as shown in Fig. 11.40.

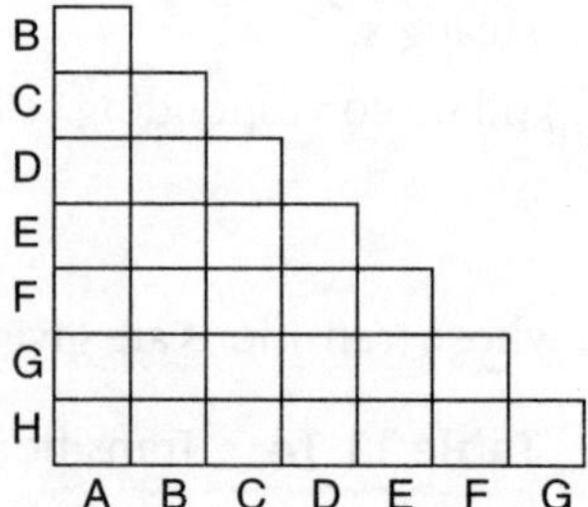

Figure 11.40 Table showing the pairs of states

The application of Step 1 of the table filling algorithm results in the marking of cells. For example, in the pair (B, C), B is a non-accepting state while C is an accepting state. Therefore the cell corresponding to (B, C) is marked with an X. This is true with other pairs of states, namely (A, C), (C, D), (C, E), (C, F), (C, G) and (C, H).The resulting entries are shown in Fig. 11.41

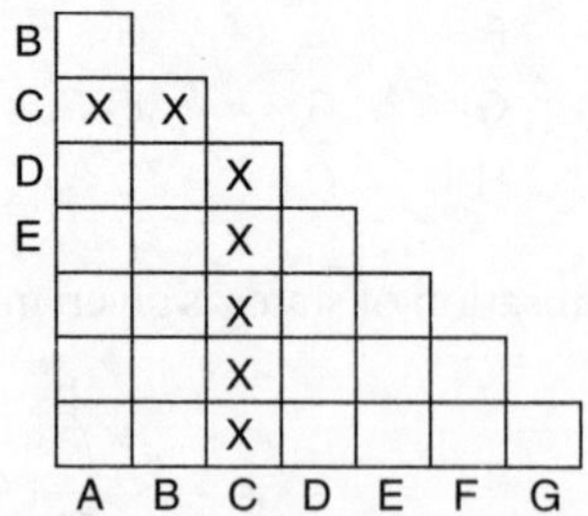

Figure 11.41 Pairs of states marked by X

Step 2 of the table filling algorithm results in differentiating between distinguishable and indistinguishable pairs of states and is shown in Fig. 11.42. Consider the pair of states (A, B). Upon the input $b \in \Sigma$, the DFA transitions to a new pair of states (F, C). However, the cell corresponding to the pair (F, C) is already marked X, hence the cell corresponding to (A, B) is also marked with X. Similarly all other pairs of states are also tested and marked. Table 11.16 gives distinguishable and indistinguishable pairs after the second iteration.

Table 11.15 Testing distinguishable and indistinguishable pairs

δ	**A**	**Mark**	**B**	**Mark**
(p, q)	***(r, s)***	***x***	***(r, s)***	**X**
(A, B)	(B, G)	–X	(F, C)	X
(A, D)	(B, C)	–X	(F, G)	----
(A, E)	(B, H)	--X	(F, F)	X
(A, F)	(B, C)	–X	(F, G)	X
(A, G)	(B, G)	--X	(F, E)	X
(A, H)	(B, G)	–X	(F, C)	X
(B, D)	(G, C)	X	(C, G)	X
(B, E)	(G, H)	X	(C, F)	----
(B, F)	(G, C)	--X	(C, G)	X
(B, G)	(G, G)	X	(C, E)	-
(B, H)	(G, G)	–	(C, C)	-
(D, E)	(C, H)		(G, F)	X
(D, F)	(C, C)		(G, G)	-
(D, G)	(C, G)		(G, E)	
(D, H)	(C, G)		(G, C)	
(E, F)	(H, C)		(F, G)	
(E, G)	(H, G)		(F, E)	
(E, H)	(H, G)		(F, C)	
(F, G)	(C, G)		(G, E)	
(F, H)	(C, G)		(G, C)	
(G, H)	(G, G)		(E, C)	

(i) δ (A, a) = B
δ (B, a) = G
(ii) δ (A, b) = F
δ (B, b) = C

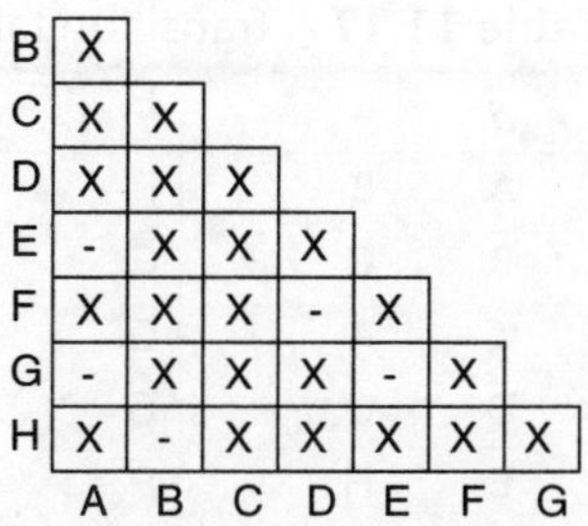

Figure 11.42 Marked table

Table 11.16 Distinguishable and indistinguishable pairs after the second iteration

δ	A	Mark	b	Mark
(p, q)	(r, s)	X	(r, s)	X
(A, E)	(B, H)	–	(F, F)	–
(A, G)	(B, G)	X	(E, E)	–
(B, H)	(G, G)	–	(C, C)	–
(D, F)	(C, C)	–	(G, G)	
E, G	(H, G)	X	(F, E)	X

The marked pairs of (AG) and (EG) are shown in Fig. 11.43

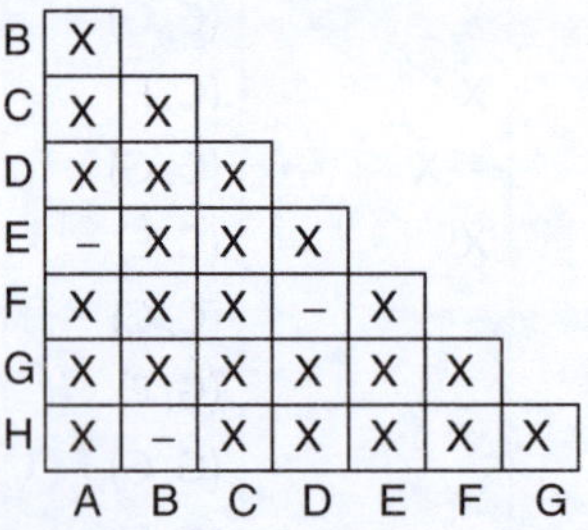

B	X						
C	X	X					
D	X	X	X				
E	–	X	X	X			
F	X	X	X	–	X		
G	X	X	X	X	X	X	
H	X	–	X	X	X	X	X
	A	B	C	D	E	F	G

Figure 11.43 Completely marked table

From Fig. 11.46 it is evident that the pairs of states (A, E), (B, H) and (D, F) are not marked and hence considered *indistinguishable.* Once distinguishable and indistinguishable pairs of states are determined, it is easy to minimise the number of states of a given DFA and accepting the same language as that of original DFA using the procedure described in Algorithm 11.2.

Algorithm 11.2: Algorithm to Minimise a DFA

Step 1: Partition the set of states into indistinguishable pairs.

Step 2: Minimise DFA states as follows:

Consider the partition $[p_1, p_2 \dots p_k]$ as one group. The $\delta(\{p_1, p_2, p_3 \dots p_k\}, a) = [r_1, r_2 \dots r_k]$ where $a \in \Sigma$. Mark an edge from the group $[p_1, p_2 \dots p_k]$ to the group $[r_1, r_2 \dots r_k]$ and label the edge with a.

Step 3: Repeat Step 2 for all the partitions and input symbols.

Step 4: The group containing the initial state of the original DFA is the initial state of the minimised DFA. The group containing the final state of the original DFA is the final state of the minimised DFA.

Example 11.21: Minimise the FA defined by Table 11.17.

Table 11.17 Transition table

δ	a	b
A	B	F
B	G	C
*C	A	C
D	C	G
E	H	F
F	C	G
G	G	E
H	G	C

Step 1: Determine the distinguishable and indistinguishable states using the table-filling algorithm (marking as x). The marked transaction table as follows.

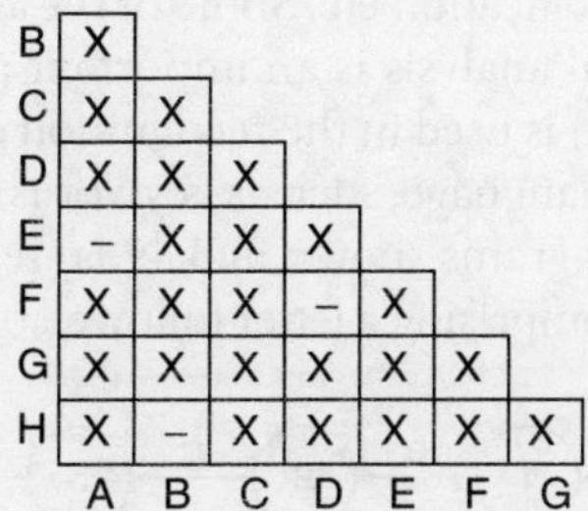

B	X						
C	X	X					
D	X	X	X				
E	–	X	X	X			
F	X	X	X	–	X		
G	X	X	X	X	X	X	
H	X	–	X	X	X	X	X
	A	B	C	D	E	F	G

Figure 11.44 Completely marked table for Example 11.23

Step 2: The pairs (A, E), (B, H) AND (D, F) are not marked and are called indistinguishable, and the states C and G are distinguishable. So the different groups that can be represented as the states of the minimised DFA are

(A, E), (B, H), C, (D, F), G.

Step 3: The transition table for the minimised DFA can be determined as follows.

	Σ	
	A	**B**
(A, E)	(B, H)	(D, F)
(B, H)	G	C
*C	(A, E)	C
(D, F)	C	G
G	G	(A, E)

Therefore, the transition diagram for the minimised DFA is shown in Fig. 11.45.

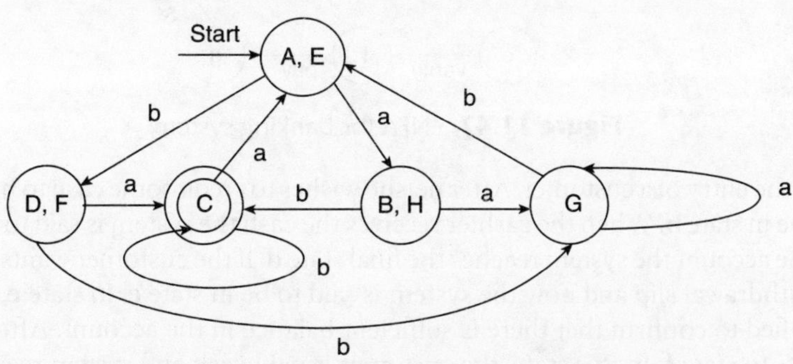

Figure 11.45 Minimised DFA for Example 11.23

11.9 APPLICATIONS OF FINITE AUTOMATA

The concept of finite automata has several applications in many areas like compiler design, special purpose hardware design, protocol specification etc. Some of the applications are described here.

(a) **Compiler Design:** Lexical analysis is an important phase in a compiler. In lexical analysis, the transition diagram (FA) is used in the recognition of tokens. The tokens are the basic constructs in a programming language, such as keywords, numbers, identifiers, operators etc. For example, the transition diagrams shown in 3.49 (a) recognises the tokens 'begin'. The typical identifiers in a language comprising a letter followed by a letter or a digit are given in 11.46 (b).

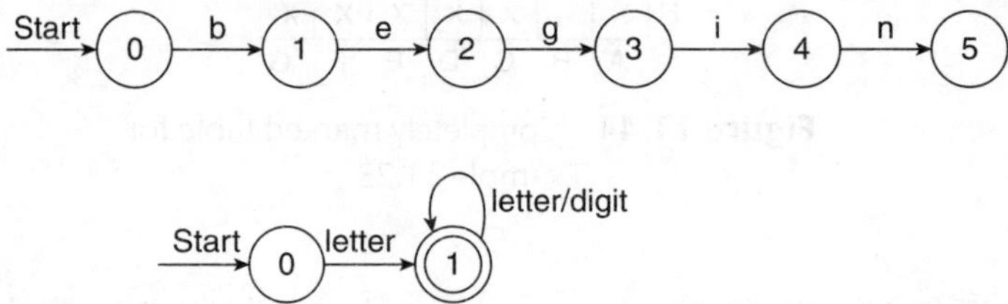

Figure 11.46 (a) NFA for keyword begin (b) NFA for identifier

The relation between an FA and regular expressions is used in many packages developed on UNIX, such as LEX. Furthermore, many UNIX tools extensively use the concept of regular expressions.

(a) **Hardware design:** In the design of computers, FA is used to design the control unit. A typical sequence of operations in a computer consists of a repetition of instructions and each instruction involves the actions fetch, decode, fetch the operand, and execute. Each state of an FA represents a specific stage of an instruction cycle.

(b) **Protocol specification:** Any system consisting of an interconnected set of subsystems requires a protocol for establishing proper coordination between the sub-systems. A protocol comprises a set of rules for the purpose of coordination and it is described by transition diagrams. An example of a banking system is shown in Fig. 11.47

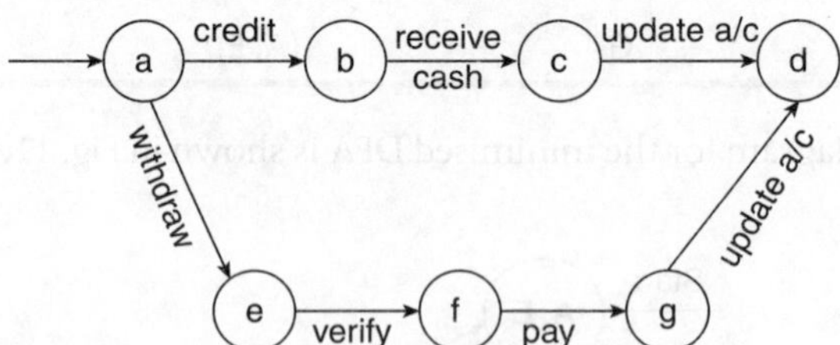

Figure 11.47 NFA for banking system

State **a** represents the entry of a customer. After he/she wishes to credit some cash to his/her account, the system is said to be in state **b**. When the cashier receives the cash the system is said to be in state **c.** After the updation of the account the system reaches the final state **d.** If the customer wants to withdraw cash, he submits the withdrawal slip and now the system is said to be in state **e**. In state **e**, the cheque/withdrawal slip is verified to confirm that there is sufficient balance in the account. After this verification, the system transits to state **f.** In this state, the customer is paid cash and system reaches a new state **g.** The customer account is finally updated and the system reaches the final state **d.**

11.10 SOLVED PROBLEMS

The following are a few examples used to enhance the understanding of the concepts discussed in this chapter.

Example 11.22: Construct an equivalent DFA for the NFA defined in the transition function given in Table 11.18.

Table 11.18 Transition table for the Example 11.24

	0	1
→p	p, q	p
q	r	r
r	s	
*s	s	s

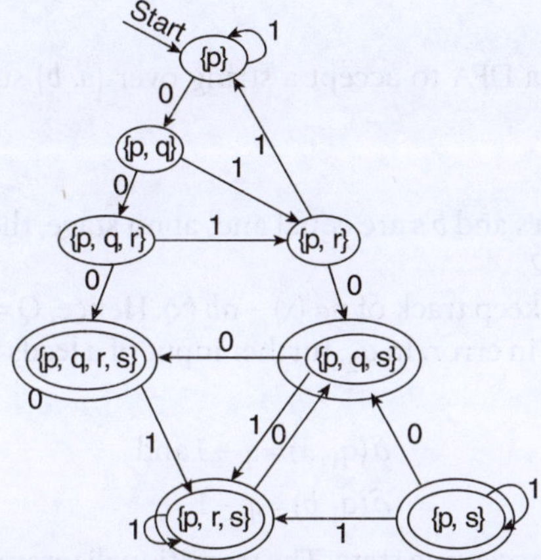

Figure 11.48 Equivalent DFA for Example 11.24

Example 11.23: Describe in English, the language accepted by the DFA given by the transition function given in Table 11.19. Prove that your answer is correct.

Table 11.19 Transition function for Example 11.25

	0	1
→*A	B	A
*B	C	A
C	C	C

The DFA accepts strings not containing two consecutive zeros. We prove this by proving the following Lemma about strings ending in A, B and C.

Lemma:

(i) A string reaching A is either ∈ or ends in I and does not contain 00 as substring.

(ii) A string reaching B ends in 0 and does not contain 00.

(iii) A string reaching C contains 00 as a substring.

Proof:

Proof is by induction on the input string w.

Basis:

W = ∈ DFA is in state A and hence satisfies the condition for A.

Induction:

Let the lemma be true for an input *w*.

Case 1: The DFA is in A.

On input I, DFA is in A and w_1 does not contain 00 and hence, satisfies the condition for A.

On input 0, DFA is in B and w_0 does not contain 00 as a substring and satisfies the condition B.

Case 2: DFA is in B.

w ends in 0 and does contain 00

On input 1, the DFA reaches A and w_1 satisfies the conditions for A.

On input 0, the DFA is in state C, w_0 contains 00 and satisfies the condition for C.

Case 3: DFA is in state C.

w contains 00 and any further input does not change this condition.

Strings reaching A and B together represent any string not containing 00 and are accepted.

Example 11.24: Construct a DFA to accept a string over {*a*, *b*} such that n*a* (*w*) = n*b* (*w*) for every prefix *x*, | n*a*(*x*) n*b*(*x*)|2.

Solution:

The condition requires that no *a*'s and *b*'s are equal and, at no stage, the difference between the number of *a*'s and number of *b*'s exceed 2.

Each state of the DFA is used to keep track of n*a* (*x*) – n*b* (*x*). Hence, $Q = \{q_i \mid -2 \le i \le 2\} \cup \{q_e\}$.

In q-2, a further input of results in error. In q_2, further input of *a* leads to an error.

In other cases

$$\delta(q_i, a) = q_i + 1 \text{ and}$$
$$\delta(q_i, b) = q_i - 1$$

q_0 is both the start state and the accepting state. The transition diagram for the DFA is given in Fig. 11.49.

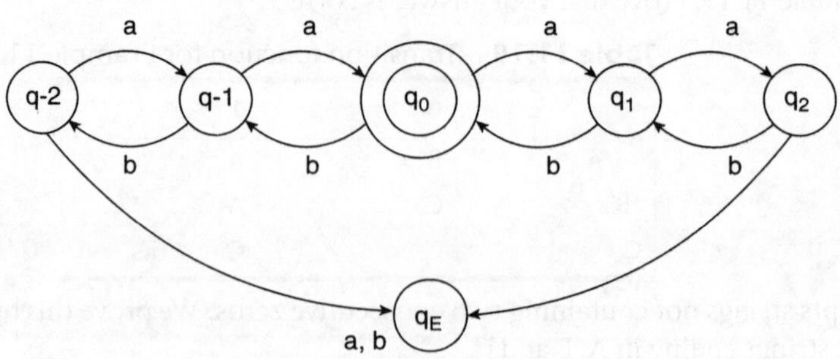

Figure 11.49 Transition diagram

Example 11.25: Minimise the states of the automata given in Table 11.20

Table 11.20 Transition table for Example 11.25

State	Input	
	a	*b*
$\rightarrow q_0$	q_0	q_3
q_1	q_2	q_5
q_2	q_3	q_4
q_3	q_0	q_5
q_4	q_0	q_6
q_5	q_1	q_4
$*q_6$	q_1	q_3

Solution:
We will construct the table for each pair of states

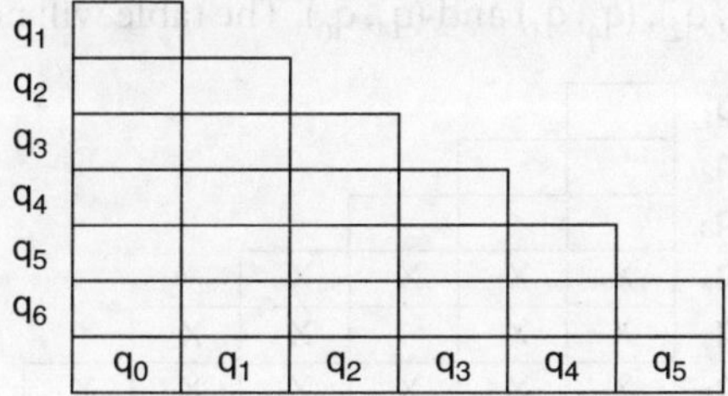

We will mark X for the pairs of final and non-final states. Hence

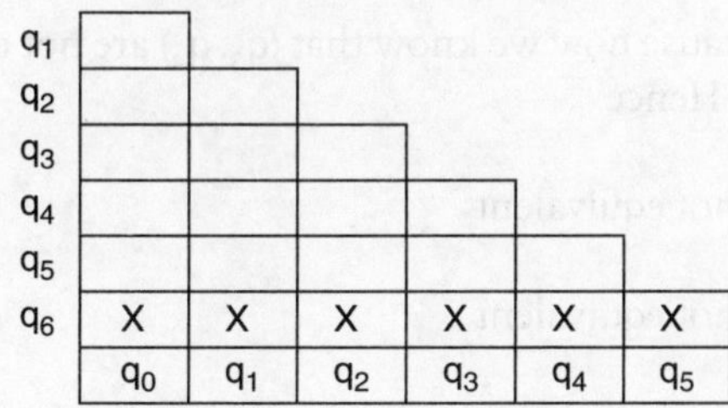

Now we will consider every pair from above table. The pairs marked by X are not equivalent.
Consider the pair (q_5, q_4),

$\delta(q_5, a) = q_1$ $\quad \delta(q_5, b) = q_4$
$\delta(q_4, a) = q_0$ $\quad \delta(q_4, b) = q_6$

As there is already an X in cell corresponding to the pair (q_4, q_6) so the pair (q_5, q_4) must be marked by X and is not equivalent.
Similarly consider

$\delta(q_5, a) = q_1$ $\quad \delta(q_5, b) = q_4$
$\delta(q_3, a) = q_0$ $\quad \delta(q_3, b) = q_5$ $\quad$ not equivalent
$\delta(q_2, a) = q_3$ $\quad \delta(q_2, b) = q_4$
$\delta(q_1, a) = q_2$ $\quad \delta(q_1, b) = q_5$ $\quad$ not equivalent
$\delta(q_0, a) = q_0$ $\quad \delta(q_0, b) = q_3$

Hence the table will be

	q_0	q_1	q_2	q_3	q_4	q_5
q_1						
q_2						
q_3						
q_4						
q_5		X		X	X	
q_6	X	X	X	X	X	X

Pair (q_5, q_2) is unknown because we don't know whether the pair (q_1, q_3) is equivalent or not. Similarly (q_0, q_1) and (q_3, q_4) are unknown for equivalence. Hence we mark X for the pair (q_5, q_0).
Now consider q_4 with all remaining states.

$\delta(q_4, a) = q_0$ $\quad \delta(q_4, b) = q_6$
$\delta(q_3, a) = q_0$ $\quad \delta(q_3, b) = q_5$ not equivalent
$\delta(q_2, a) = q_3$ $\quad \delta(q_2, b) = q_4$ not equivalent
$\delta(q_1, a) = q_2$ $\quad \delta(q_1, b) = q_5$ not equivalent
$\delta(q_0, a) = q_0$ $\quad \delta(q_0, b) = q_3$ not equivalent

Thus we will put X in (q_4, q_3), (q_4, q_2), (q_4, q_1) and (q_4, q_0). The table will now be

	q_0	q_1	q_2	q_3	q_4	q_5
q_1						
q_2						
q_3						
q_4	X	X	X	X		
q_5	X	X		X	X	
q_6	X	X	X	X	X	X

We have marked $(q_5, q_0) = X$ because now we know that (q_3, q_4) are not equivalent.
Now consider the pairs with q_3. Hence

$\delta(q_3, a) = q_0$ $\quad \delta(q_3, b) = q_5$
$\delta(q_2, a) = q_3$ $\quad \delta(q_2, b) = q_4$ not equivalent
$\delta(q_1, a) = q_2$ $\quad \delta(q_1, b) = q_5$
$\delta(q_0, a) = q_0$ $\quad \delta(q_0, b) = q_3$ not equivalent

Hence the table will be

	q_0	q_1	q_2	q_3	q_4	q_5
q_1						
q_2						
q_3	X		X			
q_4	X	X	X	X		
q_5	X	X		X	X	
q_6	X	X	X	X	X	X

The pair (q_3, q_1) is unknown for equivalence because we don't know the equivalence of the pair (q_0, q_2)
Now consider q_2 with all the remaining states.

$\delta(q_2, a) = q_3$ $\quad \delta(q_2, b) = q_4$
$\delta(q_1, a) = q_2$ $\quad \delta(q_1, b) = q_5$ (not equivalent)
$\delta(q_0, a) = q_0$ $\quad \delta(q_0, b) = q_3$ (not equivalent)

q_1						
q_2	X	X				
q_3	X	X	X			
q_4	X	X	X	X		
q_5	X	X		X	X	
q_6	X	X	X	X	X	X
	q_0	q_1	q_2	q_3	q_4	q_5

Since (q_0, q_2) is not a matching pair. The pair (q_3, q_1) is also not equivalent. Similarly the pair (q_5, q_2) is also not equivalent since the pair (q_3, q_1) is not equivalent.

Finally we check for

$\delta(q_1, a) = q_0$ $\quad\quad$ $\delta(q_1, b) = \delta(q_1, b)$

$\delta(q_0, a) = q_2$ $\quad\quad$ $\delta(q_0, b) = q_3$ not equivalent

Finally, the table will be

q_1	X					
q_2	X	X				
q_3	X	X	X			
q_4	X	X	X	X		
q_5	X	X	X	X	X	
q_6	X	X	X	X	X	X
	q_0	q_1	q_2	q_3	q_4	q_5

Thus we find no states equivalent, so we cannot remove any of the states. Hence, the given transition table itself is the minimum state finite automation.

ADDITIONAL SOLVED EXAMPLES

Example W11.1: What is a Finite State Machine (FSM)? What are the different types of FSMs

Solution:

A finite state machine is the mathematical model defined by a set of states with initial states, input alphabets and a transition function that maps the current state and the input symbol to the next state(s).

Finite state machines are classified into machines without output, which have one or more accepting states, and machines with output associated with states or transitions.

Examples of FSM without output are Deterministic Finite Automata (DFA), Non Deterministic Finite Automata (NDFA) and epsilon Non Deterministic Finite Automata ($\in$-NDFA).

Examples for FSM with output associations are Moore machines and Mealy machines.

Example W11.2: Write English descriptions for the languages generated by the RE (A|B|C|....|Z) (*a*|*b*|*c*|....|*z*)*

Solution:

All possible strings of capital and small letters

Example W11.3: Write regular expression for the following char set.

All strings of lowercase letters that begin and end in 'a'.

Solution:

letter ← {a-z}
char ← {a}
The regular expression is defined as
{char}{letter}*{char}

All strings of digits that contains no leading zeros.

first ← {1-9}
Digit ← {0-9}
The regular expression is defined as
{first}{digit}*

Example W11.4: Write the NFA for the regular expression ***(a*|b*)****. Convert the obtained NFA to DFA and minimise the states of the DFA if possible.

NFA for the Regular expression ***(a*|b*)*****; here, E means epsilon transition**

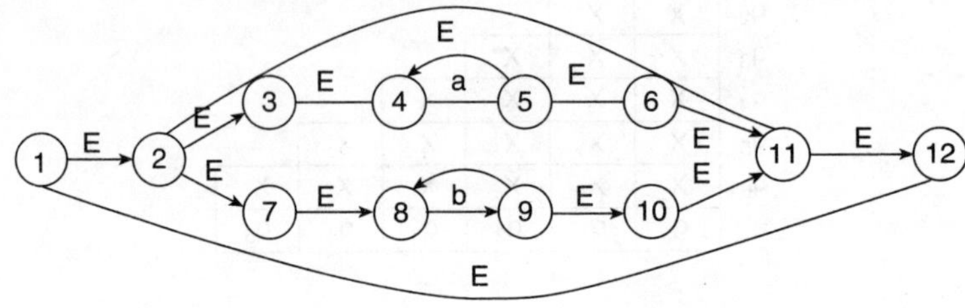

Solution:

NFA to DFA is as follows

Epsilon closure of state 1 = {1, 2, 3, 4, 6, 8, 10, 11, 12} ---- A
$\delta(A, a) = 5$

Epsilon closure of state 5 = {2, 3, 4, 5, 6, 7, 10, 11, 12} --- B
$\delta(A, b) = 9$

Epsilon closure of state 9 = {2, 3, 4, 6, 7, 8, 9, 10, 11, 12} --- C
$\delta(B, a) = 5 =$ Epsilon closure of state 5 is B
$\delta(B, b) = 9 =$ Epsilon closure of state 9 is C
$\delta(C, a) = 5 =$ Epsilon closure of state 5 is B
$\delta(C, b) = 9 =$ Epsilon closure of state 9 is C

The equivalent DFA is

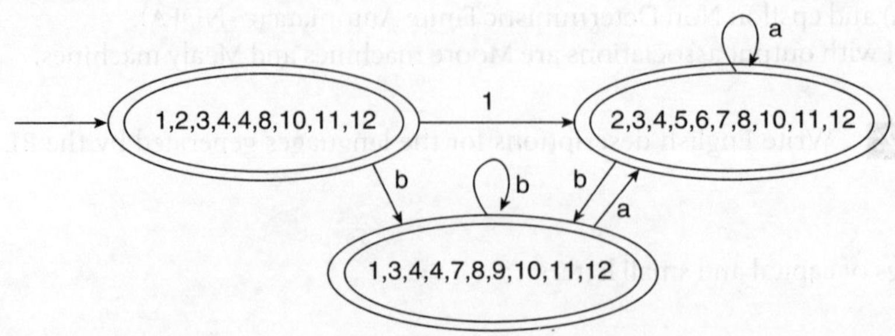

And the optimised DFA is as follows

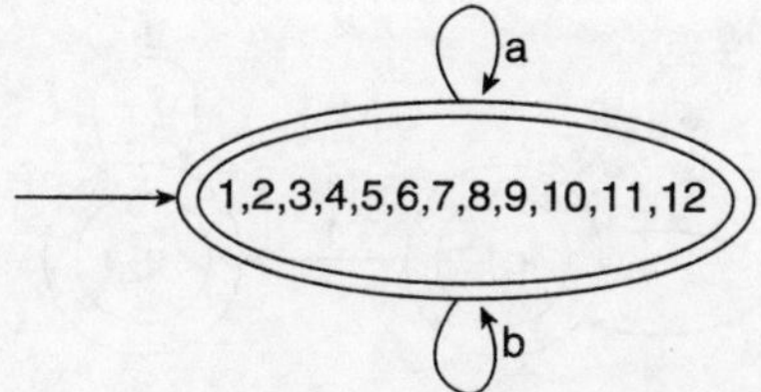

Example W11.5: Write the Epsilon NFA for the regular expression (a|b) * a (a|b) (a|b)

Solution:

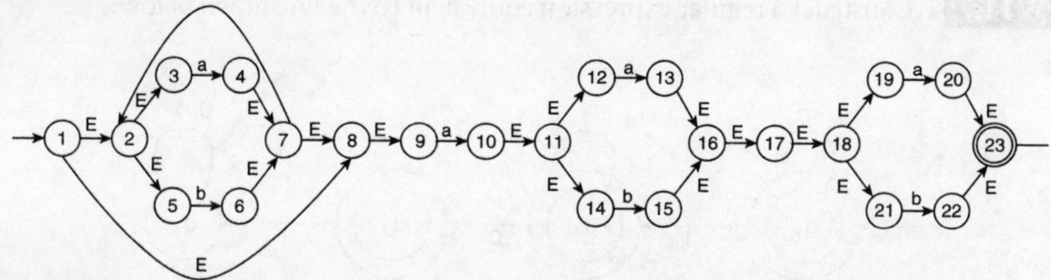

Example W11.6: Construct a regular expression equivalent to the FA shown below.

Solution:

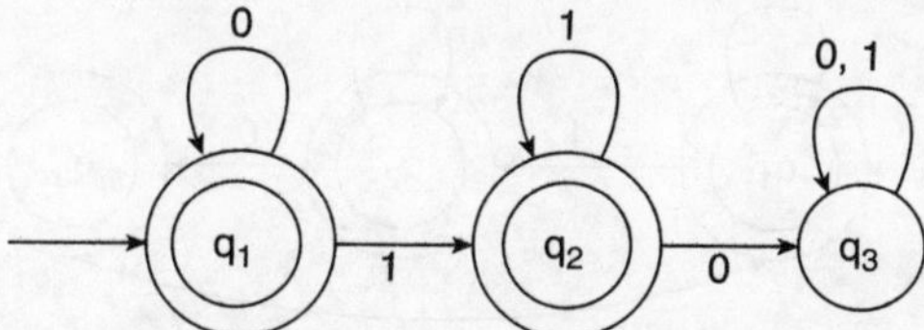

Step 1: Eliminate state q_3

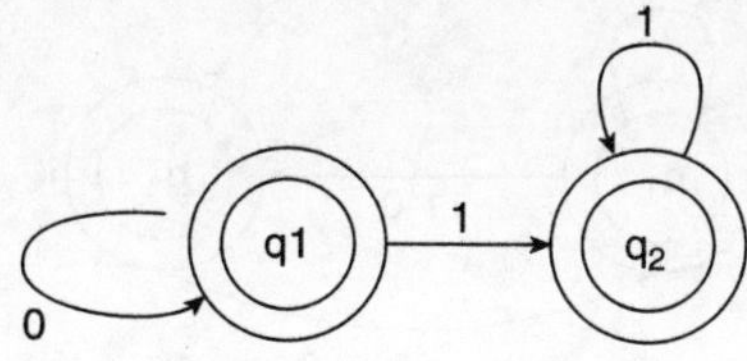

Step 2: After eliminating state q_2

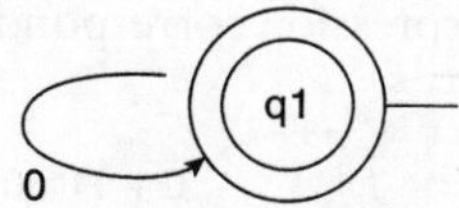

Regular expression R_1 accepting at state q_1 is 0^*

Step 3: Considering state q_2 is accepting state the automata is

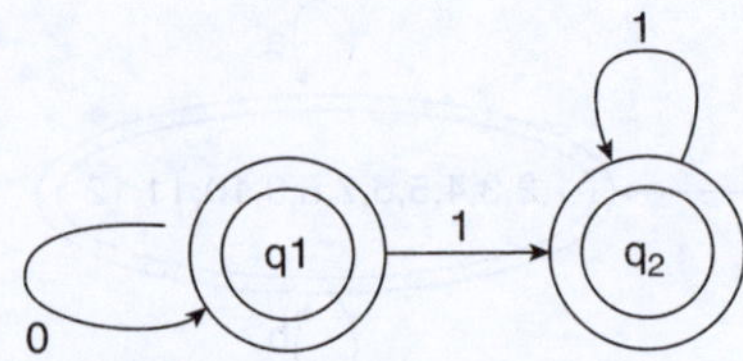

Regular expression R_2 accepting at state q_2 is $0^*[\in + 11^*]$
The final regular expression is $R_1 + R_2$ i.e., 0^*1^*

Example W11.7: Construct a regular expression equivalent to the FA shown below:

Solution:

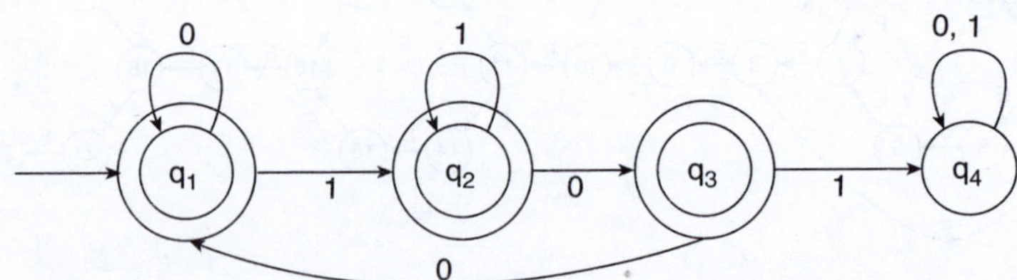

Step 1: Remove q_4 from which the final states are not reachable. The FA becomes:

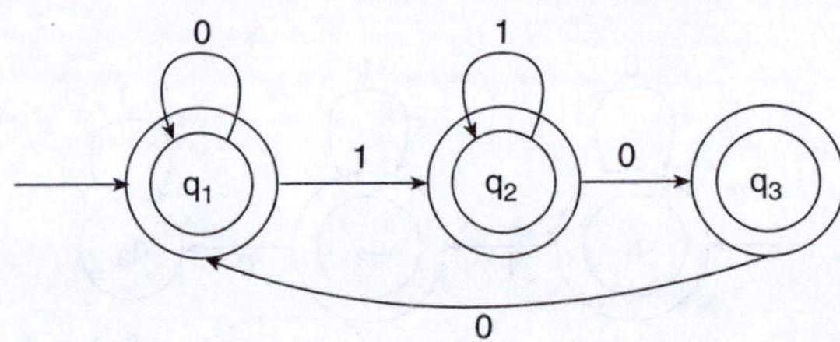

Step 2: If we eliminate q_2, we get

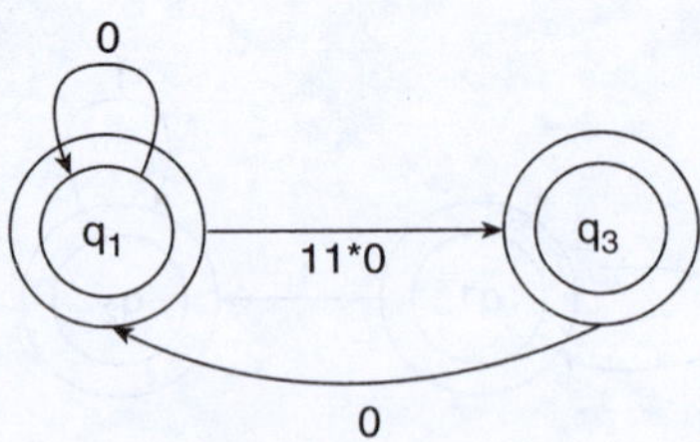

Regular expression corresponding to the final state q_3 is

$$R_1 = (0 + 11^*00)^*11^*0$$

Step 3: After eliminating state q_3, regular expression corresponding to the final state q_1 is as follows

q1 0 + 11*00

$$R_2 = (0 + 11^*00)^*$$

Step 4: After eliminating q_3 from the original finite automata, finite automata becomes

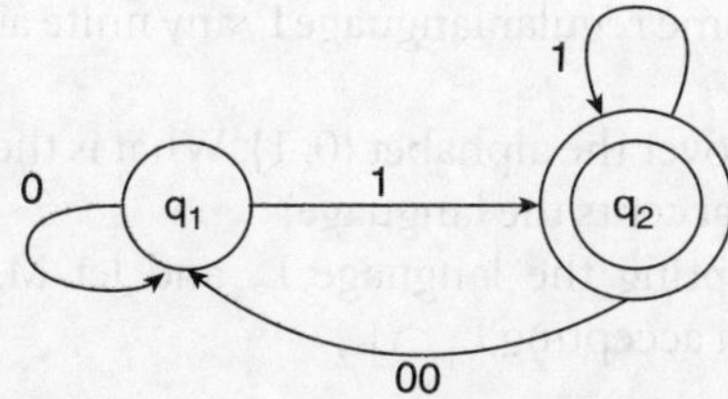

Regular expression corresponding to the final state q_2 is $R_3 = (0 + 11^*00)^*11^*$.
Final regular expression is $R = R_1 + R_2 + R_3$

$R = (0 + 11^*00)^*11^*0 + (0 + 11^*00)^* + (0 + 11^*00)^*11^*$

$= [0 + 11^*00]^*[\in + 11^* + 11^*0]$

SUMMARY

The finite state machine finds many applications in computer science. We have explained how to write a finite state machine, given the regular expression. A finite state machine recognises the regular language. We have shown the transition from a nondeterministic finite state machine to a deterministic machine. A few important theorems have also been covered. The variants of a finite state machine, such as a Melay machine etc., have been discussed. Brief accounts of some applications have been given.

EXERCISES

1. Construct a DFA to recognise decimal strings divisible by 3.
2. Construct a DFA to accept binary strings not ending in 101.
3. Let a DFA be $M = (Q, \Sigma, \delta, q_0, F)$. Show that, for any string $xy.\delta(q,xy) = \delta(\delta(q, x), y)$.
4. Construct a DFA to accept binary strings which leave a remainder of 3 after division by 7.
5. Construct a DFA which accepts binary strings containing exactly three 0's.
6. Design a DFA for accepting strings defined over {0, 1} such that their decimal values are divisible by 5.
7. Design a DFA for accepting strings defined over {*a*, *b*}, wherein every string has *a* followed by 2*b*'s.
8. A FA accepts a language L. If we add an $\in$ transition from the start state q_0 to every other state, what is the relationship between the resulting language L_1 and L.
9. Find the equivalent DFA for the NFA shown in Fig. 11.50

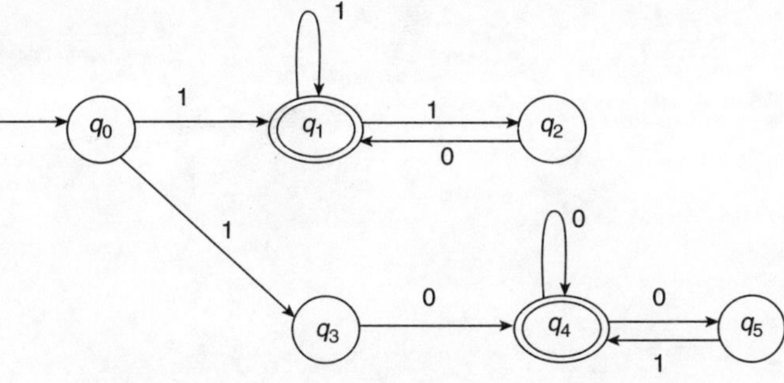

Figure 11.50 NFA for (1 + 10)* + 10 (0 + 01)*

10. Show that if for some state p, δ (p, a) = p for every a $\in \Sigma$, then δ *(p, x) = p for every $x \in \Sigma^*$.
11. Show by an example that for some regular language L, any finite automata recognising L must have more than one accepting state.
12. Let be a fixed string of length *n* over the alphabet {0, 1}. What is the minimum number of states that a finite automata can have if it accepts the language?
13. Let M_1 be the automata accepting the language L_1 and let M_2 be the automata accepting L_2 (Fig. 11.51). Draw the automata accepting $L_1 \cap L_2$

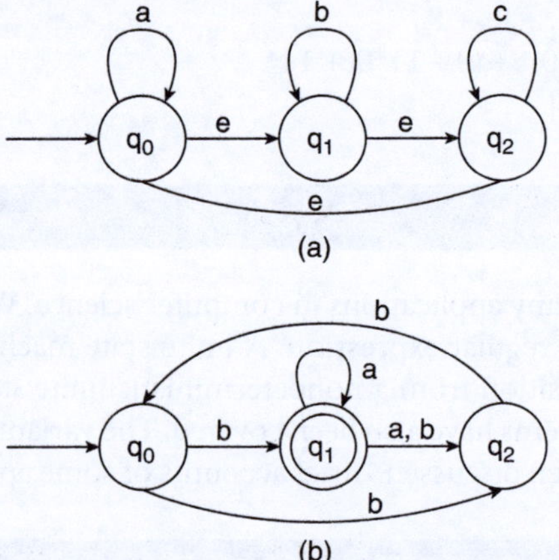

Figure 11.51 (a) Automata M_1 for L_1 (b) Automata M_2 for L_2

12 Introduction to Recurrence Relations

In Chapter 11, we have discussed the finite state machine. In this chapter, we discuss the mathematical concept called recurrence relation. The concept of recurrence relations deals with recursive definitions of mathematical functions or sequences. Solving a recurrence relation involves finding the "closed form" of a given function defined by a recurrence relation. In other words, it is to eliminate recursion from the function definition. Recurrence relations are a fundamental mathematical tool since they can be used to represent mathematical functions/sequences that cannot be easily represented non-recursively. An example is the Fibonacci sequence. We find applications of recurrence relations mainly in the design and analysis of algorithms.

12.1 CONCEPT OF RECURRENCE RELATION

A sequence is a function. Consider a function $a(n)$, which is a sequence of terms $a_0, a_1, \ldots a_{n-1}$. The term a_n depends upon the previous terms $a_0, a_1, \ldots a_{n-1}$. This type of study is called a recurrence relation or difference equations, which is considered the discrete counterpart of what is studied in differential equations.

Example 12.1: Consider the definition of 2^n. We could write the same as 2 multiplied n times, as shown in the following.

$$2 \times 2 \times 2 \times 2 \times \ldots \times 2 \text{ (}n\text{ times)} = 2^n$$

We shall write the same as follows.

$$2^1 = 2 \text{ and for } k \geq 1,\ 2^{(k+1)} = 2 \times 2^k \tag{1}$$

The statement made in (1) is called recursive definition. The same is stated in proof by the principle of mathematical induction.

Example 12.2: Consider the following recurrence relation which defines the function $f(n) = n^2$.

$$g(n) = g(n-1) + 2n - 1 \tag{2}$$

$$g(0) = 0$$

Example 12.3: Similarly, we can write the recurrence relation that defines the famous Fibonacci sequence 1, 1, 2, 3, 5, 8, 13, ...

$$\begin{aligned} f(n) &= f(n-1) + f(n-2) \\ f(1) &= 1 \\ f(0) &= 1 \end{aligned} \tag{3}$$

There are several methods for solving recurrence relations. It is mainly the iterative method that is used, which is also called the expansion, or unfolding method. The recurrence relation g(*n*), as given in Example 12.2 is solved using the iteration method as follows.

$g(n) = g(n-1) + 2n - 1$

$= [g(n-2) + 2(n-1) - 1] + 2n - 1$

/* because $g(n-1) = g(n-2) + 2(n-1) - 1$ */

$= g(n-2) + 2(n-1) + 2n - 2$

$= [g(n-3) + 2(n-2) - 1] + 2(n-1) + 2n - 2$

/* because $g(n-2) = g(n-3) + 2(n-2) - 1$ */

$= g(n-3) + 2(n-2) + 2(n-1) + 2n - 3$

.

.

.

$= g(n-i) + 2(n-i+1) + \ldots + 2n - i$

.

.

.

$= g(n-n) + 2(n-n+1) + \ldots + 2n - n$

$= 0 + 2 + 4 + \ldots + 2n - n$

/* because $g(0) = 0$ */

$= 2 + 4 + \ldots + 2n - n$

$= 2*n*(n+1)/2 - n$

/* using arithmetic progression formula $1 + \ldots + n = n(n+1)/2$ */

$= n^2$

Consider the following recurrence relations, given in (4), that have more than one variable. In some applications, we may consider recurrence relations with two or more variables. The famous Ackermann's function is one such example and is a recurrence relation in two variables.

$$\begin{aligned} &T(m, n) = 2*T(m/2, n/2) + m*n, \ (m > 1), \text{ and } (n > 1) \\ &T(m, n) = n, \ \text{ if } (m = 1) \\ &T(m, n) = m, \ \text{ if } (n = 1) \end{aligned} \tag{4}$$

We can solve this recurrence using the iteration method as follows. Assume that ($m <= n$). Then, we can write the relation (4) as follows.

$$\begin{aligned}
T(m, n) &= 2*T(m/2, n/2) + m*n \\
&= 2^2*T(m/2^2, n/2^2) + 2*(m*n/4) + m*n \\
&= 2^2*T(m/2^2, n/2^2) + m*n/2 + m*n \\
&= 2^3*T(m/2^3, n/2^3) + m*n/2^2 + m*n/2 + m*n \\
&\;\;\vdots \\
&= 2^i*T(m/2^i, n/2^i) + m*n/2^{(i-1)} + ... + m*n/2^2 + m*n/2 + m*n
\end{aligned}$$

Let $k = \log\_2\, m$. Then, we have

$$\begin{aligned}
T(m, n) &= 2^k*T(m/2\hat{}k, n/2\hat{}k) + m*n/2\hat{}(k-1) + ... + m*n/2\hat{}2 + m*n/2 + m*n \\
&= m*T(m/m, n/2\hat{}k) + m*n/2\hat{}(k-1) + ... + m*n/2\hat{}2 + m*n/2 + m*n \\
&= m*T(1, n/2\hat{}k) + m*n/2\hat{}(k-1) + ... + m*n/2\hat{}2 + m*n/2 + m*n \\
&= m*n/2\hat{}k + m*n/2\hat{}(k-1) + ... + m*n/2\hat{}2 + m*n/2 + m*n \\
&= m*n*(2-1/2\hat{}k) \\
&= \theta(m*n)
\end{aligned}$$

12.2 GENERAL SOLUTION

The general solution to the recurrence relation is given as under.

$a_{n+1} = da_n$, where $(n \geq 0)$, d is a constant and $a_0 = A$ is unique and is given by $a_n = Ad^n$ $(n \geq 0)$ and defines a discrete function whose domain is the set of non-negative integers. Given a particular recurrence relation and certain initial conditions, we are supposed to devise a formula for the nth term. Consider the recurrence relation of the form

$$a_n = ra_{n-1} + sa_{n-2} + f(n)$$

where, r and s are constants and f(n) is some function of n. This type of recurrence relation is called a second order linear recurrence relation with constant coefficients. If f(n) = 0, the recurrence relation is called homogeneous. The recurrence relation is called second order because the nth term is defined as the function of the previous two terms. It is also called linear because the terms a_{n-1} and a_{n-2} are of the first power.

Consider r = 1 and s = 1 with f(n) = 0. The recurrence relation $a_n = a_{n-1} + a_{n-2}$ represents the Fibonacci sequence. Similarly, in $a_n = 5\,a_{n-1} - 6\,a_{n-2} + n$ the value of r = 5, s = –6 and f (n) = n. Another example is $a_n = 3\,a_{n-1}$, which is a homogeneous recurrence relation with $r = 3$ and $s = 0$.

Example 12.4: Solve the recurrence relation $a_{n+1} = 3\,a_n$, where $n \geq 0$ and $a_0 = 5$.

Solution: A few terms of this sequence are as follows

$$\begin{aligned}
a_0 &= 5 \\
a_1 &= 3\,a_0 = 3(5) \\
a_2 &= 3\,a_1 = 3(3\,a_0) = 3^2(5) \\
&\;\;\vdots
\end{aligned}$$

Thus, for $n \geq 0$, $a_n = 3^n$ (5) and is called the general solution of the given recurrence relation. Any term can be obtained using the general solution.

Example 12.5: Consider the following two recurrence relations

(i) $a_n = 5a_{n-1} - 3a_{n-3}$

(ii) $a_n = a_{n-1}\, a_{n-2} + n^2$

The relation in (i) is not of the second order and the relation in (ii) is not linear because of the product $a_{n-1}\, a_{n-2}$.

The homogeneous recurrence relation $a_n = ra_{n-1} + sa_{n-2}$, can be rewritten as follows,

$$a_n - ra_{n-1} - sa_{n-2} = 0$$

The quadratic polynomial $x^2 - rx - s$ is associated with the above and is called the characteristic polynomial of the recurrence relation. Its roots are called the characteristic roots of the recurrence relation. For example, the recurrence relation $a_n = 5a_{n-1} - 6a_{n-2}$ has the characteristic equation $x^2 - 5x + 6$ and the characteristic roots are 2 and 3.

If x_1 and x_2 are the roots of the characteristic polynomial $x^2 - rx - s$, then the solution of the recurrence relation $a_n = ra_{n-1} + sa_{n-2}$, for ($n \geq 2$) is given as follows.

$$a_n = \{c_1 x_1{}^n + c_2 x_2{}^n\} \quad \text{if } x_1 \neq x_2 \text{ (for complex roots as well, i.e., for } x = a + ib)$$

$$(c_1 + c_2 n)\, x^n \quad \text{if } x_1 = x_2 = x$$

Here, c_1 and c_2 are arbitrary constants determined by initial conditions.

Note: The solution obtained in the above process is known as solution due to a homogeneous equation.

Example 12.6: Solve the recurrence relation $a_n = 5\, a_{n-1} - 6\, a_{n-2}$ for ($n \geq 2$), given $a_0 = -3$ and $a_1 = -2$.

Solution: The characteristic polynomial $x^2 - 5x + 6$ has distinct roots $x_1 = 2$ and $x_2 = 3$. The solution is $a_n = c_1\,(2^n) + c_2\,(3^n)$. Since $a_0 = -3$, we can write $c_1\,(2^0) + c_2\,(3^0) = -3$. Similarly, $a_1 = -2$, so we can write $c_1\,(2^1) + c_2\,(3^1) = -2$. We get the following equations in c_1 and c_2.

$$c_1 + c_2 = -3.$$

$$2c_1 + 3c_2 = -2$$

Solving the above equations, we get $c_1 = -7$ and $c_2 = 4$. Therefore the solution is $a_n = -7\,(2^n) + 4\,(3^n)$.

Example 12.7: A bank pays 6% interest on savings, compounding the interest monthly. If Shankar deposits Rs 1000 on the first day of May, how much will this deposit be worth after a year?

Solution: The annual interest is 6%, so the monthly interest rate is 6%/12 = 0.5% = 0.005. For $0 \leq n \leq 12$, let P_n denote the value of Shankar's deposit after n months. We can write the relation as follows.

$$P_{n+1} = P_n + 0.005P_n$$

Here the term $0.005P_n$ is the interest earned on P_n during the month (n+1), for $0 \leq n \leq 11$ and P_0 = Rs 1000. The relation $P_{n+1} = (1 + 0.005)P_n$ with P_0 = Rs 1000 has the following solution:

$$P_n = P_0\,(1.005)^n = 1000(1.005)^n.$$

The deposit of Shankar at the end of one year is Rs 1061.68.

12.3 MORE APPLICATIONS OF RECURRENCE

Many algorithms, particularly divide-and-conquer algorithms, and time complexities are modelled by recurrence relations. The recurrence relations are equations which are defined in their terms. A recurrence relation is also mathematical induction. Matrix multiplication is $O(n^3)$ while Strassen's multiplication algorithm is $O(n^{2.81})$ and has used a divide and conquer strategy. Triangulation is an important step in geometric algorithms. Given a polygon in the plane, add diagonals such that each face is a triangle. None of the diagonals are allowed to cross. Merge-sort is an example of a recurrence relation.

Example 12.8: A particle is moving in the horizontal direction. The distance it travels in each second is equal to 3 times the distance it traveled in the previous second. If a_n denotes the position of the particle in the nth second, determine a_n, given that $a_0 = 3$, $a_3 = 10$.

Solution: Let a_r, a_{r+1}, a_{r+2} be the positions of the particle in the nth, $(n+1)^{th}$, and $(n+2)^{th}$ seconds of its motion.
Then by the problem definition,

$$a_{r+2} - a_{r+1} = 3(a_{r+1} - a_r)$$

On simplification we get

$$a_{r+2} - 4a_{r+1} + 3a_{r+1} + 3a_r = 0$$

The characteristic equation is given by

$$m^2 - 4m + 3 = 0$$
$$(m-3)(m-1) = 0,\ m = 1, 3$$

The solution is given by $a_n = c_1\,(1)^n + c_2\,(3)^n$.
Using the initial conditions to determine the arbitrary constants,

$$a_0 = 3,\quad 3 = c_1 + c_2$$

$$a_3 = 10,\ 10 = c_1 + c_2\,(3)^3 \text{ or } 10 = c_1 + c_2\,27$$

Solving, we get $c_1 = 71/26$, $c_2 = 7/26$.
The solution is given by $a_n = 71/26\,(1)^n + 7/26\,(3)^n$ which gives the distance traveled in the nth second.
Examples of homogeneous recurrence relations:

Example 12.9: Solve the following recurrence relation.

$$a_n - 5a_{n-1} + 6a_{n-2} = 0$$

The characteristic equation is given by

$$x^2 - 5x + 6 = 0$$

$$(x-2)(x-3) = 0$$

$$x = 2, 3$$

The homogeneous solution of the equation is given by

$$a_n = A\,(2)^n + B\,(3)^n.$$

Example 12.10: Solve the recurrence relation

$$a_n - 8a_{n-1} + 16a_{n-2} = 0$$

The characteristic equation is given by

$$x^2 - 8x + 16 = 0$$
$$(x-4)^2 = 0$$
$$x = 4, 4$$

The homogeneous solution of the equation is given by

$$a_n = (A + Bn)\,(4)^n.$$

Example 12.11: Solve the recurrence relation

$$a_n + 3a_{n-1} + 4a_{n-2} = 0$$

The characteristic equation is given by

$$x^2 + 3x + 4 = 0$$

$$x = \frac{-3 \pm \sqrt{9-16}}{2} = \frac{-3}{2} \pm i\frac{\sqrt{7}}{2}$$. The roots are complex conjugate.

The homogeneous solution of the equation is given by

$$a_n = A\left(\frac{-3+i\sqrt{7}}{2}\right)^n + B\left(\frac{-3-i\sqrt{7}}{2}\right)^n$$

12.4 NON-HOMOGENEOUS RECURRENCE RELATIONS

The general form of linear non-homogeneous recurrence relations is given by

$$u_n + c_1u_{n-1} + c_2u_{n-2} + \ldots + c_ku_{n-k} = g(n)\ ;\ n > k \qquad \ldots\text{(I)}$$

Where $c_1, c_2,$ ---- c_k are constants.
The corresponding homogeneous recurrence relation is,

$$u_n = c_1u_{n-1} + c_2u_{n-2} + \ldots + c_k u_{n-k} \qquad \ldots\text{(II)}$$

The general solution to a non-homogeneous linear recurrence relation takes the form of a particular solution added to an arbitrary linear combination of solutions to the associated homogeneous linear recurrence relation. Once the homogeneous equation has been solved, it is relatively easy to find the general solution to the non-homogeneous equation.

12.4.1 Method of Undetermined Co-efficient

This method is useful in finding the particular solution of non-homogeneous recurrence relations, when the right-hand side function $g(n)$ of equation (I) is non-zero, i.e., $g(n) \neq 0$. The method of undetermined coefficient is not a general method. It produces particular solutions to the nth order equations of the form (I) for which c_1, c_2,----- are constants and $g(n)$ is the sum of terms of the form $c^n p(n)$, where c is a constant and $p(n)$ is a polynomial in n. Let us discuss the various cases and finally the general case.

1. To find a particular solution to $a_{n+2} + a_{n+1} + a_n = n$. ...(1)
The right hand side term is a linear equation in the form $a_n = An + B$ indicates the nature of solution which also will be of linear form, where A and B are undetermined coefficients.
Take the particular solution $a_n = An + B$ and substitute in equation (1).

$$a_{n+2} = A(n+2) + B,\ a_{n+1} = A(n+1) + B \text{ and } a_n = An + B$$

$$A(n+2) + B + A(n+1) + B + An + B = n$$

$$3An + 3(A + B) = n$$

Comparing coefficients of n and the constants, on both sides,

$$3A = 1,\ 3(A + B) = 0$$

Which gives $A = 1/3, B = -1/3$.

The particular solution $a_n = \frac{n}{3} - \frac{1}{3} = \frac{(n-1)}{3}$

2. To find a particular solution to $a_{n+3} - a_{n+2} + a_{n+1} - a_n = 2^n n^2$. ...(2)
A solution of the form $a_n = 2^n (An^2 + Bn + C)$ is used.
Now $a_{n+3} = 2^{n+3} (An^2 + Bn + C)$, $a_{n+2} = 2^{n+2} (An^2 + Bn + C)$

$$a_{n+1} = 2^{n+1} (An^2 + Bn + C),\ a_n = 2^n (An^2 + Bn + C)$$

Using all these in equation (2),

$$2^{n+3} (An^2 + Bn + C) + 2^{n+2} (An^2 + Bn + C) + 2^{n+1} (An^2 + Bn + C) + 2^n (An^2 + Bn + C) = 2^n n^2.$$

Simplifying

$$2^n [5An^2 + (36A + 5B)n + (58A + 18B + 5C)] = 2^n n^2.$$

$$[5An^2 + (36A + 5B)n + (58A + 18B + 5C)] = n^2.$$

Comparing the coefficients of n^2, n, and the constant terms on both sides.

$$5A = 1,\ 36A + 5B = 0,\ 58A + 18B + 5C = 0.$$

$$A = 1/5,\ B = -36/25,\ C = 358/125.$$

The particular solution is $a_n = 2^n \left[\frac{1}{5}n^2 - \frac{36}{25}n + \frac{358}{125}\right]$

3. To find a particular solution to $a_n - 5a_{n-1} + a_{n-2} = 1$. ...(3)
The RHS term is a constant.
The particular solution is obtained by $a_n = p$ (a constant).
Substituting this in equation (3), we get
$p - 5p + 6p = 1$, which can be further simplified to
$2p = 1, p = ½$.
The particular solution is given by $a_n = ½$

4. To find the particular solution to the recurrence relation
   $a_{n+2} - 3a_{n+1} + 2a_n = \beta^n$ where β is some constant.
   The particular solution is obtained in the form of $a_n = p\beta^n$
   Using in the given equation,
   $$p[\beta^{n+2} - 3\beta^{n+1} + 2\beta^n] = p\beta^n$$
   $$p[\beta^2 - 3\beta + 2] = 1$$
   Using this in a trial solution
   $$a_n = \frac{\beta^n}{(\beta-1)(\beta-2)} \qquad \beta \neq 1, 2$$
   We have discussed the nature of solutions to be implemented based on RHS terms. On the other hand if the RHS is of a mixed type, the trial solution is obtained in such a way that it represents all types of solutions and without a repetition of constants.

The following rules help to find particular solutions in all possible cases.

1. If $g(n) = (b_0 + b_1 n + \ldots + b_m n^m)k^n$, k is not a root of the characteristic equation, the particular solution obtained is of the form $a_n = (d_0 + d_1 n + \ldots + d_m n^m)k^n$.
2. If $g(n) = (b_0 + b_1 n + \ldots + b_m n^m)1^n$ and 1 is not a root of the homogeneous recurrence relation, the particular solution is obtained in the form $a_n = (d_0 + d_1 n + \ldots + d_m n^m)$
3. If $g(n) = k^n$ and k – is not a root of a characteristic equation, the particular solution is obtained in the form $a_n = d_0 k^n$
4. If $g(n) = (b_0 + b_1 n + \ldots + b_m n^m)k^n$ where k is a root of a characteristic equation repeated l times, the particular solution is obtained in the form $a_n = n^l(d_0 + d_1 n + \ldots + d_m n^m)k^n$
5. If $g(n) = (b_0 + b_1 n + \ldots + b_m n^m)1^n$ and 1 is a root of a characteristic equation repeated l times, the particular solution is obtained in the form $a_n = n^l\ (d_0 + d_1 n + \ldots + d_m n^m)$
6. If $g(n) = k^n$, where k is a root of a characteristic equation repeated l times, the particular solution is obtained in the form $a_n = d_0 n^l k^n$.

Example 12.12: Find the general solution for

$$a_n + 3a_{n-1} - 10\, a_{n-2} = n^2 + n + 1. \qquad \ldots\text{(I)}$$

The characteristic equation is

$$r^2 + 3r - 10 = 0$$
$$r(r+5) - 2\,(r+5) = 0$$
$$r = 2, -5.$$

The solution for the homogeneous equation, $a_{nh} = c_1\,(2)^n + c_2\,(-5)^n$.
Let the particular solution be $a_{np} = A + Bn + Cn^2$.
Substituting the particular solution in the recurrence relation (I)
$A + Bn + C\,n^2 + 3\,[A + B(n-1) + C(n-1)^2] - 10\,[A + B\,(n-2) + C\,(n-2)^2] = n^2 + n + 1.$

Rearranging the terms,

$-6A + 17B - 37C + (-6B + 34C)n - 6Cn^2 = 1 + n + n^2$

Equating the coefficients of powers of n, we get

$-6A + 17B - 37C = 1$; coefficient of n^0

$-6B + 34C = 1$; coefficient of n^1

$-6C = 1$; coefficient of n^2

Solving we get $C = \frac{-1}{6}, B = \frac{-10}{9}, A = \frac{-247}{108}$

The particular solution is given by $a_{np} = \frac{-247}{108} + B\frac{-10}{9}n + \frac{-1}{6}n^2$

Hence the general solution is given by

$$a_n = a_{nh} + a_{np} = A + Bn + Cn^2 - \frac{247}{108} + B\frac{-10}{9}n + \frac{-1}{6}n^2.$$

Example 12.13: Find the complete solution of the recurrence relation

$$a_{n+2} - 4a_{n+1} - 4a_n = 3n + 2^n \quad \text{...(I)}$$

The solution for the homogeneous equation is given by

$$r^2 - 4r + 4 = 0 \quad (r-2)^2 = 0,\ r = 2, 2$$

The solution for the homogeneous equation $a_{nh} = (c_1 + c_2 n)(2)^n$

To find particular solution, the RHS of (I) is of mixed type. The trial solution is taken as $a_{np} = A + Bn + Cn^2\, 2^n$.This is because 2^n is the solution recurrence relation (obtained from the homogeneous solution) and $n\,2^n$ is also solution. Hence the trial solution is as given above.

Using the particular solution in equation (I)

$$[A + B(n+2) + C(n+2)^2 2^{n+2}] - 4[A + B(n+1) + C(n+1)^2 2^{n+1}] + 4[A + Bn + Cn^2 2^n] = 3n + 2^n$$

Simplifying, we get

$$(A - 2B) + Bn + 8C\, 2^n = 3n + 2^n$$

Comparing the coefficients on both sides, we get

$$A = 6, B = 3, C = 1/8.$$

Hence the particular solution is given by

$$a_{np} = 6 + 3n + 1/8\, n^2\, 2^n$$

The general solution is given by $= a_n = a_{nh} + a_{np}$

$$= (c_1 + c_2 n)(2)^n + 6 + 3n + 1/8\, n^2\, 2^n.$$

Example 12.14: The Tower of Hanoi is a neat little puzzle invented by the French mathematician Édouard Lucas in 1883. We are given a tower of *r* disks, initially stacked in decreasing size on one of three pegs. The objective is to transfer an entire tower to one of the other pegs, moving only one disk at a time and never moving a larger one onto a smaller one. What is the minimum number of moves required?

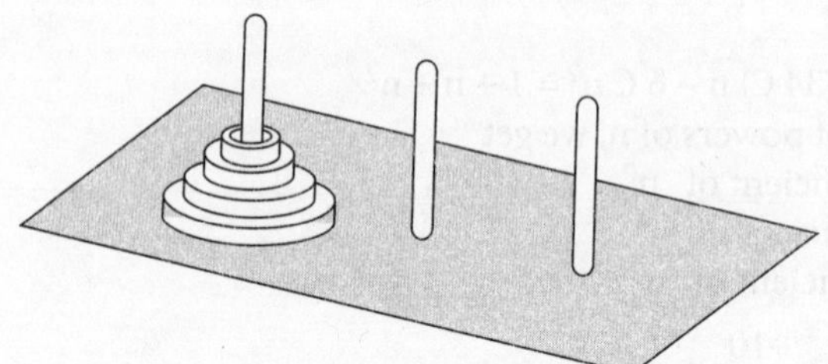

Figure 12.1 Tower of Hanoi Problem

Solution: Let T_n be the minimum number of moves that will transfer n disks from one peg to another under Lucas's rules. It is advantageous to look at small cases first. It is easy to see that $T_1 = 1$ and $T_2 = 3$. Here is one of the methods to transfer n ($n \geq 2$) disks: We first transfer the $(n-1)$ smallest to a different peg (requiring at least T_{n-1} moves), then move the largest (requiring one move), and finally transfer the $(n-1)$ smallest one back onto the largest (requiring at least another T_{n-1} moves). It follows that $T_n \leq 2T_{n-1} + 1$. We now show that there can be no shorter way. No matter what method we use to transfer the disks, at some point we must move the largest disk. When we do, then the other $(n-1)$ smaller disks must be on a single peg, and it must have taken at least T_{n-1} moves to put them there. After moving the largest disk, we must transfer the $(n-1)$ smaller disks (which must again be on a single peg) back onto the largest; this too requires at least T_{n-1} moves. Hence $T_n \geq 2T_{n-1} + 1$. Thus, we have the recurrence relation $T_n = 2T_{n-1} + 1$ with initial condition $T_1 = 1$. We shall later show that $T_n = 2n - 1$ for $n = 1, 2, ...$

ADDITIONAL SOLVED EXAMPLES

Example W12.1: Solve the following recurrence relation.

$$a_n - 7a_{n-2} + 6\,a_{n-3} = 0$$

with initial conditions $a_0 = 8$, $a_1 = 6$, and $a_2 = 22$. The characteristic equation of the given recurrence relation is as given under.

$$m^3 - 7m + 6 = 0$$

Solution:

By inspection, $m = 1$ is a root. Using synthetic division method, we can write as

$$\begin{array}{r|rrrr} m-1 & 1 & 0 & -7 & 6 \\ & & -1 & 1 & -6 \\ \hline & 1 & 1 & -6 & 0 \end{array}$$

$\Rightarrow m^2 + m - 6 = 0$. $m = 3$ and -2. Therefore, the roots are $m = 1, -2$, and 3.

The general homogeneous solution to the recurrence relation is given as under.

$$a_n = A_1(1)^n - A_2(2)^n + A_3(3)^n$$

where A_1, A_2 and A_3 are arbitrary constants. To find these arbitrary constants, we have to use the given conditions and arrive at the system of three linear equations with three unknowns.

Putting $n = 0, 1, 2$ in the above homogeneous solution, we can write as under.

$a_0 = A_1 + A_2 + A_3$ or $A_1 + A_2 + A_3 = 8$

$a_1 = A_1 + 2A_2 - 3A_3$ or $A_1 + 2A_2 - 3A_3 = 6$

$a_2 = A_1 + 4A_2 + 9A_3$ or $A_1 + 4A_2 + 9A_3 = 22$

Solving above system using Gauss elimination technique

$$\begin{bmatrix} 1 & 1 & 1 \\ 1 & 2 & -3 \\ 1 & 4 & 9 \end{bmatrix} \begin{bmatrix} A_1 \\ A_2 \\ A_3 \end{bmatrix} = \begin{bmatrix} 8 \\ 6 \\ 22 \end{bmatrix}$$

The augmented matrix is given by

$$\approx \begin{bmatrix} 1 & 1 & 1 & 8 \\ 1 & 2 & -3 & 6 \\ 1 & 4 & 9 & 22 \end{bmatrix}$$

Using row operations, $R_2^1 = R_2 - R_1$: $R_3^1 = R_3 - R_1$

$$\approx \begin{bmatrix} 1 & 1 & 1 & 8 \\ 0 & 1 & -4 & -2 \\ 0 & 3 & 8 & 14 \end{bmatrix}$$

$R_3^1 = R_3 - 3R_2$ gives

$$\approx \begin{bmatrix} 1 & 1 & 1 & 8 \\ 0 & 1 & -4 & -2 \\ 0 & 0 & 20 & 20 \end{bmatrix}$$

We write the matrix in the form of simultaneous equations, as under

$$A_1 + A_2 + A_3 = 8$$

$$4A_2 + A_3 = -2$$

$$20A_3 = 20$$

Solving again using back substitution, we get the values $A_3 = 1, A_2 = 2, A_1 = 5$.
$A_3 = 1, A_2 = 2, A_1 = 5$.
Therefore, the solution is given by

$$a_n = 5 + 2(2)^n + (-3)^n$$

or

$$a_n = 5 + 2^{n+1} + (-3)^n.$$

Example W12.2: Elaborate the method of generating functions to solve non-homogeneous recurrence relations.

Solution:

The method is based on the concept that the recurrence relation is transformed from the sequence into an algebraic expression.

1. The sequence of repeated constant numbers such as 1, 1, 1, ___ has the generating function.

$$S(1,z)=1+z+z_2^2+----+--=\sum_{n=0}^{\infty} z^n=\frac{1}{z}=(1-z)^{-1}$$

2. The sequence a^n ($a \neq 0$ and is constant) has the generating function.

$$S(a,z)=1+az+a^2z^2+----+--=\sum_{n=0}^{\infty} a^n z_n^n=\frac{1}{1-az}=(1-az)^{-1}$$

3. The sequence of binomial coefficients ${}^nC_0, {}^nC_1, {}^nC_2,$ ___ ${}^nC_n$ has the generating function

$$S(a_n,z)={}^nC_0+{}^nC_1 z+{}^nC_2 z^2+--+{}^nC_n z^n+—=\sum_{k=0}^{\infty} {}^nC_k z^k=(1+z)^n.$$

4. The sequence $a_n=7.3^n, n \geq 0$ has the generating function $S(a,z)=\frac{7}{1-3z}$ which is the sum to infinity terms of a GP with $A=7$ (the first term) and $r=3z$ (the common ratio).
5. The sequence $a_n=5(-1)^n$ has the generating function $S(a,z)=\frac{5}{1+z}$.

The following are the sequences and their corresponding Generating Functions:

Sl no.	General Sequence	Corresponding Generating function
1	$a_k=1$	$S(a,z)=(1-z)^{-1}$
2	$a_k=k+1$	$S(a,z)=(1-z)^{-2}$
3	$a_k=k$	$S(a,z)=\frac{z}{(1-z)^2}$
4	$a_k=ba^k$ (a, b constants)	$S(a,z)=\frac{b}{1-az}$
5	$a_k=k(k+1)$	$S(a,z)=\frac{2z}{(1-z)^3}$
6	$a_k=A^k$ (A - constant)	$S(a,z)=(1-Az)^{-1}$
7	$a_k={}^nC_k B^{n-k} A^k$ $(k=0,1,2,\ldots,n), k>n$	$S(a,z)=(B+Az)^n$
8	$a_k=b\,k\,a^k$ (a, b constants)	$S(a,z)=\frac{abz}{(1-az)^2}$
9	$a_k=\frac{1}{k!}$	$S(a,z)=e^z$
10	$a_k=a_{k+n}$	$\frac{S(a,z)-a_0-a_1z\ldots a_{n-1}z^{n-1}}{z^n}$

Following are some of the observations:

(i) If the generating function for the sequence whose general term a_n is known, then the generating function for the sequence a_{n+1} and a_{n+2} can be obtained.

Consider

$$S(z) = a_0 + a_1 z + a_2 z^2 + \ldots a_r z^r + \ldots$$

$$S(z) - a_0 = a_1 z + a_2 z^2 + --- + a_r z^r + --$$

$$\frac{S(z) - a_0}{z} = a_1 + a_2 z^2 + \ldots + a_r z^{r-1} + a_{r+1} z^r + \ldots$$

Which is the generating function whose general term is a_{n+1}.

Similarly,

$$S(z) - a_0 - a_1 z = a_2 z^2 + \ldots + a_r z_r^r + \ldots$$

$$\frac{S(z) - a_0 - a_1 z}{z^2} = a_2 + a_3 z + \ldots + a_r z^{r-2} + a_{r+1} z^{r-1} + \ldots$$

Which is the generating function whose general term is a_{n+2}

(ii) $S(z) = \sum_{k=0}^{\infty} a_k z^k$ generates the sequence $(0, a_0, a_1, \ldots)$

$z. S(z)$ generates the sequence $(0, a_0, a_1, a_2, \ldots)$

$z^2 S(z)$ generates the sequence $(0, 0, a_0, a_1, a_2, \ldots)$

In general, $z^n S(z)$ generates the sequence $(0, 0, ----0, a_0, a_1, a_2, \ldots)$ Which contains n-zeroes before a_0.

For example,

(a) $\frac{1}{1-x} = \sum_{k=0}^{\infty} x^k$ generates 1,1,1, … i.e. $a_n = 1, n \geq 0$.

(b) $\frac{x}{1-x} = \sum_{k=0}^{\infty} x^{k+1} = \sum_{k=1}^{\infty} x^k$ generates the sequence 0,1,1,1, …

(c) $\frac{x^2}{1-x} = \sum_{k=0}^{\infty} x^{k+2} = \sum_{k=2}^{\infty} x^k$ generates the sequence 0,0,1,1, …

(iii) The division of the sequence by z, shifts the sequence to left, and it generates the sequence $(a_1, a_2, a_3, \ldots)$.

$$\frac{S(z) - a_0}{z} = \sum_{n=1}^{\infty} a_n z^{n-1}.$$

Similarly,

$\frac{S(z) - a_0 - a_1 z}{z^2} = \sum_{n=2}^{\infty} a_n z^{n-2}$ generates the sequence $(a_2, a_3, \ldots)$.

In general

$\frac{S(z) - a_0 - a_1 z - a_2 z^2 \ldots a_z z^k}{z^k} = \sum_{n=k}^{\infty} a_n z^{n-k}$ generates the sequence $(a_k, a_{k-1}, \ldots)$.

Example W12.3: Solve the following recurrence relation by the method of generating functions.

$$a_r - 7a_{r-1} + 10a_{r-2} = 3^r, \ r \geq 2.$$

Given the conditions $a_0 = 0, \ a_1 = 1.$

Solution:

The given equation is $a_r - 7a_{r-1} + 10a_{r-2} = 3^r, \ r \geq 2.$ ...(i)

Multiply both sides by z^r and sum each term from $r = 2$ to ∞, we get,

$$\sum_{r=2}^{\infty} a_r z_r - 7\sum_{r=2}^{\infty} a_{r-1} z^r + 10\sum_{r=2}^{\infty} a_{r-2} z^r = \sum_{r=2}^{\infty} 3^r z^r, \ r \geq 2.$$

$$= \frac{3^2 z^2}{1-3z}$$

(RHS is a GP with $a = 3^2 z^2$ and $r = 3z$ and sum to infinite terms of a GP $= \frac{a}{1-r}$)

Now solve the first summation for $r = 0, 1$.

Let $\sum_{r=0}^{\infty} a_r z_r = A(z)$

$$[A(z) - a_0 - a_1 z] - 7z[A(z) - a_0] + 10z^2 A(z) = \frac{3^2 z^2}{1-3z}$$

$$A(z)\left[1 - 7z + 10z^2\right] = a_0 + a_1 z - 7a_0 z + \frac{9z^2}{1-3z}$$

$$(1-5z)(1-2z)A(z) = z + \frac{9z^2}{1-3z}, \text{ where } a_0 = 0, a_1 = 1.$$

$$(1-5z)(1-2z)A(z) = \frac{(1-3z)^z + 9z^2}{1-3z} = \frac{z + 6z^2}{1-3z}$$

$$A(z) = \frac{z + 6z^2}{(1-3z)(1-5z)(1-2z)} \text{ using partial fractions.} \quad \text{...(ii)}$$

$$\frac{z + 6z^2}{(1-3z)(1-5z)(1-2z)} = \frac{A}{1-3z} + \frac{B}{1-5z} + \frac{C}{1-2z}. \ A = -9/2, \ B = 11/6, \ C = 8/3.$$

$$\frac{z + 6z^2}{(1-3z)(1-5z)(1-2z)} = \frac{-9}{2(1-3z)} + \frac{11}{6(1-5z)} + \frac{8}{3(1-2z)}$$

Now Equation (ii) becomes.

$$A(z) = -\frac{9}{2}(1-3z)^{-1} + \frac{11}{6}(1-5z)^{-1} + \frac{8}{3}(1-2z)^{-1}$$

The general solution is obtained by finding the coefficient of z^r in $A(z)$.

$$a_r = -\frac{9}{2} \text{ coeff of } z^r \text{ in } (1-3z)^{-1} + \frac{11}{6} \text{coeff of } z_r \text{ in } (1-5z)^{-1} + \frac{8}{3} \text{coeff of } z_r \text{ in} (1-2z)^{-1}$$

$$\left[\text{coeff of } z^r \text{ in } (1-az)^{-1} = n - 1 + {}^r C_r a_r\right]$$

$$a_r = -\frac{9}{2}\left[\left({}^{1-1+r}C_r\right)(3)^r\right] + \frac{11}{6}\left[\left({}^{1-1+r}C_r\right)(5)^r_r\right] + \frac{8}{3}\left[\left({}^{1-1+r}C_r\right)(2)^r\right]$$

Example W12.4: Solve the following recurrence relation

$$a_{r+2} - 3a_{r+1} + 2a_r = 0$$

by the method of generating functions with $a_0 = 2, a_1 = 3$.

Solution:

Let $\sum_{r=0}^{\infty} a_r z^r = A(z)$, be the generating function for the given recurrence relation.

Multiply each term of recurrence relation with z^n and finding the sun from $r=0$ to ∞.

$$\sum_{r=0}^{\infty} a_{r+2}z^r - 3\sum_{r=0}^{\infty} a_{r+1}z^r + 2\sum_{r=0}^{\infty} a_r z^r = 0,$$

Solving the summation,

$$(a_2 + a_3 z + a_4 z^2 + \ldots) - 3(a_1 + a_2 z + a_3 z^2 + \ldots) + 2(a_0 + a_1 z + a_2 z^2 + \ldots) = 0$$

$$\frac{A(z) - a_0 - a_1 z}{z^2} - 3\left[\frac{A(z) - a_0}{z}\right] + 2\left[a_0 + a_1 z + a_2 z^2 + \ldots\right] = 0$$

putting $a_0 = 2,\ a_1 = 3$ and solving the summation for A(z), we get

$$\frac{A(z) - 2 - 3z}{z^2} - 3\left[\frac{A(z) - 2}{z}\right] + 2A(z) = 0$$

$$A(z)\left[\frac{1}{z^2} - \frac{3}{z} + 2\right] - \frac{2}{z^2} - \frac{3}{z} + \frac{6}{z} = 0$$

$$A(z) = \frac{2 - 3z}{2z^2 - 3z + 1} = \frac{2 - 3z}{(1-z)(1-2z)}$$

Using partial fractions.

$$A(z) = \frac{2 - 3z}{2z^2 - 3z + 1} = \frac{2 - 3z}{(1-z)(1-2z)} = \frac{A}{1-z} + \frac{B}{1-2z}$$

Solving, we get A = 1, B = 1.

$$A(z) = \frac{1}{1-z} + \frac{1}{1-2z} = \sum_{r=0}^{\infty} z^r + \sum_{r=0}^{\infty} (2z)^r$$

$$A(z) = \sum_{r=0}^{\infty} (1 + 2^r) z^r.$$

The solution $a_r = 1 + 2^r$.

Example W12.5: Find the generating functions for the following sequences

(i) 0, 1, −2, 4, −8 and (ii) 2, −2, 2, −2.

Solution:

(i) Now $A(z) = 0 + z - 2z_2^2 + 4z^3 - 8z^4 + \ldots$

$$= z - 2z^2 + 4z_3^3 - 8z^4 + \ldots$$

$$= z\left[1-(2z)+(2z)^2-(2z)^3+\ldots\right]$$

This is a GP with $a = 1$ and $r = 2z$.The sum infinity is given by

$$A(z) = \frac{z}{1-2z} \quad |2z| < 1$$

(ii) Now, $A(z) = 2 - 2z + 2z^2 - 2z^3 + \ldots$

$$= 2(1 - z + z^2 - z^3 + \ldots)$$

$$A(z) = \frac{z}{1+z} \quad |-z| < 1$$

SUMMARY

Recurrence relations are the extensions of recursive functions and are useful mathematical concepts in computer science in the analysis of algorithms. Divide and conquer strategies are analysed based on recurrence relations. Some of the best-known difference equations have their origins in the attempt to model population dynamics. The Fibonacci numbers that were once used as a model for the growth of a rabbit population is a case in point. In digital signal processing, recurrence relations are used to model closed-loop feedback systems, where outputs at one time become inputs for a future time. This principle is applied in infinite impulse response (IIR) digital filters. Recurrence relations, especially linear recurrence relations, are used extensively in both theoretical and empirical economics. Recurrence relations are also used in testing economic theories from the evidences because these functions have analytic continuation property.

EXERCISES

1. Find the general solutions of the following non-homogeneous recurrence relations
   (a) $a_n + 5a_{n-1} + 6a_{n-2} = 3n^2 - 2n + 1$
   (b) $a_{n+2} - 4a_{n+1} + 4a_n = n^2 2^n$
   (c) $a_n - 5a_{n-1} + 6a_{n-2} = 2$ with conditions $a_0 = 1$, $a_1 = 1$ and $n \geq 2$
   (d) $a_n - 7a_{n-1} + 10a_{n-2} = n4^n$
2. Solve the following homogeneous recurrence relations.
   (a) $a_n - 7a_{n-1} + 10a_{n-2} = 0$ Satisfying $a_0 = 0$, $a_1 = 6$.
   (b) $a_{n+2} - 3a_{n+1} + 2a_n = 0$.
   (c) $a_{n+2} - 2a_{n+1} + 4a_n = 0$
   (d) $a_n - 8a_{n-1} + 16a_{n-2} = 0$ with $a_2 = 16$, $a_3 = 8$.
   (d) $a_n - 5a_{n-1} + 6a_{n-2} = 0$ with $a_0 = a_1 = 3$

13 Introduction to Graph Theory

In Chapter 12, we discussed recurrence relations. A pictorial representation of the Fibonacci series is considered a graph representation. A graph is a discrete mathematical structure. In this chapter, we introduce the basics of graph theory, which is a full-fledged sub discipline in mathematics that finds innumerable applications in computer science, physics, chemistry, sociology, computational linguistics and many other domains.

13.1 HISTORY OF GRAPH THEORY

Many concrete and practical problems can be simplified and solved with graphs. A multitude of puzzles are solved using graphs. Algorithms that use graph representation are called Graph algorithms. One classical problem in literature is the Konigsberg Bridge Problem. Figure 13.1 shows a diagrammatic representation of the Konigsberg bridges. There are two banks and two islands formed by the flow of the Pregel River in Konigsberg, the capital of East Prussia. These land regions are connected by seven bridges, as shown in Fig. 13.1. The land masses are labeled A, B, C and D. People wondered if it were possible to start from one of the land masses, walk over each of the seven bridges exactly once, and return to the starting point without getting wet.

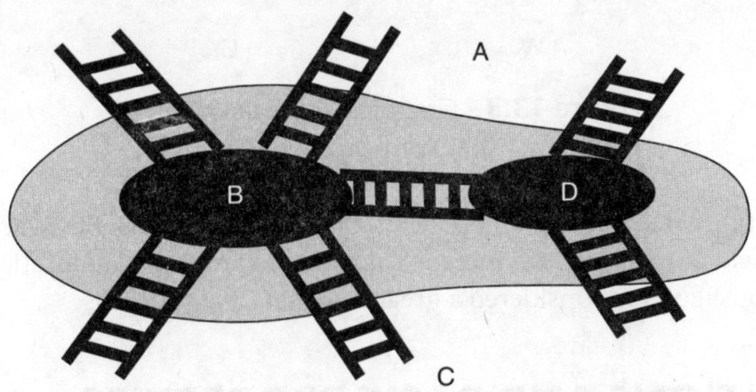

Figure 13.1 Bridges of Konigsberg

Such real world problems are considered puzzles and are often tackled with mathematics. The Konigsberg problem was first tackled by the great Swiss mathematician Leonhard Euler (1707–1783) and solved with some imagination in 1736. It is because of his imaginative pictorial solution to the Konigsberg problem that Euler is considered to be the father of modern-day graph theory. He represented the Konigsberg problem as a graph. The land masses are shown as vertices and the bridges are shown as edges. The equivalent graph of the problem is given in Fig. 13.2. The real world problem of Konigsberg is restated as having to take a pen or pencil, start from any vertex, and go through all the edges without lifting the pen to reach the starting vertex. If it is done then there is a solution, otherwise not. It has been seen that the Kongsberg has no solution.

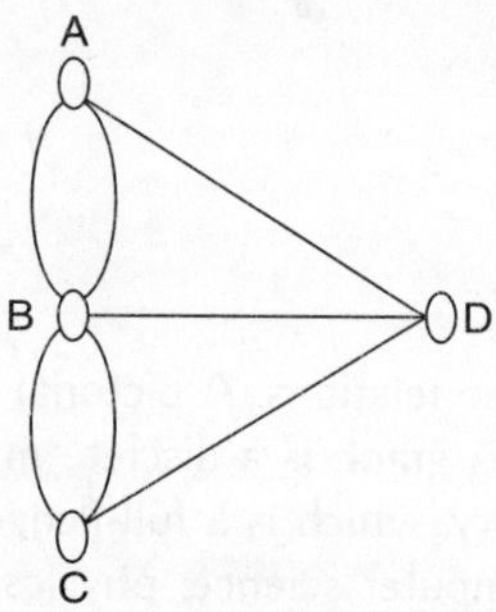

Figure 13.2 Graphical representation of the Konigsberg problem

Another classical problem in computer science is known as the *utility problem*. In this problem, three utilities, namely gas, electricity and water, need to be connected to three houses without crossing one another. The graphical representation of this problem is given in Fig. 13.3. The utilities are shown as vertices and the connections as edges on a plane. A solution has not been found for this problem.

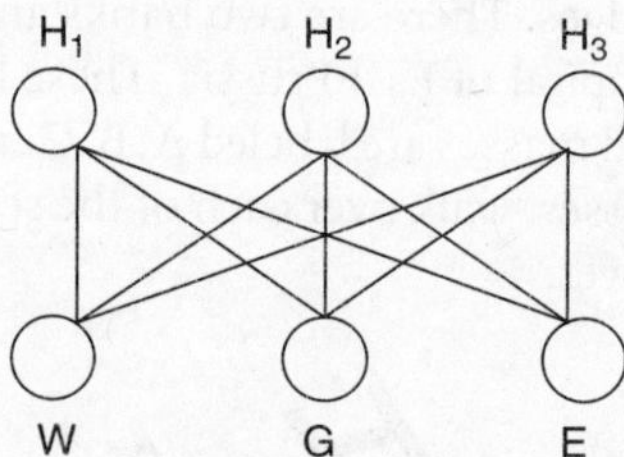

Figure 13.3 Graphical representation of the utility problem

Euler started his scientific career soon after the death of the great scientist Sir Isaac Newton. Some of his contributions to the field of mathematics include calculus of variations, quadratic equations, Euclidean space and conic sections. Euler is considered a great designer of algorithms.

13.2 DEFINITIONS AND BASIC PROPERTIES

A graph is a tuple G = (V, E) where V is a finite set, called the set of vertices of G and E is a finite set, called the set of edges of G. The edge set is also expressed as $E \subseteq V \times V$. The graphs are finite when the vertex and edge sets are finite. Otherwise, the graphs are called infinite. Graphs are also classified as simple graphs and multigraphs. A simple graph is one that has no self loops or parallel edges. A graph with parallel edges is called a multigraph. The maximum number of parallel edges represents the multiplicity of a graph. Consider the graph in Fig. 13.4. The multiplicity of the graph is three, as there are three parallel edges between the vertices B and C. The edge (C, E) is called a pendant edge and the vertex E is called a pendant vertex. At the vertex A there is self loop. The vertices which are directly connected by edges are called adjacent vertices. The vertices B and D are adjacent to A whereas the vertices C and E are not adjacent. Similarly, the vertices C, D and A are adjacent to vertex B. The vertex B is not adjacent to E.

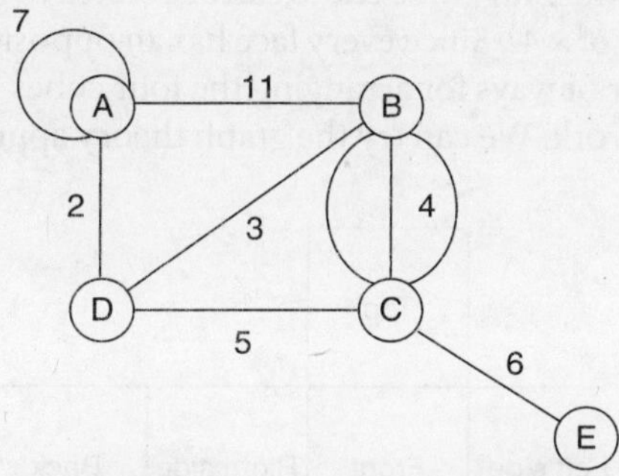

Figure 13.4 Some example graph

13.2.1 Sub-graph and Spanning Sub-graph

Consider the two graphs G = (V, E) and G_1= (V_1, E_1) in Fig. 13.5(a) and (b). The graph G_1= (V_1, E_1) is said to be a subgraph of G = (V, E), when $V_1 \subseteq V$ and $E_1 \subseteq E$ but $V_1 \neq Ø$. The sets V = {A, B, C, D} and E = {1, 2, 3, 4, 5, 6}. The sets V_1 = {A, B, C} and E_1 = {1, 4, 6}.

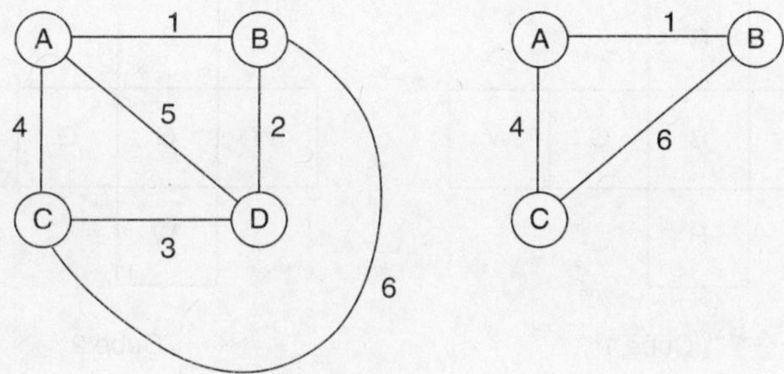

Figure 13.5 Example graph (a) G = (V, E) (b) G_1= (V_1, E_1)

The graph G_1= (V_1, E_1) is called a spanning subgraph of G when G_1 is a subgraph of G and V_1 = V. The graph G_1 shown in Fig. 13.6 is the spanning subgraph.

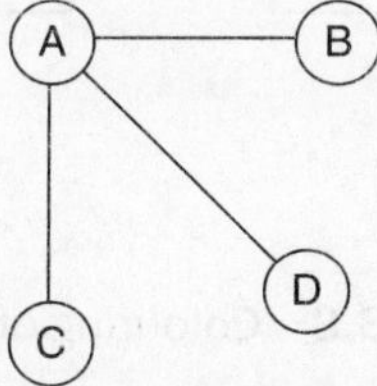

Figure 13.6 Spanning Sub-graph G_1 = (V_1, E_1) of G = (V, E) given in Fig.13.5 (a)

13.2.2 Game of Instant Insanity

There has, for a long time, been a popular game called Instant Insanity. It is also known as the Rubik's Cube game. The unfolding of a cube is shown in Fig. 13.7. The game consists of four cubes, each of whose six faces is in one of the four colours red, blue, green and white, as shown in Fig. 13.8. The game is to stack the cubes one over the other such that each of the four colours appears on each side of the resulting column. Each cube can be placed in six ways. After placing the cubes one over the others, leaving the first one as it is, the remaining three cubes can be rotated to get colours in the four directions. The number of ways to do this is $6^4 \times 4^3$. Since every face has an opposite face, this number is divided by two. Therefore, the total number of ways for arranging the four cubes is $(½) \times 6^4 \times 4^3$, which is equal to 41,472. Trial and error will not work. We can try the graph theory approach.

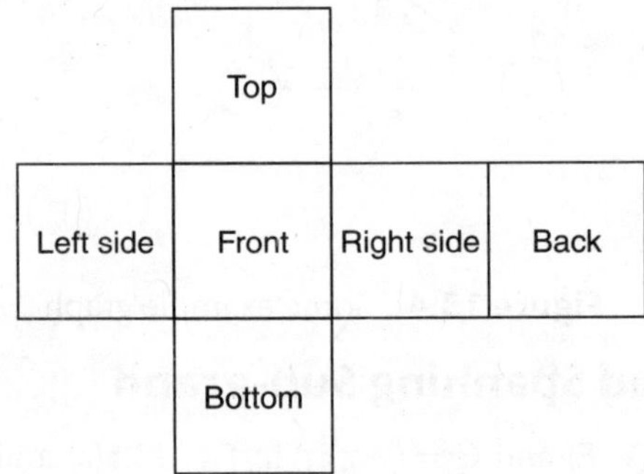

Figure 13.7 Unfolding of the cube and its sides

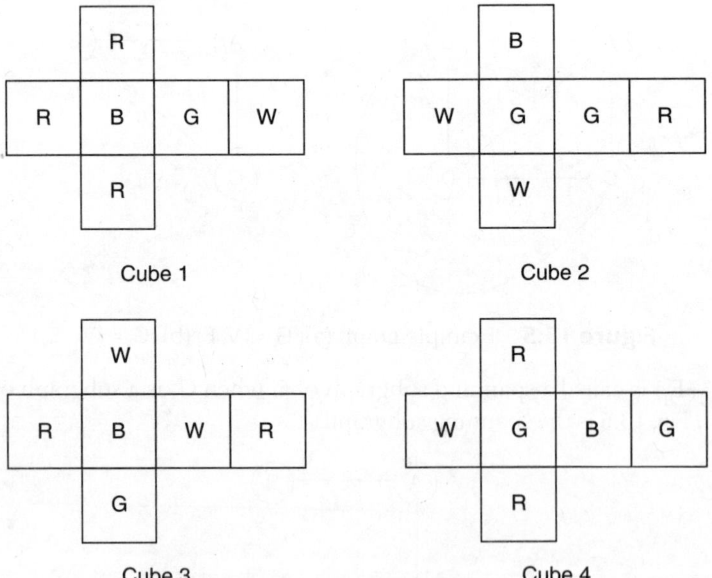

Figure 13.8 Colouring of four cubes

The four colours, R, G, B and W are represented as four vertices. The edges indicate the colours on opposite sides of the cube. In the case of cube 1, the three edges indicate three pairs of opposite colours, RR, RG and BW, as shown in Fig. 13.9.

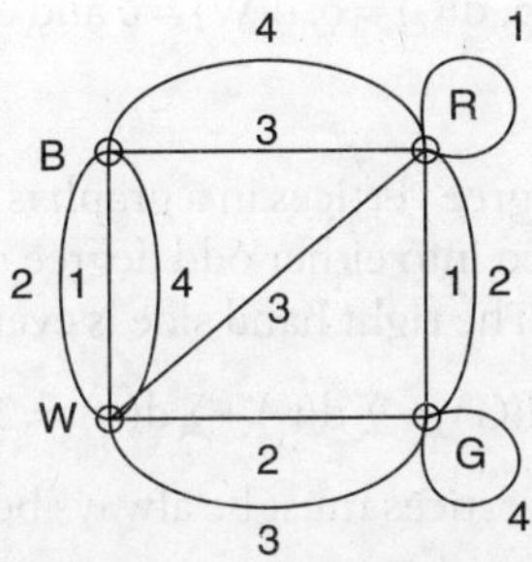

Figure 13.9 Graph representation of game of instant insanity

In the graph shown in Fig. 13.9, we try to find two edge-disjoint sub-graphs, as shown in Fig. 13.10, which represent the front-back and left-right sides of the stack. The solution to this game is given in Fig. 13.11.

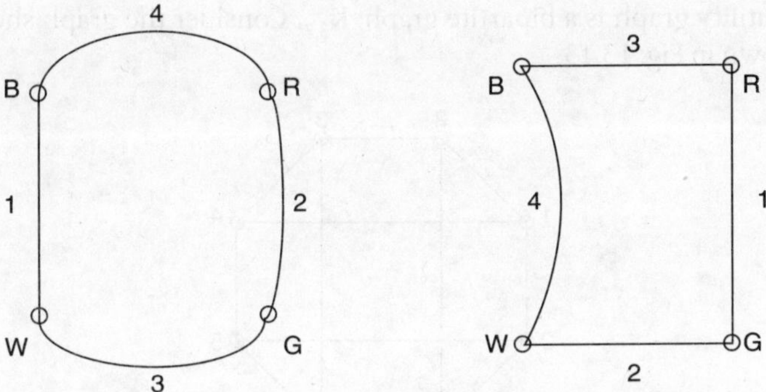

Figure 13.10 Two edge-disjoint sub-graphs of graph shown in Fig. 13.9

	Front	Back	Right	Left
Cube 4	B	R	W	B
Cube 3	G	W	B	R
Cube 2	R	G	G	W
Cube 1	W	B	R	G

Figure 13.11 Solution to the game of instant insanity

Proposition 1: The total degree of a given graph G = (V, E) is always equal to twice the number of edges.

$$\sum d(v_i) = 2|E|$$

Since each edge in a graph is incident on two end vertices, the total degree is equal to twice the number of edges. The self loop contributes two to the degree of the graph. Consider the graph given in Fig. 13.8, where the total degree is 24.

$$d(B) = 5, d(G) = 6, d(W) = 6 \text{ and } d(R) = 7$$

The number of edges is 24/2 = 12.
Proposition 2: The number of odd degree vertices in a graph is always even.
The vertices in a given graph are divided into either odd degree or even degree vertices. The addition of vertices of even degree is always even. The right hand side is even.

$$\sum d(G) = \sum d(v_i) + \sum d(v_j) = 2|E|$$

Therefore, the number of odd degree vertices must be always be even, to make this partial sum even.

13.2.3 Bipartite Graph

A graph G = (V, E) is called a bipartite graph if the vertex set V is divided into two disjoint subsets V_1 and V_2 called partition sets, such that every edge has one end vertex in V_1 and the other vertex in V_2. The complete bipartite graph is a bipartite graph in which every vertex in V_1 is connected to every vertex inV_2. The complete bipartite graph on the partition sets of *m* vertices and *n* vertices is denoted as $K_{m,n}$. The example of the utility graph is a bipartite graph, $K_{3,3}$. Consider the graph shown in Fig. 13.12. It's bipartite graph is shown in Fig. 13.13.

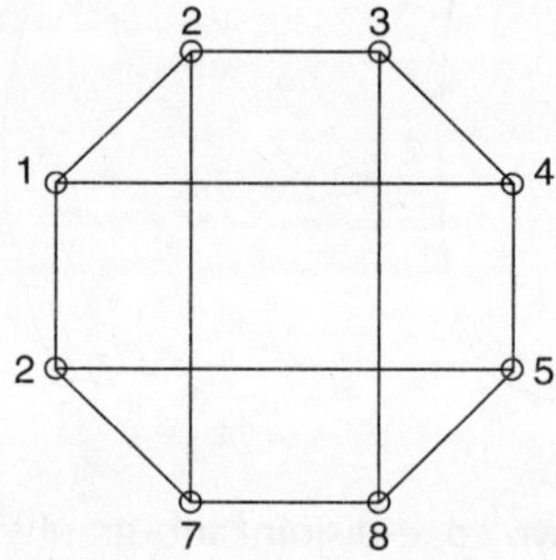

Figure 13.12 Sample graph

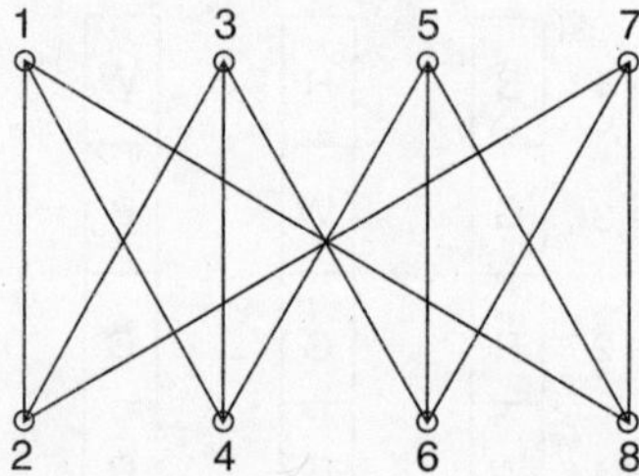

Figure 13.13 Bipartite graph of graph in Fig.13.12

Example 13. 1: How many edges does the graph $K_{3,6}$ contain?

Solution:
The complete bipartite graph has three vertices of degree six and six vertices of degree three.
Total degree of the graph = 6×3 + 3×6 = 36 = 2 |E|
Number of edges is = 18

13.3 GRAPH ISOMORPHISM

The two graphs $G_1 = (V_1, E_1)$ and $G_2 = (V_2, E_2)$ are called isomorphic graphs when there exists a bijective (one-to-one) function (f) between V_1 and V_2 such that the edge relationship is maintained. f: $V_1 \rightarrow V_2$. The two graphs are essentially the same graphs but labelled or drawn differently.

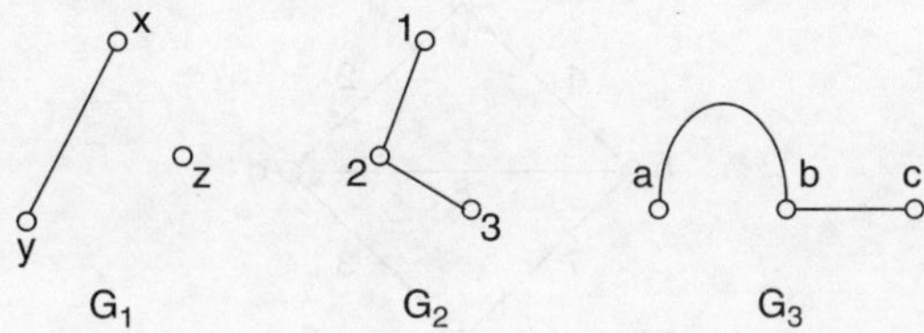

Figure 13.14 Examples of graphs

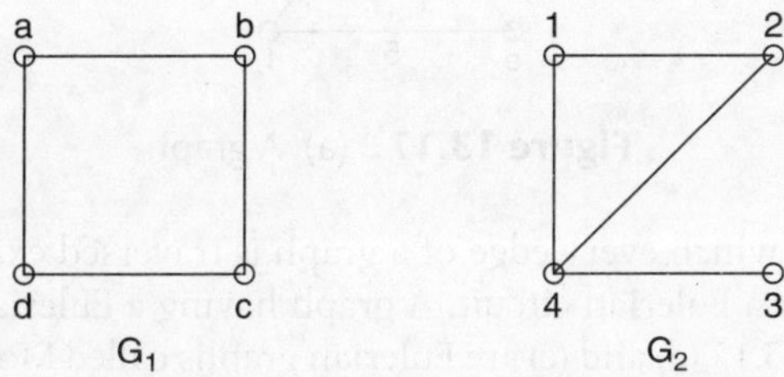

Figure 13.15 Non-isomorphic graphs

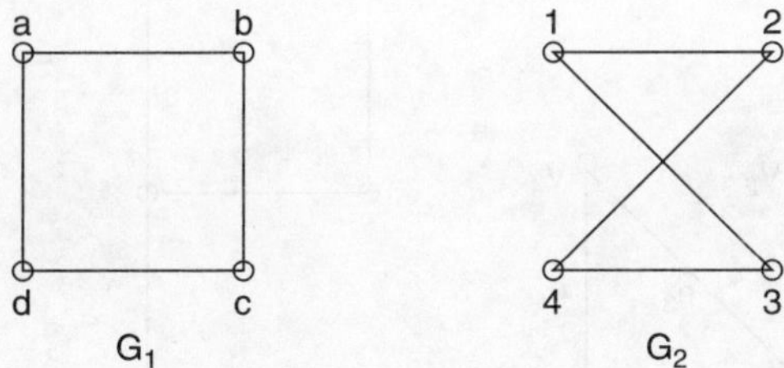

Figure 13.16 Isomorphic graphs

The graphs G_2 and G_3 given in Fig. 13.14 are isomorphic. The mapping is f(1) = a, f(2) = b and f(3) = c. G_1 is not isomorphic to G_2 and G_3. In Fig. 13.15, neither of the graphs are isomorphic. However, the graphs shown in Fig. 13.16 are isomorphic. The mappings are f(1) = a, f(2) = b f(3) = d and f(4) = c.

13.4 WALK, PATHS, CYCLES AND CIRCUITS

Consider the graph given in Fig. 13.17. An alternating sequence of vertices and edges, starting with a vertex and ending with a vertex is called a walk. The sequence **a2d3c4f** is a walk, wherein the starting and ending vertices are different, and is called an open walk. The sequence **a2d8b1a** is called a closed walk as the starting and ending vertices are same. The number of edges traversed in a walk is called the length of the walk. The restricted walks are given different names, namely, trail and path. A trail is walk wherein no edge is traversed more than once. A closed trail is called a circuit. The sequence **a2d8b1a** is called a closed walk, and also qualifies to be a trail and a circuit. The walk wherein no vertex is traversed more than once is called a path. If the end vertices in the path are different then the path is called an open path. If the end vertices in the path are the same then it is called a cycle. The sequence **b7c3d8b** is a cycle of length three. The sequence **c6e5f4c7b8d3c** is not a cycle because the vertex c is repeated.

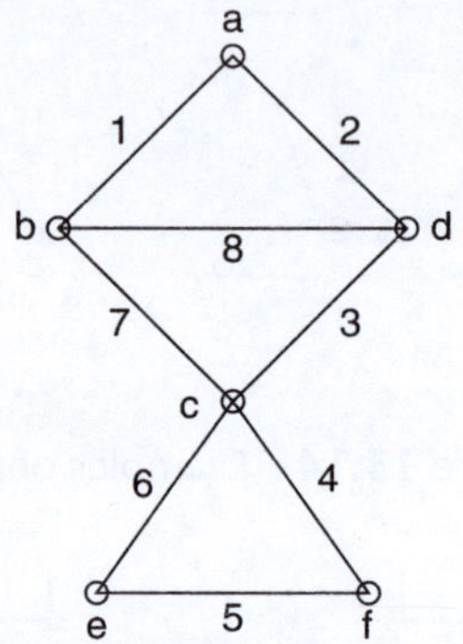

Figure 13.17 (a) A graph

An Eulerian circuit is one in which every edge of a graph is traversed exactly once. In the Konigsberg problem, we tried to look for a Eulerian circuit. A graph having a Eulerian line or a Eulerian circuit is called a Eulerian graph. Fig. 13.17 (d) and (e) are Eulerian graphs called Mohammed's Scimitars and Star of David respectively.

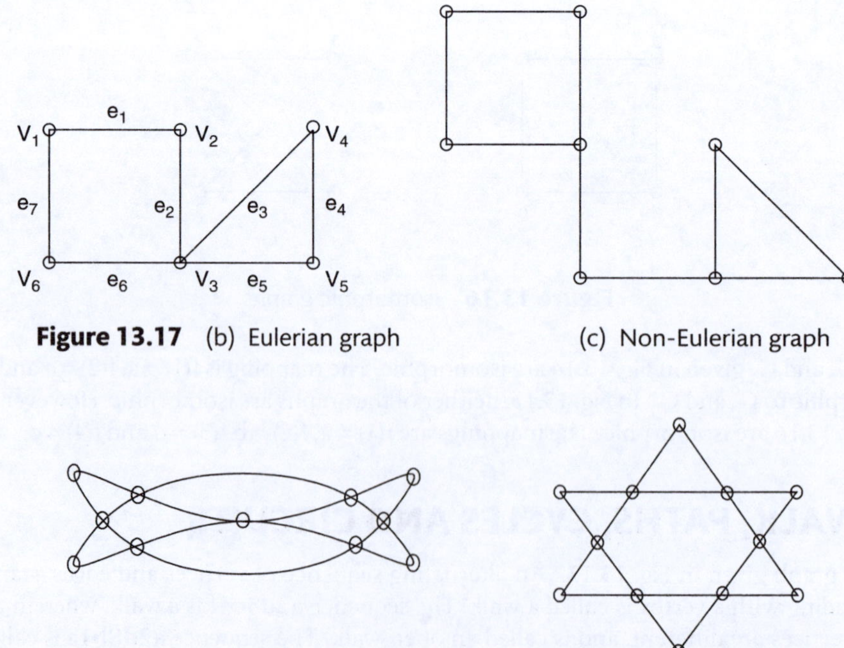

Figure 13.17 (b) Eulerian graph (c) Non-Eulerian graph

Figure 13.17 (d) Mohammed's scimitars (e) Star of David

13.5 HAMILTONIAN CYCLES

A Hamiltonian cycle in a graph is a cycle that contains every vertex of the graph exactly once, except the end vertices. The graph containing a Hamiltonian cycle is called a Hamiltonian graph. Some Hamiltonian graphs are shown in Fig. 13.18 (a) (b) and (c).

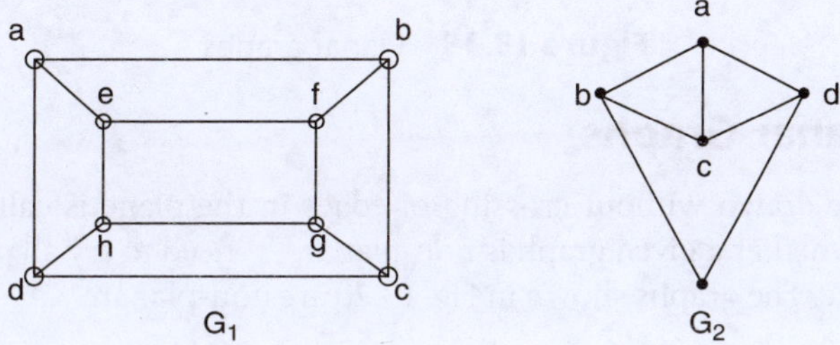

Figure 13.18 (a) Hamiltonian graphs

In some books, the term Hamiltonian circuit is used synonymously; a Hamiltonian circuit is a circuit in which every vertex is visited exactly once, except the first and the last. Obviously, such a circuit is a cycle too. Similarly, a Hamiltonian path is a path in which every vertex is visited exactly once.

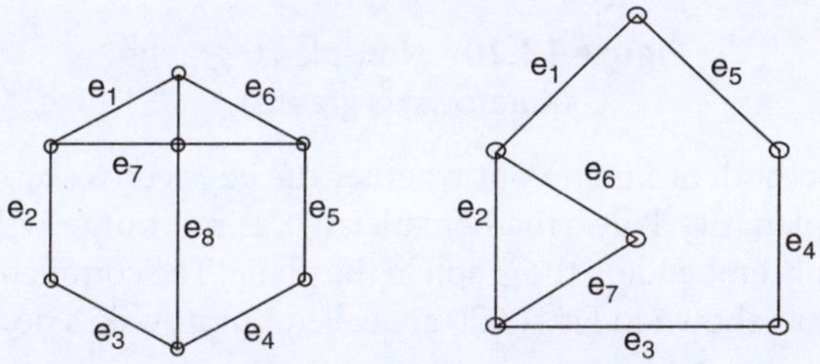

Figure 13.18 (b) and (c) Hamiltonian graphs

Proposition 3: A simple connected complete graph of N vertices will have (N–1)/2 Hamiltonian circuits.

We know that a complete graph on N vertices has N (N–1)/2 edges. A circuit with N vertices will have N edges. Therefore, the complete graph with N vertices will have (N–1)/2 Hamiltonian circuits.

The Hamiltonian graphs take their name from Sir William Rowan Hamilton (1805–1865), a contemporary and personal friend of William Wordsworth.

13.6 PLANAR GRAPHS

The graph G is said to be planar if it can be drawn on a plane without an intersection of the edges other than at the end vertices. The graph is said to be embedded in the plane. The graphs shown in Fig. 13.19 are some examples of planar graphs. A disconnected graph is said to be planar if its connected components are planar.

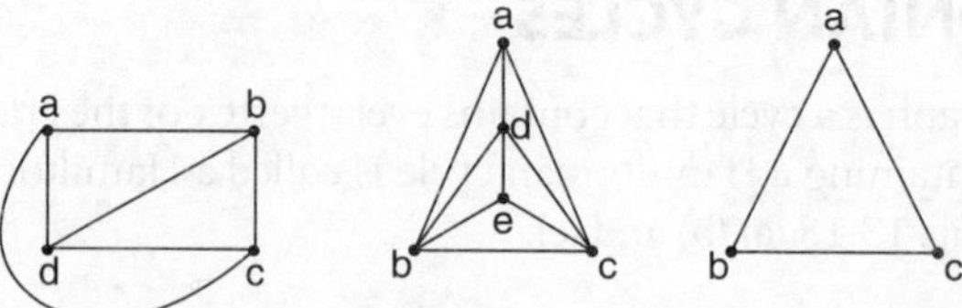

Figure 13.19 Planar graphs

13.6.1 Non-planar Graphs

A graph that cannot be drawn without crossing of edges in the plane is called a non-planar graph. In order to determine whether a given graph is non-planar, we need to try all possible ways of embedding the graph in a plane. The graphs shown in Fig. 13.20 are non-planar.

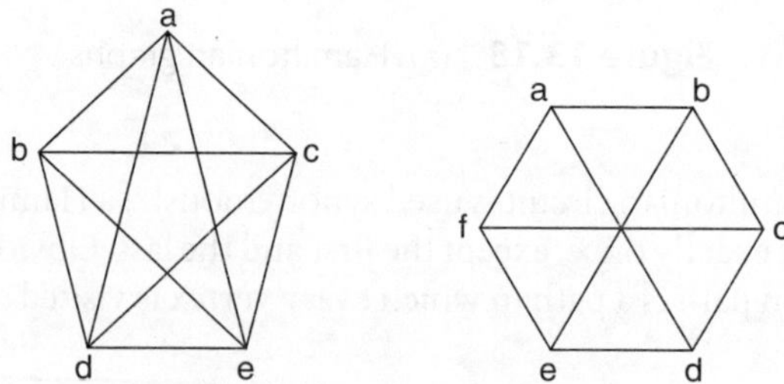

Figure 13.20 Non-planar graphs (Kuratowski's graphs)

But this trial-and-error method, of finding out whether the edges cross by drawing the graph is not a good way of determining planarity. Polish mathematician, Kasimir Kuratowski, has shown that it can be determined without actually embedding the graph in the plane. The complete graph on five vertices (K_5) and the utility graph ($K_{3,3}$) as shown in Fig. 13.20 are called Kuratowski's non-planar graphs (Also refer Fig. 13.3).

13.7 DETECTION OF PLANARITY

In order to detect planarity, the following steps need to be followed to simplify the given graph and check the resultant one.

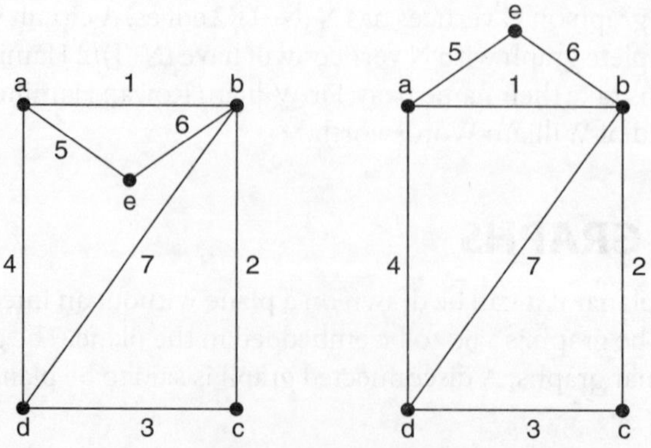

Figure 13.21 Planar graphs

Step 1: If the graph G is disconnected, and has connected components G_1, G_2 … Gn, it is planar when its components are planar.
Step 2: Self loops do not affect planarity. Therefore, remove all self loops from the graph.
Step 3: Parallel edges do not affect the planarity. Therefore, remove all self loops except one.
Step 4: The edges in series share a vertex of degree two and these edges do not affect planarity. Therefore, eliminate all edges in series.

The Steps 3 and 4 are applied repeatedly till we get one of the following graphs, which are planar. If we get one of them, declare the given graph as planar. Otherwise, declare it as nonplanar. The resulting graph has to be one amongst the following.

(i) a single edge or
(ii) a complete graph on four vertices or
(iii) a nonseparable simple graph with $n \geq 5$ and $e \geq 7$.

The aforesaid steps are carried out on the graphs shown in Fig. 13.21. Both the graphs show alternate embedding in the plane.

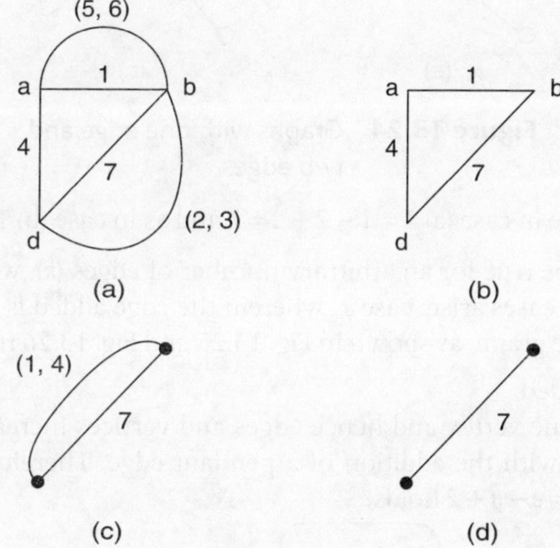

Figure 13.22 Series-parallel reduction

The series-parallel reductions carried out are shown in Fig. 13.22 (a), (b), (c) and (d). The graph in Fig. 13.22 (d) is a single edge and hence the graph shown in Fig. 13.21 is planar and this is evident from its embedding.

13.8 EULER'S FORMULA (EULER'S POLYHEDRAL FORMULA)

A plane is divided into regions or faces by embedding a graph, as shown in Fig. 13.23, and these regions or faces are numbered from 1 to 6. It is observed from Fig. 13.23 that the graph has a minimum of a self loop or parallel edges or a triangle. Euler's formula relates the number of edges (e), number of vertices (n) and the number of regions or faces (r or f), as follows:

$$f = e - n + 2$$

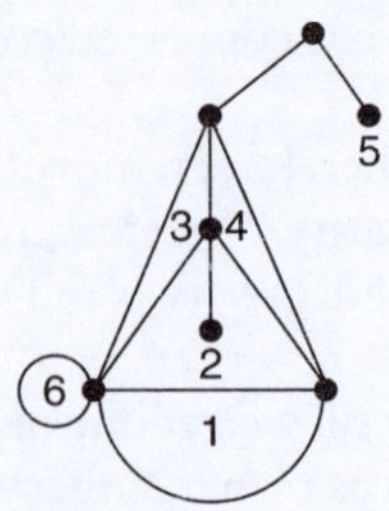

Figure 13.23 Regions in a graph

The formula can be proved using mathematical induction on the number of edges (e), a method which consists of basis and induction steps.

Basis: When $e = 1$ or $e = 2$, represent the graphs as shown in Fig. 13.24 and trivially the plane is divided into one region.

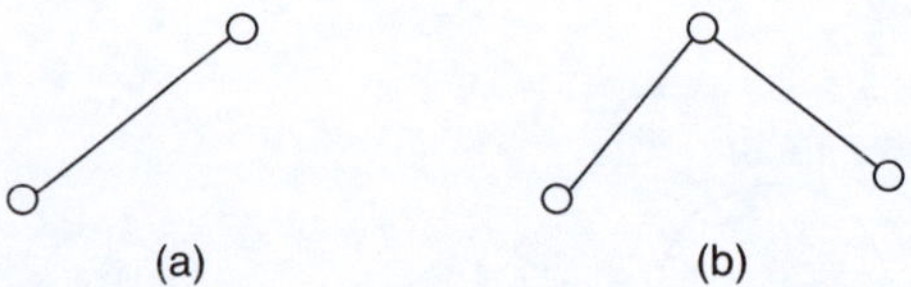

Figure 13.24 Graphs with one edge and two edges

The formula holds good, since in case (a) $1 = 1 - 2 + 2 = 1$ and as in case (b) $1 = 2 - 3 + 2 = 1$.

Induction: Let the formula be true for an arbitrary number of edges (k), where $k \geq 2$ and try to prove it for $(k + 1)$ edges. Here, two cases arise: case 1, wherein the edge added is a pendant edge and case 2, wherein the edge is within the graph, as shown in Fig. 13.25 and Fig. 13.26 respectively.

Case 1: A pendant edge is added.
By adding an edge, we add one vertex and hence edges and vertices increase by one. The number of regions does not get altered with the addition of a pendant edge. Therefore, the term $e - n$ remains altered. Hence, the formula $f = e - n + 2$ holds.

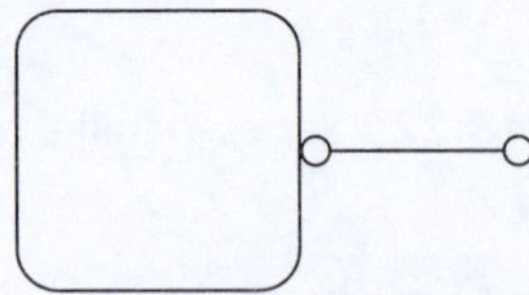

Figure 13.25 Graph with a pendant edge added

Case 2: The edge is added between any two vertices within the graph.

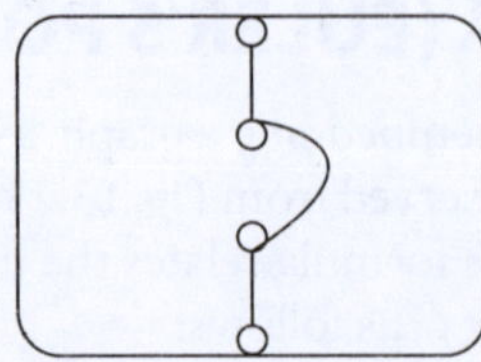

Figure 13.26 Graph with an edge other than a pendant edge added

If this additional edge is added, the number of edges is increased by one, the vertices remain the same and the number of regions is increased by one. Thereby the formula $f = e - n + 2$ holds. Thus the formula holds for a connected graph with planar embedding.

13.9 GRAPH COLOURING

Graph colouring implies painting the vertices of a graph with colours, not necessarily with distinct colours. It is also called the vertex colouring of a graph. A vertex colouring is called proper colouring if no adjacent vertices receive the same colour. The graph G with N vertices is said to be of the order N. One can use a maximum of N colours for proper colouring of a given graph. It is really interesting to use the minimum number of colours for proper colouring of a given graph. The minimum number of colours required for proper colouring of a given graph G is called the chromatic number of G and is denoted as k(G). The graph given in Fig. 13.27 has four distinct colours. The colours used are white (W), red (R), green (G) and blue (B). However, the graph is 3-chromatic, as shown in Fig. 13.28 and only three colours are used. Figure 13.28(a) gives coloured regions and the equivalent graph.

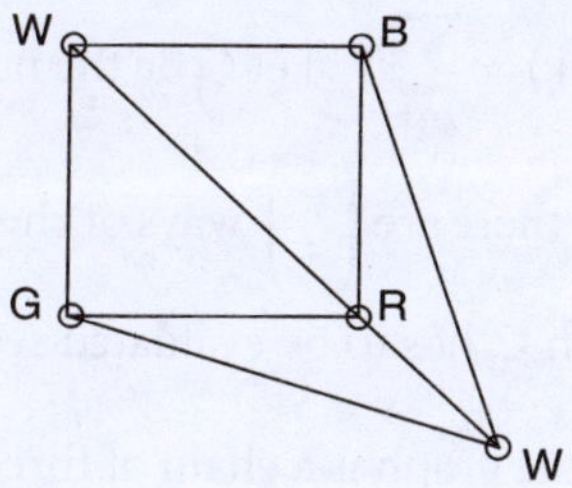

Figure 13.27 Four colours used in a graph

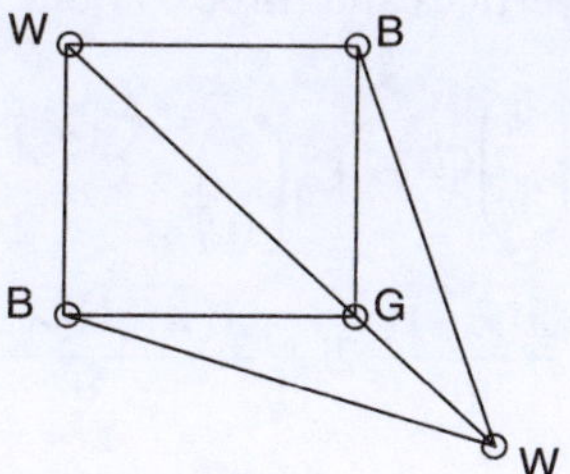

Figure 13.28 A 3-chromatic graph

The following are some observations on the chromatic number of a graph.

(i) The chromatic number of an isolated vertex is one. The chromatic number of an edge graph is two.
(ii) The chromatic number of a complete graph K_n is n.
(iii) The cnumber of a path is two.
(iv) The chromatic number of a cycle is three if n is odd and two when n is even.
(v) The chromatic number of a binary tree is two.

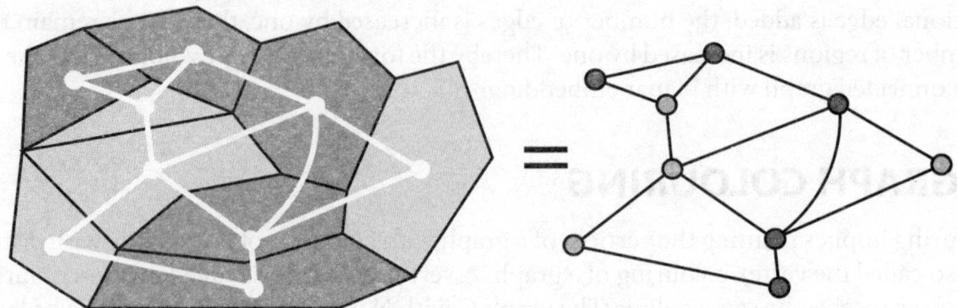

Figure 13.28 (a) Coloured regions and the equivalent graph

13.9.1 Chromatic Polynomial

We know that the vertices of a graph can be properly coloured in many ways using a large number of colours. In how many ways can a graph on n vertices be properly coloured with at the most λ colours?

Let a_k be the number of ways in which a graph can be coloured with k colours. The total number of ways of colouring is a polynomial $P(G, k) = \sum_{k=1}^{\lambda} ak$. Let C_i be the number of ways of colouring the graph with i distinct colours. Given λ colours, there are $\binom{\lambda}{i}$ ways of choosing i colours from λ colours, where $\lambda \le n$. Hence, $P(G, \lambda) = \sum_{i=1}^{n} \binom{\lambda}{i} C_i$. Each C_i has to be evaluated separately for each graph. For example, consider graphs on three vertices (i) One graph is a chain of three vertices and (ii) another is a triangle and write polynomials.

(i) Chain of vertices: We find that this graph cannot be coloured with one colour. Therefore, $C_1 = 0$. The end vertices can be coloured with one colour and the middle vertex with another colour. Therefore, $C_2 = 2$. With three vertices and three colours, there are $3! = 6$ ways of colouring and $C_3 = 6$.

$$P(G, \lambda) = \sum_{i=1}^{n} \binom{\lambda}{i} Ci = C_0 \binom{\lambda}{0} + C_2 \binom{\lambda}{2} + C_3 \binom{\lambda}{3}$$

$$= 0 + \frac{\lambda(\lambda-1)}{2!} 2! + \frac{\lambda(\lambda-1)(\lambda-2)}{3!} 3!$$

$$= \lambda(\lambda-1) + \lambda(\lambda-1)(\lambda-2)$$

$$= \lambda(\lambda-1)^2$$

(ii) Triangle: It is not possible to colour with one and two colours. Therefore, $C_1 = 0$ and $C_2 = 0$.

$$P(G, \lambda) = \sum_{i=1}^{n} \binom{\lambda}{i} Ci = C_0 \binom{\lambda}{0} + C_2 \binom{\lambda}{2} + C_3 \binom{\lambda}{3}$$

$$= 0 + 0 + \frac{\lambda(\lambda-1)(\lambda-2)}{3!} 3!$$

$$= \lambda(\lambda-1)(\lambda-2)$$

13.9.2 Decomposition Theorem for Chromatic Polynomials

If G = (V, E) is a connected graph and $e \in$ E then

$$P(G_e, \lambda) = P(G, \lambda) + P(G_{e'}, \lambda)$$

Proof: Let $e = \{a, b\}$. The number of ways to properly colour the vertices in G_e with at the most λ colours is $P(G_e, \lambda)$. Those colourings where a and b have different colours form proper colourings of G. The colourings of G_e are not proper colourings when the end vertices of the edges a and b have the same colour. But each of these colourings correspond to proper colouring of $G_{e'}$. This partition of $P(G_e, \lambda)$ which is a proper colouring of G_e into two disjoint sub-graphs results in the equation $P(G_e, \lambda) = P(G, \lambda) + P(G_{e'}, \lambda)$

For example, consider the graph G given as follows.

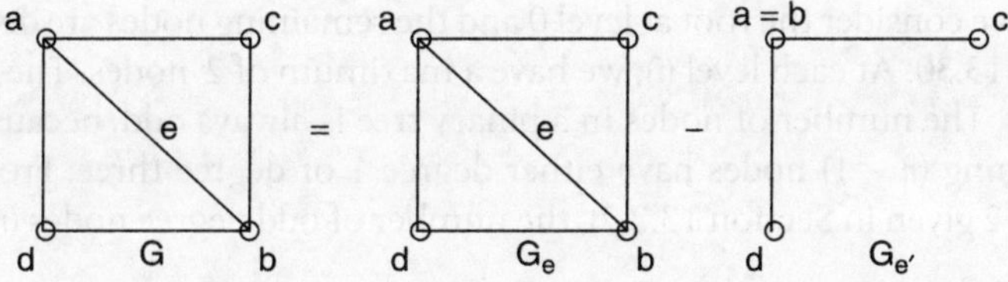

13.10 BASICS OF TREES

A tree is a connected graph which is acyclic in nature. In short, a tree is a connected acyclic graph. In fact a connected graph is a tree if there exists a unique path between every pair of vertices. The vertices are called nodes and one special node is designated as the root. The nodes with degree one are called leaf nodes and nodes other than these two types are called intermediate nodes. (See Fig. 13.29).

The root can have any number of nodes as descendants, called children. If every node in the tree except the leaf nodes has a maximum of m descendants, then it is called an m-ary tree. If $m = 2$, then it is called a binary tree. When all the nodes except leaves have exactly two children, it is called complete or full binary tree.

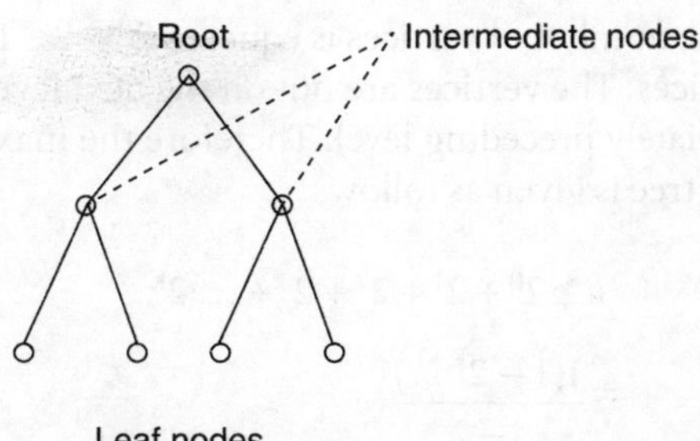

Figure 13.29 Basic terms used in defining trees

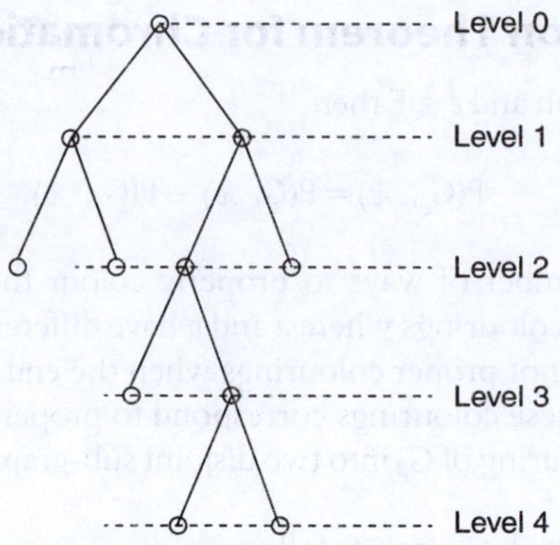

Figure 13.30 Levels in a binary tree

In case of binary tree, we consider the root at level 0 and the remaining nodes are distributed at different levels as shown in Fig. 13.30. At each level (i), we have a maximum of 2^i nodes. The maximum level in a tree is called its height. The number of nodes in a binary tree is always odd, because the root has even degree and the remaining ($n - 1$) nodes have either degree 1 or degree three. From the handshaking principle (Proposition 2 given in Section 13.2.2), the number of odd degree nodes ($n - 1$) is always even. Therefore, n is odd.

13.10.1 Some Properties of Trees

(i) Counting the number of vertices of degree three in a binary tree on N vertices and having K number of pendant vertices, we find that the tree has K pendant vertices having degree one. There is a root with degree two. The number of remaining vertices of degree three is (N – K – 1).

(ii) In a binary tree T on N vertices, the number of pendant vertices is of degree equal to (N + 1)/2. We shall assume that the number of pendant vertices is K and the degree is one. There are (N – K – 1) vertices of degree three. One vertex is of degree two. The tree has (N – 1) edges. We know that

The sum of degrees of vertices = 2 × Number of edges

$$K \times 1 + 1 \times 2 + 3 \times (N - K - 1) = 2(N - 1)$$

$$K + 2 + 3N - 3K - 3 = 2N - 2$$

$$-2K = -N - 1 \text{ or } 2K = N + 1$$

Therefore, $K = (N + 1)/2$

(iii) In a binary tree of level K, the number of vertices is equal to $2^{(K+1)} - 1$. This is because, at each level i we have at the most 2^i vertices. The vertices are put on the next level only when the required vertices are filled at the immediately preceding level. Therefore the maximum number of vertices (n) possible for a K level binary tree is given as follows.

$$n \geq 2^0 + 2^1 + 2^2 + 2^3 + \ldots 2^K$$

$$= \frac{1(1 - 2^{K+1})}{1 - 2}$$

$$= 2^{K+1} - 1$$

since the right hand side is a geometric progression with the first term 1 and a common ratio 2.

Further, we can say that the maximum possible level of a binary tree on $(2n+1)$ vertices is n. Let K be the height of a binary tree with $(2n+1)$ vertices. To get the maximum possible level, we must get exactly two vertices at each level except the root vertex. Thus, out of $(2n+1)$ vertices, one is the root vertex and the remaining $2n$ vertices are kept exactly at n levels. Therefore, the maximum height is $n = \mathrm{K}$. It can also be shown that $2 + 2 + 2 + \ldots + 2$ (K times) $\geq 2n$. That is $2\mathrm{K} \geq 2n$ and hence $\mathrm{K} \geq n$. Thus, the maximum value of K is n. Similarly, the minimum height of a binary tree on $(2n-1)$ vertices is $\log_2 n$. This is because, $(2n-1) \leq 2^0 + 2^1 + 2^2 + 2^3 + \ldots 2^{\mathrm{K}} = 2^{\mathrm{K}+1} - 1$ and $n \leq 2^{\mathrm{K}}$. Thus, $\log_2 n \leq \mathrm{K}$ or $\mathrm{K} \geq \log_2 n$.

13.11 DIRECTED, WEIGHTED AND DIRECTED ACYCLIC GRAPHS

Directed graphs are also common in mathematics and computer science. A graph is said to be directed when it has arrow marks on the edges as shown in Fig. 13.31. The undirected edge has arrows in both directions and hence has been omitted in the representations. In directed graphs we consider both in-degree and out-degree for a vertex, and the total in-degree is equal to the out-degree across the vertices. A directed acyclic graph, also called a DAG, is a directed graph with no directed cycles. This means that it is not possible to start at some vertex (v) and follow a sequence of directed edges and eventually loop back to (v). A tree is an example of a DAG.

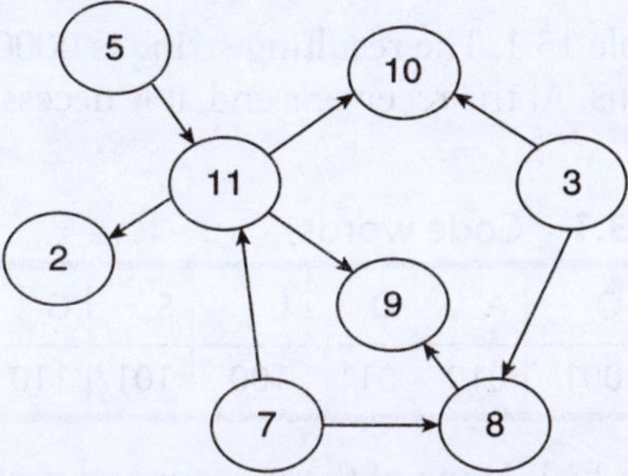

Figure 13.31 Directed graph and also directed acyclic graph

A weighted graph is a directed graph with edges having labels and these labels indicate certain parameters in the real world. Figure 13.32 shows an example of a directed graph, wherein the nodes indicate cities and the labels on the edges may indicate the distances or costs in moving from one city to another.

Figure 13.32 Example of a directed graph

13.11.1 Prefix Codes

One of the applications of trees is found in coding theory. The messages are coded and transmitted over the network and at the receiving end the messages are decoded. Consider a set of symbols called the alphabet. On this alphabet the codes are defined. Codes are nothing but sequences of symbols from the alphabet. Consider an alphabet = {0, 1} and the code words as binary sequences. Consider the message "ROAD IS GOOD". To generate the code words, we can use a code tree as shown in Fig. 13.33.

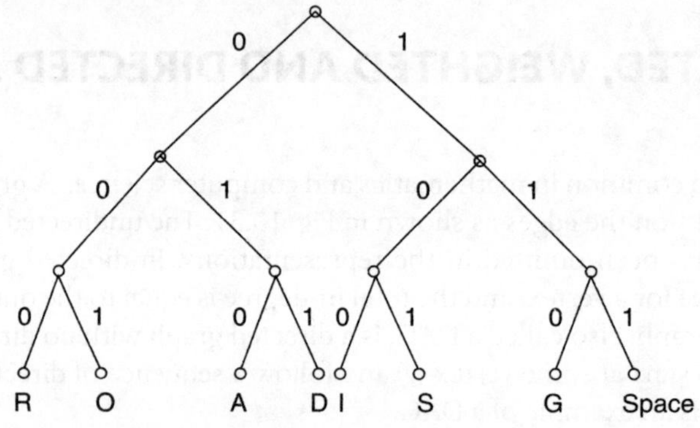

Figure 13.33 Coding tree

The code word set is given in Table 13.1. The resulting string is 000001010011111100101111110001001011. The coding is unambiguous. At the receiver's end, it is necessary to know the codes and decode the coded string.

Table 13.1 Code words

R	O	A	D	I	S	G	Blank
000	001	010	011	100	101	110	111

Consider the codes given in Table 13.1. None of them occur as a prefix to other code words. This is the main property of prefix codes.

13.11.1.1 *Weighted path length*

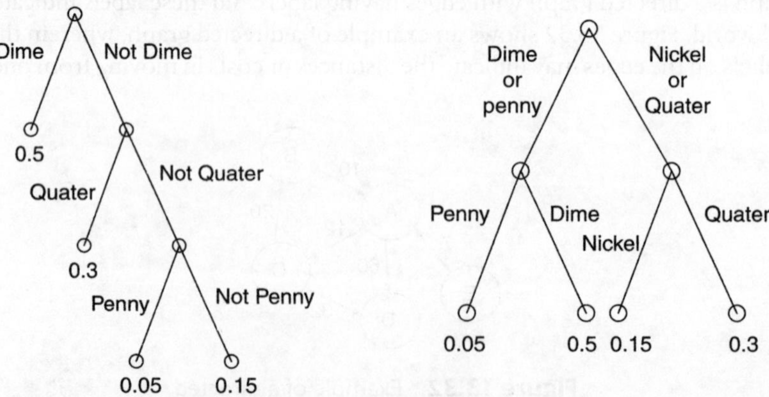

Figure 13.34 Weighted path length

In some applications, every pendant vertex v_i of a binary tree is associated with a positive real number w_i. We need to construct the binary tree such that $\sum w_i l_i$ is minimum, where l_i is the level of the pendant vertex v_i. The sum is taken over all pendant vertices.

Consider a coke vending machine and two decision trees as shown in Fig. 13.34. It has to identify through a sequence of tests what coins have been entered into the machine. Assume that only pennies, dimes, nickels and quarters are allowed and their respective probabilities as 0.05, 0.5, 0.15 and 0.3. The two possible trees are as shown in the figure. During each test, the coins are partitioned into two complementary sets. The first tree gives you the total weight as 1.7 and the second gives 2.0. Therefore, the first is the minimum weighted tree.

13.11.2 Spanning Trees and Minimal Spanning Trees

A spanning tree is a sub-graph, also called a spanning sub-graph, of an undirected graph G that includes all of the vertices of G and is a tree. In general, a graph may have several spanning trees, as shown in Fig. 13.35 and it gives the count of trees of a complete graph on four vertices. A graph that is not connected will not contain a spanning tree.

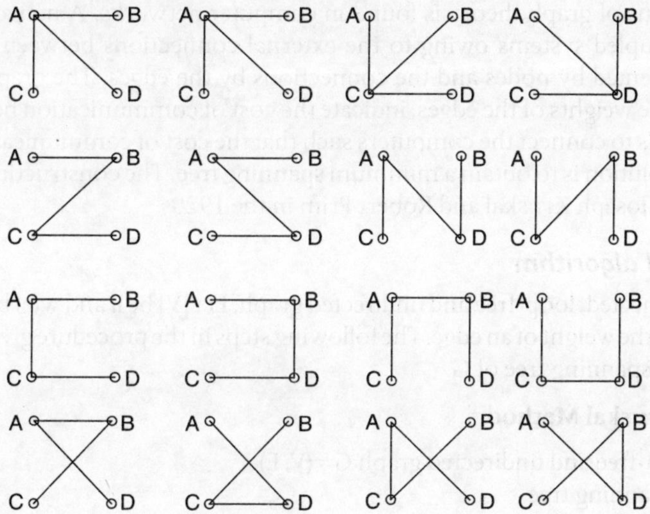

Figure 13.35 Spanning trees

13.11.2.1 Counting trees

The individual graphs in Fig. 13.35 indicate the possible counting trees of a complete graph on four vertices.

The number of labelled trees for a connected graph with *n* vertices and $(n \geq 2)$ is $n^{(n-2)}$

13.11.2.2 Minimum spanning tree

A given graph can have many different spanning trees. We can also assign weights to the edges, resulting in weighted graph. We can obtain the total weight of a spanning tree by computing the sum of the weights of the edges in that spanning tree in moving from one node, called the source node, to another node, called the destination node. A minimum spanning tree (MST) or minimum weight spanning tree is a spanning tree with weight less than or equal to the weight of every other spanning trees. More generally, any undirected graph (not necessarily connected) has minimum weight. One typical spanning tree is shown in Fig. 13.36.

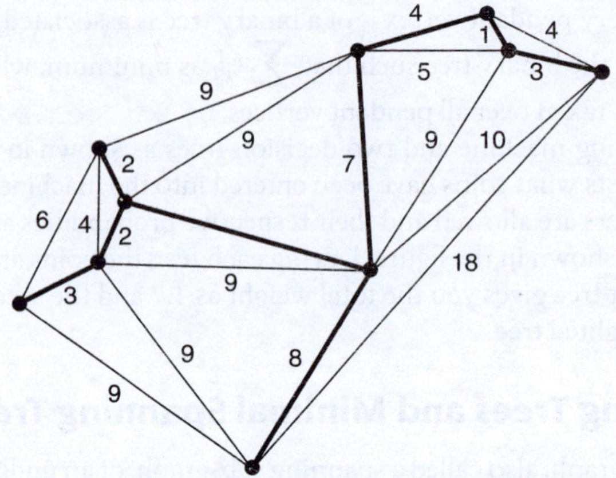

Figure 13.36 Minimum spanning tree

13.11.3 Kruskal's and Prim's Algorithm

One of the applications of graph theory is found in computer networks. A network of computers are termed as loosely coupled systems owing to the external connections between the computers. The computers are represented by nodes and the connections by the edges. The graph is labelled and the labels on the edges, the weights of the edges, indicate the cost of communication between the two computers. The objective is to connect the computers such that the cost of communication between them is the minimum. The solution is to obtain a minimum spanning tree. The construction of such a spanning tree was proposed by Joseph Kruskal and Robert Prim in the 1920s.

13.11.3.1 Kruskal algorithm

Let G = (V, E) be a connected, loop-free and undirected graph. Let |V| be *n* and w(e) be a non-negative real number representing the weight of an edge. The following steps in the procedure given in Algorithm 13.1, result in the minimal spanning tree of G.

Algorithm 13.1: Kruskal Method

Input: Weighted, loop-free and undirected graph G= (V, E)
Output: Minimum spanning tree

```
Start
  Step1: Initialise the counter (i) to 1 and select the edge e_1 from
         G such that weight w(e_1) is the least.
  Step2 : For ( 1≤i≤(n-1))and if the edges e_1, e_2, e_3, ..., e_i are
          already selected, then select the edge e_{i+1} from the
          remaining edges from G, such that
     Step 2.1: w(e_{i+1}) is as small as possible.
     Step 2.2: The subgraph formed by e_1, e_2, e_3, ..., e_i, e_{i+1} does
     not contain any cycle.
  Step3: Set i = i + 1.
     Step 3.1: If i = (n - 1) then the subgraph of G formed by e_1,
               e_2, e_3, ..., e_i, e_{i+1} is connected and it is the min-
               imal spanning subgraph.
     Step 3.2: If (i < (n - 1) then goto Step 2.
End
```

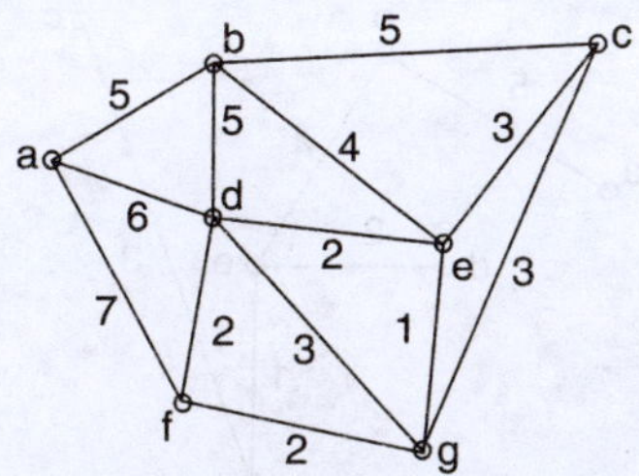

Figure 13.37 (Kruskal)

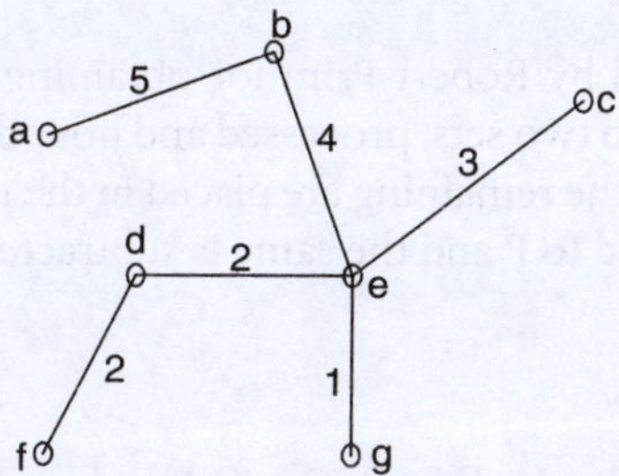

Figure 13.38 Minimal spanning tree (Kruskal)

Consider the graph given in Fig. 13.37 and apply Kruskal's algorithm to find the minimal spanning tree which is given in Fig. 13.38. Following is the detailed application of the method.

Iteration 0: Initialise *i* to 1 and the tree T to { }. Amongst all the edges, the edge {e, g} with the smallest weight is 1. Insert the edge in T and T = {{e, g}}.

Iteration 1: Amongst the remaining edges in G, we find three edges {d, f}, {d, e}, and {f, g} with weight 2. Let us select {d, f} and insert this edge in T. Now T = {{e, g}, {d, f}}, which is a forest. Increment *i* by 1 and we have *i* = 2. Since (*i* < 6), we go to Step 2.

Iteration 2: Now we will add the edge {d, e} to T and increment the value of *i* by 1. The value of *i* becomes 3 and T = {{e, g}, {d, f}, {d, e}}. Since (*i* < 6) we go to Step 2.

Iteration 3: Now we will add the edge {d, f} with weight 2 to T but it results in a cycle and the definition of tree is defeated. Therefore, the edges having the next smallest weight, that is, 3, namely {c, e}, {c, g} and {d, g} are considered. However, the addition of edge {d, g} results in a cycle and hence the edges {c, e}, and {c, g} are considered as these do not result in cycles. Let us add {c, e} to T, which gives us T = {{e, g}, {d, f}, {d, e},{c, e}}, and increment *i* by 1 resulting in *i* = 4. Since (*i* < 6), we go to Step 2.

Iteration 4: Now, we will add to T, the edges having weight 4. The only edge is {b, e} and addition to T does not result in a cycle. Increment the value of *i* by 1. The value of *i* becomes 5 and T = {{e, g}, {d, f}, {d, e}, {c, e}, {b, e}}. Since (*i* < 6), we go to Step 2.

Iteration 5: Similar to the previous iterations, the edge {a, b} of weight 5 is added to T and T = {{e, g}, {d, f}, {d, e}, {c, e}, {b, e}, {a, b}}. Increment the value of *i* by 1. The value of *i* becomes 6, which is same as the number of vertices in G. Therefore, it is an optimal tree having weight = 17, as given in Fig. 13.38.

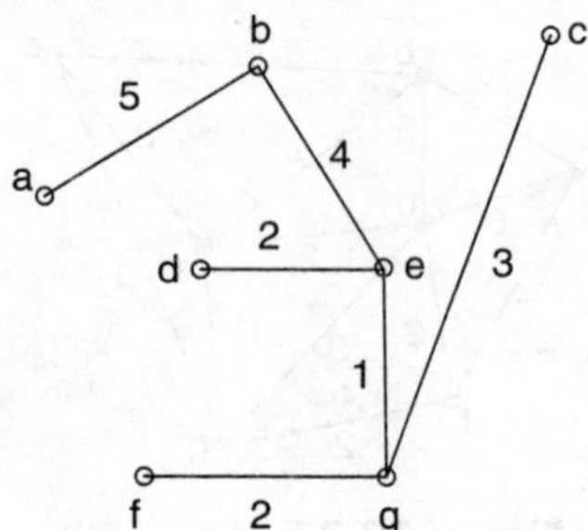

Figure 13.39 Minimal spanning tree (Prim's)

13.11.3.2 Prim's algorithm

This is another method developed by Robert Prim for obtaining the minimal spanning tree. In this method the vertices are divided into two sets, processed and not processed. To begin with, one vertex is placed in the processed set (P) and the remaining are placed in the not processed set(NP). Every time we get the smallest vertex, it gets added to P and the same is subtracted from (NP). The procedure is given in Algorithm 13.2.

Algorithm 13.2: Prim's Method

Input: Weighted, loop-free and undirected graph G = (V, E), where |V| = n
Output: Minimum spanning tree

```
Start
   Step1: Let the counter i = 1. Consider an arbitrary vertex v1∈ V
          and place it in the set P.
          Let NP = V – {v1} whose and T = { }.
   Step 2:   For (1 ≤ i ≤ n) do
      Step 2.1: Add to T, the weight of an edge (x, y) whose vertex
                x ∈ P, vertex y ∈ NP and that has
            minimal weight. Add the vertex y to P and delete it from NP
   Step3: Increment the counter i
      Step 3.1: If (i = n) then the set P contains n vertices and T
                contains (n – 1) edges
                representing a minimal spanning tree.
                 Step 3.2: If (i < n) then go to Step 2.
End
```

Consider the graph G = (V, E) as given in Fig. 13.37 and let us apply Prim's method to find the minimal spanning tree.

Iteration 0: P = {a}, NP = {b, c, d, e, f, g} and T = { }, i = 1

Iteration 1: P = {a, b}, NP = {c, d, e, f, g} and T = {(a, b) } and i = 2

Iteration 2: P = {a, b, e}, NP = {c, d, f, g} and T = {(a, b), (b, e) } and i = 3

Iteration 3: P = {a, b, e, g}, NP = {c, d, f } and T = {(a, b), (b, e), (e, g)} and i = 4

Iteration 4: P = {a, b, e, g, d}, NP = {c, f } and T = {(a, b), (b, e), (e, g), (d, e)} and i = 5

Iteration 5: P = {a, b, e, g, d, f}, NP = {c} and T = {(a, b), (b, e), (e, g), (d, e), (f, g)} and i = 6

Iteration 6: P = {a, b, e, g, d, f, c}, NP = { } and T = {(a, b), (b, e), (e, g), (d, e), (f, g), (c, g)} and i = 7. The resulting tree is a minimal spanning tree with weight = 17 as given in Fig. 13.39.

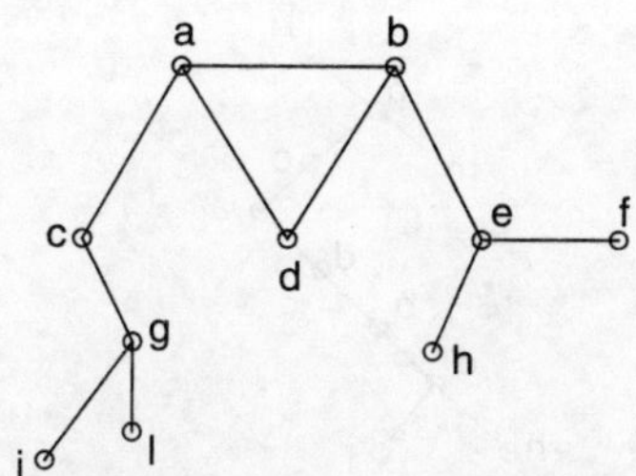

Figure 13.40 Some graph

13.11.4 DFS and BFS Algorithms

Let G = (V, E) be a connected, loop-free and undirected graph as shown in Fig. 13.40. The popular methods of searching for vertices in a connected, loop-free and undirected graph include breadth first search (BFS) and depth first search (DFS).Both ultimately result in spanning trees. The term backracking in data structures is evolved from one of these searching techniques.

13.11.4.1 Depth-First search

Algorithm 13.3: DFS Method

Input: Connected and undirected graph G = (V, E)

Output: Depth-first spanning tree T of G for a given order.

Start

Step 1: Assign v_1 to the variable v and initialise T as the tree consisting of just this one vertex. The vertex v_1 will be the root of the resulting spanning tree. Visit the vertex v_1.

Step 2: Select the smallest subscript i, for $(2 \leq i \leq n)$, such that $\{v, v_1\} \varepsilon$ E and v_i is not already visited.

Step 2.1 If no such subscript is found, then go to Step 3, Otherwise, perform the following.

(i) Attach the edge $\{v, v_i\}$ to the tree T and visit v_i

(ii) Assign v_i to v; and

(iii) Return to Step (2).

Step 3: If $(v = v_1)$, the tree T is the rooted ordered spanning tree for the order specified.

Step 4: For $(v \neq v_1)$, backtrack from v to its parent u in T.

Step 4.1 Then assign u to v and return to Step (2).

End

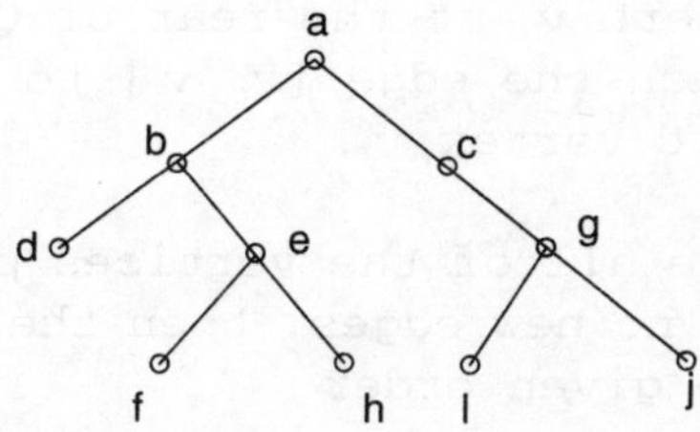

Figure 13.41 DFS tree of Fig. 13.40

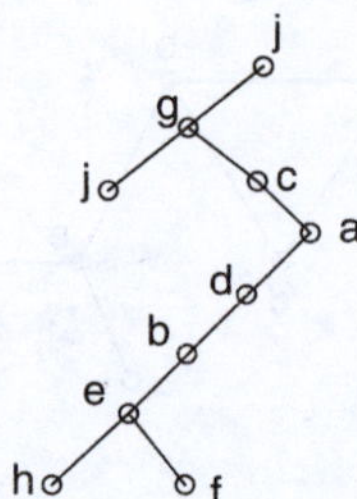

Figure 13.42 DFS Tree of Fig. 13.40

Consider the graph given in Fig. 13.40 and the vertex order a, b, c, d, e, f, g, h, i, j for the depth-first search method. Let the vertex 'a' be assigned to v and T = {(a, b)}. Now assign b to v and return to Step 2. At v = b, we find that the vertex d is not visited. Add the edge {b, d} to T and assign d to v and return to Step 2. At this time, no new vertex is available that has not been visited and we go to Step 3. Since v = d, not a, we go to Step 4. Now we backtrack from d and assign vertex b to v. We return to Step 2 and add edge {b, e} to T. Continuing this way we next add the edges {e, f} and {e, h}.

We assign the vertex h to v. We backtrack from h to e to b to a. Now we add the edge {a, c} to T. Continuing in a similar manner, other edges, namely, {c, g}, {g, i} and {g, j} are added to T. Now all the vertices of G have been visited. We backtrack from j to g to c to a. We go to Step 2 and then Step 3 where the method terminates. The tree (T) obtained is given in Fig. 13.41. The depth-first tree for the vertex order j, i, h, g, f, e, d, c, b, a. is given in Fig. 13.42.

13.11.4.2 Breadth-first (BFS) method

In this method of searching vertices, we designate one vertex as the root and reach or fan out all the vertices adjacent to the root, or children of this root. Next we consider one of these vertices, not already visited, and fan out to its adjacent vertices. This process is continued until all the vertices are visited. The process is presented as Algorithm 13.4

Algorithm 13.4: BFS Method

Input: Connected and undirected graph G = (V, E)

Output: Breadth-first spanning tree T of G for a given order.

```
Start
  Step 1: Insert vertex v1 at the rear of the queue Q, which is initially empty.
      Step 1.1: Initialise T as a tree formed by this one vertex v1,
                which is the root of the final breadth-first spanning tree T.
      Step 1.2: Now visit v1.
  Step 2: While the queue Q is not empty do
      Step 2.1: Delete the vertex v from the front of the queue Q.
      Step 2.2: Now estimate the vertices vi (for 2 ≤ i ≤ n) that are
                adjacent to v in the specified order.
      Step 2.3: If vi has not been visited, perform the following:
                (i) Insert vi at the rear of Q;
               (ii) Attach the edge [v, vi] to T; and
              (iii) Visit vertex vi.
      While-end
   Step 3: If we examine all of the vertices previously in the queue
           Q and obtain no new edges, then the tree T is the spanning
           tree for the given order.
End
```

Consider the graph given in Fig. 13.40 and apply Algorithm 13.3 with an order a, b, c, d, e, f, g, h, i, j on the vertices. We shall start with the vertex a by inserting it in queue Q. The vertex 'a' forms the root of the tree T.

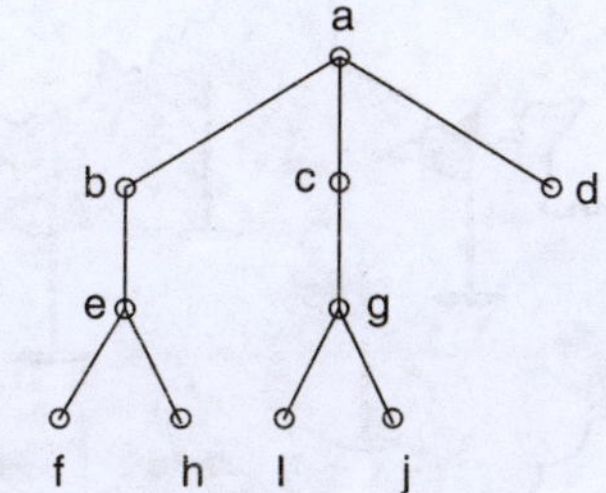

Figure 13.43 BFS Tree of Fig. 13.40

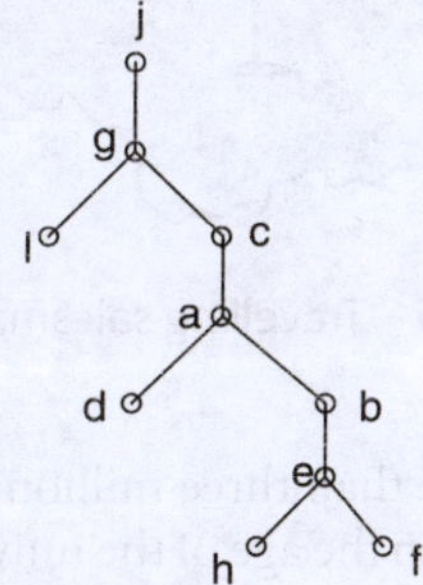

Figure 13.44 BFS Tree for the vertex order j, i, h, g, f, e, d, c, b, a

As per Step 2, the Q is not empty and we shall delete the vertex a by visiting the adjacent vertices b, c, d, as these are not previously visited. Add the edges (a, b), (a, c) and (a, d) to T. In Step 3, we add these vertices b, c, d to the rear end of Q and return to Step 2.

Now, each of these vertices is deleted by visiting the adjacent vertices e and g which have not previously been visited. Thus, add the edges (b, e) and (c, g) to T. From Step 2, add the vertices e and g to Q and return to Step 2. Similarly, continuing in the same way, we get the edges (e, f), (e, h), (g, i), (g, j) added to T. The vertices are added to Q. We delete these vertices and do not find new vertices. The tree T is as shown in Fig. 13.43 and is the minimal spanning tree for the order prescribed. Fig. 13.44 gives the breadth-first tree for the vertex order j, i, h, g, f, e, d, c, b, a.

13.12 CERTAIN APPLICATIONS OF GRAPH THEORY

In this section we have given certain applications of graph theory, namely, the travelling salesman problem, intelligent transportation systems, social networks and circuit layout.

13.12.1 Travelling Salesman Problem

Many a time in real world situations, we try to find the shortest possible routes. One such scenario is for a salesman. This is called the travelling salesman problem: A salesman has to visit a number of cities to deliver items. What is the shortest route that connects all the cities? Of course we need to know how many cities there are and what the distances between any two cities are. But even with all that information the Traveling Salesman problem is still very difficult. We could just try every possible path, but with many cities this becomes very impractical. This is depicted in Fig. 13.45.

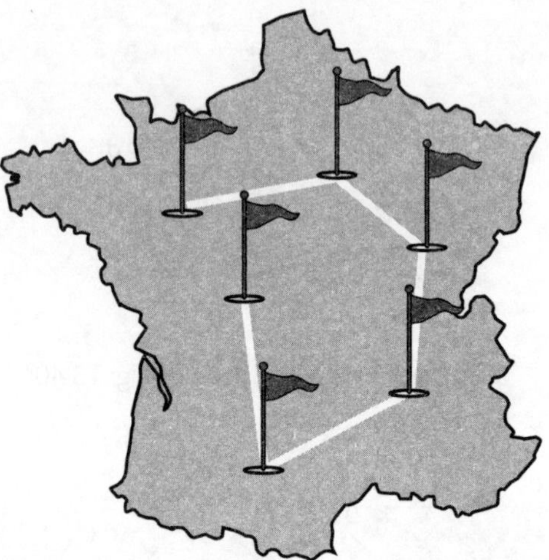

Figure 13.45 Travelling salesman problem

If we have ten cities to visit there are more than three million possible orders to visit them in! Testing 100 cities in that way would take longer than the age of the universe! Visiting 100 cities does not happen very often in real life but it does happen if you plan airline networks or want to optimise production lines in factories. There are many other important applications which you can read about in the rest of the chapter. Unfortunately mathematicians have not been able to find a way to solve the travelling salesman problem in a much more efficient way. Problems like this, which take a very long time to solve using a computer, are called NP-hard (**N**on-deterministic **P**olynomial-time hard). If we are connecting n cities there are $n!$ (n-factorial) different routes (this is explained in the article on Combinatorics)., the most efficient computer algorithm takes of the order of 2^n steps. This is not very useful since the number of steps increases exponentially with n. Instead there are approximate algorithms to solve the travelling salesman and other similar problem. These are much faster and usually give very good results, but sometimes they miss the best solutions.

13.12.2 Intelligent Transportation Systems

Everything in our world is linked: cities are linked by street, rail and flight networks. Pages on the internet are linked by hyperlinks. The different components of an electric circuit or computer chip are connected and paths of disease outbreaks also form a network. Scientists, engineers and many others want to analyse, understand and optimise these networks. And this can be done using graph theory. A typical situation is given in Fig. 13.46. These **Intelligent Transportation Systems** could work by collecting location data from smart-phones of motorists and telling them where and how fast to drive in order to reduce overall congestion. Graph theory is already utilised in flight networks. Airlines want to connect countless cities in the most efficient way, moving the most passengers with the fewest possible trips: a problem very similar to that of the travelling salesman. At the same time, air traffic controllers need to make sure hundreds of planes are at the right place at the right time and don't crash: an enormous task that would be almost impossible without computers and graph theory.

Figure 13.46 Transportation problem

13.12.3 Circuit Layouts

One area where speed and the best connections are of crucial importance is the design of computer chips. Integrated circuits (ICs) consist of millions of transistors which need to be connected. Although the distances are only a few millimetres, it is important to optimise these countless connections to improve the performance of the chip (Fig. 13.47). Graph theory also plays an important role in the evolution of animals and languages, crowd control and in checking the spread of diseases.

Figure 13.47 Circuit layout

13.12.4 Social Networks

The internet is one of the largest graphs in life. In recent years, there has been another important use of graph theory: the internet. Every page in the internet could be a vertex in a graph, and whenever there is a link between two pages, there is an edge between the corresponding vertices.

Figure 13.48 Internet is a graph

Figures 13.48 and 13.49 represent a social network. The resulting graph is of course *very* large. Early web search engines had a very big problem: they could search the web for a particular keyword, but they couldn't determine whether a page was "good" or just spam. If you searched for 'London', you might get hundreds of websites of small shops in London, or people who lived there, before you got the official london.gov website. Google found a solution to this: any page that is very good will have many other pages linked to it. Pages that are rarely visited, or not very interesting, will be very "lonely" in the internet graph with only few other pages linking to it. This gives a way to *rank* websites and allows Google to display the best results at the beginning

Figure 13.49 Everyone is connected via the internet

There is a digital graph of which most of us are also a part: Facebook. All the users form vertices and whenever two users are friends they are linked by an edge. Graph theory can help web developers improve the performance of social networking sites, and it can help us understand Facebook better. If we were to choose two users of Facebook completely at random, how many of these friendship edges do I need to get from one to the other? 10? 100? Surprisingly the answer, on average, is only six. On an average, you are less than six friends away from *any* other Facebook user: that's globalisation! Experiments have been conducted to test this and the longest distance ever found on Facebook was only 12. The number of **degrees of separation** has inspired countless writers and film makers. How many degrees are you away from the Queen? Or the American president? Nobody knows the exact answer, but it is probably a lot less than you think!

ADDITIONAL SOLVED EXAMPLES

Example W13.1: How many vertices are there in a graph with six edges with each vertex having degree two?

Solution:
We know that the total degree of G = (V, E) is equal to twice the number of edges.
Total degree of graph = 2E = 2 × 6 = 12
Let the cardinality of V is |V| = n and the total degree of G = 2|V|
Therefore, 2|V| = 12 and |V| = 6, the number of vertices in the graph.

Example W13.2: Write down the chromatic polynomial for a complete graph on n vertices.

Solution:
Since a minimum of n colours are required for the proper colouring of the complete graph K_n on n vertices, $C_i = 0$ for $i = 1$ to $(n-1)$. Since the graph contains n vertices, n distinct colours can be assigned in $n!$ ways. Thus $Cn = n!$.

$$P(G, \lambda) = \sum_{i=1}^{n} \binom{\lambda}{i} C_i = C_0\binom{\lambda}{0} + C_2\binom{\lambda}{2} + C_3\binom{\lambda}{3}$$

$$= Cn\binom{\lambda}{n} = \lambda(\lambda-1)(\lambda-2)\ \ldots\ (\lambda-n+1)$$

Example W13.3: Show that the chromatic polynomial for a tree on n vertices is given by $P(G, \lambda) = \lambda(\lambda-1)^{n-1}$.

Solution:
Let graph G be a tree on n vertices. We use mathematical induction on number of vertices.
Basis: If n = 1 then G contains only one vertex, which can be coloured with λ colours in λ distinct ways. $\lambda(\lambda-1)^0 = \lambda$. Hence the result holds good.

If $(n = 2)$ then the graph is an edge. We require a minimum of two colours to properly colour the graph. Hence, $C_1 = 0$ and the edge can be coloured in two distinct ways. $C_2 = 2$. The resulting polynomial is as follows.

$$P(G, \lambda) = \lambda(\lambda-1)^{2-1} = \lambda(\lambda-1) = \frac{\lambda(\lambda-1)}{2!} \times 2! \quad \text{Hence, the result holds good.}$$

Induction: We shall assume that the result hold good for an arbitrary positive integer k $(k < n)$ and prove it for $(k + 1)$.

$$P(G, \lambda) = \lambda(\lambda-1)^{k-1}$$

Now, add one pendant vertex. Its adjacent vertex has some colour and the pendant vertex can be coloured in $(\lambda-1)$ ways. Therefore, together the graph with $k + 1$ vertices is

$$\lambda(\lambda-1)^{k-1}(\lambda-1) = \lambda(\lambda-1)^k \quad \text{Hence, the result holds good for any } n.$$

Example W13.4: Obtain the chromatic polynomial for the following graph G and the edge marked by e.

Solution:
From the following figure, it follows that $P(G_{e}, \lambda) = \lambda(\lambda-1)^3$ and $P(G_{e'}, \lambda) = K_3$.

a b e = a b − a b = d
c d c c
P(G, λ) P(G_e, λ)

$P(G_{e'}, \lambda) = \lambda(\lambda-1)(\lambda-2)$. Therefore, P(G, λ) is given as follows.

$$\begin{aligned} P(G, \lambda) &= \lambda(\lambda-1)^3 - \lambda(\lambda-1)(\lambda-2) \\ &= \lambda(\lambda-1)[(\lambda-1)^2 - (\lambda-2)] \\ &= \lambda(\lambda-1)[\lambda^2 - 3\lambda + 3] \\ &= \lambda^4 - 4\lambda^3 + 6\lambda^2 - 3\lambda \end{aligned}$$

Example W13.5: A tree with more than one vertex has at least two leaves, and at least two vertices of degree one. Explain.

Solution:
Suppose T is a tree with n vertices $v_1, v_2, \ldots v_n$ and $(n-1)$ edges. We know that the total degree is $\sum_{i=1}^{n} v_i = 2n-2$. We already know that some vertex, say, v_i, has degree one. If the remaining vertices are of degree two or more, then the total degree is at least $1 + 2(n-1) = 2n-1$. This contradicts the handshaking theorem. Hence, there has to be another vertex of degree one in addition to v_i.

SUMMARY

Graph theory finds many applications in Engineering in general and Computer Science in particular. Graph Theory is a research area in mathematics as well as in Computer science. It is a full semester course and the coverage is limited. The content will help the readers to attempt reading books on the topic. The basic knowledge of graph theory is being presented in this chapter and the topics covered help the readers to read material connected with the topic.

EXERCISES

1. Prove that the sum of the degrees of the vertices of any finite graph is even.
2. Show that every simple graph has two vertices of the same degree.
3. Show that if n people attend a party and some shake hands with others (but not with themselves), then at the end, there are at least two people who have shaken hands with the same number of people.

4. Prove that a complete graph with n vertices contains $n(n-1)/2$ edges.
5. Prove that a finite graph is bipartite if and only if it contains no cycles of odd length.
6. Show that if every component of a graph is bipartite, then the graph is bipartite.
7. Prove that if u is a vertex of odd degree in a graph, then there exists a path from u to another vertex v of the graph where v also has odd degree.
8. If the distance $d(u, v)$ between two vertices u and v that can be connected by a path in a graph is defined to be the length of the shortest path connecting them, then prove that the distance function satisfies the triangle inequality: $d(u, v) + d(v, w) \geq d(u, w)$.
9. What is the minimum number of spanning trees in a connected graph with n nodes?
10. Prove that "If G is a (multi)graph with minimal degree 2 then G contains a cycle".
11. The figure on the left is a graph colouring, but the figure on the right is not. Why not?

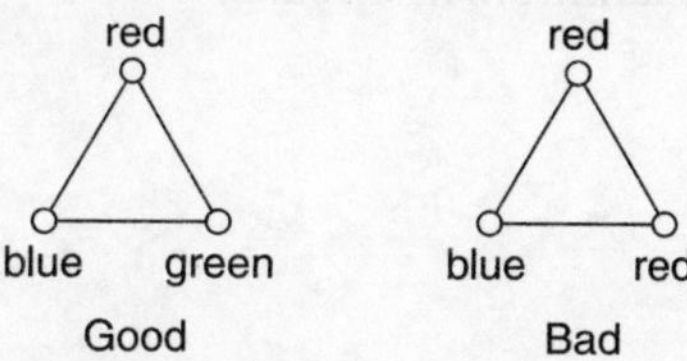

12. Find a graph colouring of

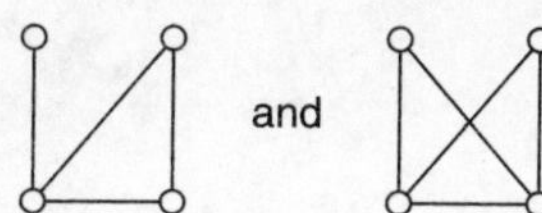

What is the least number of colours you can use?

13. What is the chromatic number of K_n?
14. List the applications of graphs.
15. What are the degrees and neighbourhoods of the vertices in the graphs G and H?

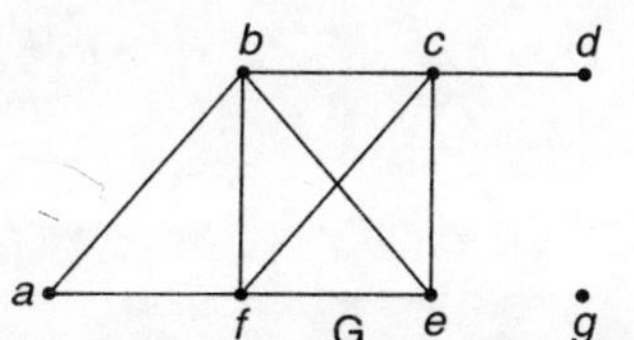

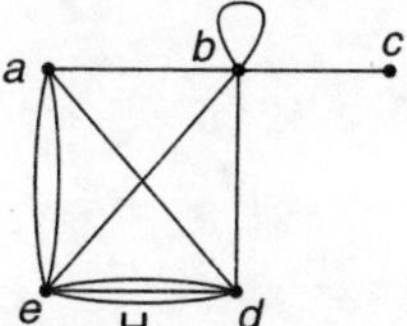

16. How many edges are there in a graph with 10 vertices, each having degree six?
17. If a graph has five vertices, can each vertex have degree 3?

18. Show that C_6 is bipartite.
19. When does a bipartite graph have a perfect matching?
20. Show that the graphs G = (V, E) and H = (W, F) are isomorphic.

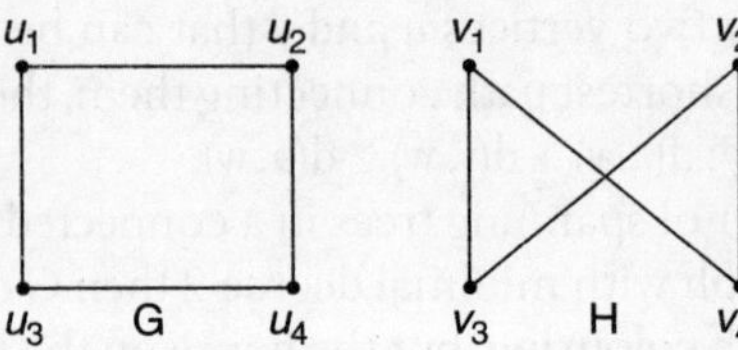

21. Define Hamiltonian Paths and Hamiltonian Circuits.

Bibliography

Apostol TM, *Calculus Volume1: One Variable Calculus with an Introduction to Linear Algebra,* Wiley India, New Delhi, 2007.

Bender CM and Orszag SA, *Advanced Mathematical Methods for Scientists and Engineers,* McGraw–Hill, 1978.

Burnside W, *Theory of Groups,* Dover, New York, 1955.

Deo N, *Graph Theory with Applications to Engineering and Computer Sciences,* Prentice Hall of India, New Delhi, 2001.

Grimaldi RP and Ramana BV, *Discrete and Combinatorial Mathematics: An Applied Introduction,* Pearson, 2002.

Hankerson DR, Hoffmann DG, Leonard DA, Linder CC, Phelps KT, Roder CA and Wall JR, *Coding Theory and Cryptography,* Marcel Dekker, New York, 2006.

Herstein IN, *Topics in Algebra,* Wiley Eastern, New Delhi, 1975.

Hopcroft JE and Ullman JD, *Introduction to Automata Theory, Languages and Computation,* Narosa Publishing House, New Delhi, 1967.

Hoperott JE and Ullman JD, *Introduction to Automata Theory Languages and Computation,* Narosa Publishing House, New Delhi, 1998.

Koshy T, *Discrete Mathematics with Applications,* Elsevier, 2004.

Lipschutz S and Lipson ML, *Theory and Problems of Discrete Mathematics,* Tata McGraw–Hill, New Delhi, 1999.

Liu CL, *Elements of Discrete Mathematics,* Tata McGraw–Hill, New Delhi. 2000.

Milton JS and Arnold JC, *Introduction to Probability and Statistics,* Tata–McGraw Hill, New Delhi, 2007.

Mott JL, Kandel A, and Baker TP, *Discrete Mathematics for Computer Scientists and Mathematicians,* Prentice Hall of India, 2005.

Mott SL, *Discrete Mathematics for Computer Scientists and Mathematicians,* Prentice Hall of India, New Delhi, 2005.

Peebles PZ Jr., *Probability Random Variables and Random Signal Principles,* Tata–McGraw Hill, New Delhi, 2002.

Rosen KH, *Discrete Mathematics and Its Applications with Combinatorics and Graph Theory,* Tata McGraw–Hill, New Delhi, 2002.

Stewart J, *Essential Calculus Early Transcendental,* McMaster University, Cengage Learning, 2007.

Truss JK, *Discrete Mathematics for Computer Scientists,* Addison–Wesley, Harlow, England, 2000.

Wilson RJ, *Introduction to Graph Theory,* 4e, Addison–Wesley, Reading Mass, 1996.

Yedidyah L, Moshe JA and Tenenbeum AM, *Data Structures Using C and C++,* Prentice Hall of India, 1999.

Index

A

Addition principle 64
Adjacency matrix 106
Algebraic structure 143
Algebraic topology 184
Algorithm 32, 327
  BFS 349, 350
  DFS 349
  Kruskal's 346
  minimize DFA 298, 299
  Prim's 346, 348
Anti-symmetry 82, 107
Application of FA 300
Arithmetic mean 44
Arithmetic progression 46
Asymmetric relation 81, 82, 107
Automata minimisation 294

B

Baye's theorem 245
Bernoulli's trials 248
Big(O) 128, 129, 131
Big Omega - Ω 130
Big theta - Θ 130
Binary operation 145
Binary properties 79
Binary relation 76
Binomial distribution 249
Binomial theorem 227
Bipartite graph 332
Boolean algebra 21, 108
Bubble sort 134

C

Cardinality 53, 54, 56, 64, 73, 77
Cartesian product of two sets 76, 77
Chromatic polynomial 340
Chromatic decomposition theorem 341
Circuit layout 353
CNF 21, 22
Co-domain 111
Coding, channel 191
Coding, source 191
Coding theory 189, 191
Collinear 226
Combinations 225
  complementary 236
  with repetition 232
Complex number 152
Concatenation 69
Conclusion 8
Congruence modulo *n* 173
Connective 7, 8, 9
Consequence 8
Contradiction 5, 11, 18
Contra-positive 14
Converse 14
Conversion cycles and circuits 333
  Moore's machine – Meelay's machine 292
  ∈-NFA to DFA 287
  ∈-NFA to NFA 286
  NFA to DFA 281, 283
Cosets 174
  leader 204
  left, right 174
Counting 217
Cryptography 184, 189
Cube roots of unity 154
Cyclic groups 167, 169

D

Data processing 191
De-Morgan's law 11, 12
Decoder 193
Decoding 195
Detection of planarity 336
DFA 269, 302, 309
Digraph 92, 104, 105, 108

Digraph transitive closure 110
Domain 111, 112

E

∈-NFA 285
Encoder 193
Encoding 207
Equivalence relation 82, 83, 108
Equivalent, NFA and DFA 282
Error coding 191
  correction 192, 195
  detection 192
Euler formula 337
Events 240
  mutually exclusive 241
Exponential distribution 256

F

Factorial function 119
Fibonacci series 39
Fourth root of unity 154
FSM 270–273
Functions 111
  algebraic 117
  bijective 116, 119
  composition 121, 122, 125, 129, 132
  decoding 196, 210
  encoding 194, 196, 201, 202, 205
  exponential 118
  growth 111, 128
  hashing 118, 141
  identity 126
  into 115
  inverse 119, 120
  invertible 126, 127
  many- to-one 127
  one-one 115, 119, 125
  onto 115, 125
  rational 118
  transcendental 118
  constant 115
  range 111

G

Game of instant insanity 330, 331
GCD 100
Generator 169, 170
Generator matrix 197, 205, 209
Geometric mean 44
Geometric progression 37
GLB 100
Graph 32
  colouring 339
  directed, weighted and directed acyclic 343
  isomorphism 333
  non-planar 336
  planar 335
  theory 327
Group 143, 148
  Abelian 152, 155, 156, 158, 167
  abstract 141
  code 192, 203, 209
  cyclic 167
  dihedral 166
  finite and infinite 148
  homomorphism 180
  isomorphic 183
  matrix 144
  permutation 144
  semi 160
  symmetric 162, 163
  topological and algebraic 145
  transformation 144

H

Haar measures 184, 189
Hamiltonian cycles 335
Hamming 191, 213
  additional parity check 208
  code 208
  metric 193
Hasse diagram 96, 97, 99, 100, 108

I

Identity relation 79
Implication 8, 9

Independence 243
Informal logic 29
Intelligent transport system 352
Inverse 14, 90
Inverse relation 79
Irreflexivity 80
Isomorphism 177, 181, 182

J

Jokes 45

K

Konigsberg bridge 327
  graphical representation of 328
Kleene's closure 69, 95

L

Lagrange's theorem 175
Lattice 100, 108
  complemented 101
  distributive 101, 103
  properties 100
Laws of logic 11
LCM 100
Logic 5
Logical AND – OR 88
Logically equivalent 11–13, 17
Loop invariant 40, 50, 51
Lower bound 99
LUB 100, 110

M

Mathematical induction 32
Matrix representation of relations 88
Maximal, minimal elements 98, 110
Modal logic 29
Monoid 160
Moore and Mealy machine 290, 291

N

NDFA 274, 276–280
NFA 275, 309
Non-planar graphs 336
Normal distribution 258

O

Operations on relations 85
Output associated finite automata 289

P

Paradoxes 45
Parity check 192, 197
  Even 194
  generator matrix 196
  matrix 197, 198, 200, 210, 212, 213
  Odd 195
Partial order 95
Partition 82–84
Pascal's triangle 231
  properties 231
Paths 333
Permutations 161
  circular 223
  composition 161
Poisson distribution 252
POSET 95, 97–100, 108–110
  graphical representation of 96
Power set 97, 100, 102
Predicate logic 5, 23, 54
Prefix codes 344
Principle of duality 12, 59, 103
Probability 238
  addition theorem 242
  classical approach 239
  conditional 243
  continuous distribution 253
  models 238
  personal approach 239
  relative frequency approach 239
Proposition 6–9
Pseudo code 106

Q

Quantifiers 23
  existential 23
  universal 23

R

Random variables 248
Range 80

RDBMS 76
Recurrence relation 311
  general solution 313
  generating function 321, 322, 324, 325
  method of undetermined coefficient 317
  non-homogeneous 316
Reflexive relation 81
Reflexivity 80
Regular expression 69, 70, 306–308
Regular grammar 69, 71
Regular language 69
Relations 76
  composition 86, 94
  graphical representation 92
Residue classes – addition, multiplication 173
Rule of inference 15
Rules – sum and product 217
Russel's paradox 55

S

Sample space 240
Sequences 67
Social networks 353
Sorting 133
  bubble 134
  merge 136
  quick 135
  selection 136
Square roots of unity 154
Strings 68
Sub-graph, spanning 324
Subgroups 171, 174
Symbolic logic 5
Symmetric relation 81, 107
  closure 87
  graphical representation 92

T

Tautology 11, 15
Temporal logic 29
Testing equivalence of states 294
Tower of Hanoi 320
Transitive
Transitive relation 82, 93
Trees 341
  counting 345
  properties 342
  spanning, minimal spanning 345, 347
Trichotomy 133

U

Universal relation 77
Universal upper bound 102

V

Venn diagrams 62, 73
Void relation 79

W

Walk, paths, cycles and circuits 333
Warshall's algorithm 104, 106